CO_2 Capture and Conversion Processes: Recent Trends and Future Perspectives

CO_2 Capture and Conversion Processes: Recent Trends and Future Perspectives

Guest Editors

Georgios Bampos
Georgios Karanikolos

MDPI

Basel • Beijing • Wuhan • Barcelona • Belgrade • Novi Sad • Cluj • Manchester

Guest Editors

Georgios Bampos
Department of Chemical Engineering
University of Patras
Patras
Greece

Georgios Karanikolos
Department of Chemical Engineering
University of Patras
Patras
Greece

Editorial Office
MDPI AG
Grosspeteranlage 5
4052 Basel, Switzerland

This is a reprint of the Special Issue, published open access by the journal *Processes* (ISSN 2227-9717), freely accessible at: https://www.mdpi.com/journal/processes/special_issues/844481IQ59.

For citation purposes, cite each article independently as indicated on the article page online and as indicated below:

Lastname, A.A.; Lastname, B.B. Article Title. *Journal Name* **Year**, *Volume Number*, Page Range.

ISBN 978-3-7258-7805-5 (Hbk)
ISBN 978-3-7258-7806-2 (PDF)
https://doi.org/10.3390/books978-3-7258-7806-2

Contents

Preface

This Reprint brings together selected contributions on carbon dioxide capture, utilization, storage, and conversion, reflecting the growing scientific and industrial need for effective carbon management strategies. Its scope covers experimental, analytical, and process-oriented studies, including absorption, adsorption, geological storage, chemical looping, blue hydrogen assessment, monitoring methods, and catalytic CO_2 conversion.

The aim of this Reprint is to present recent progress in technologies and methodologies that can support greenhouse gas mitigation and the transition toward lower carbon energy and chemical processes. The motivation for compiling these works lies in the urgent need to better understand, evaluate, and improve practical solutions for reducing CO_2 emissions while maintaining industrial performance and energy security.

This Reprint is addressed to researchers, engineers, graduate students, and professionals working in chemical engineering, environmental science, energy systems, process optimization, and carbon capture and utilization. It is also intended to support readers seeking an overview of current research directions and applied challenges in the development of scalable CO_2 management technologies.

Georgios Bampos and Georgios Karanikolos
Guest Editors

processes

MDPI

Editorial

Editorial for the Special Issue "CO_2 Capture and Conversion Processes: Recent Trends and Future Perspectives"

Georgios Bampos [1,2,*] and Georgios N. Karanikolos [1,2,*]

1 Department of Chemical Engineering, University of Patras, 26504 Patras, Greece
2 Institute of Chemical Engineering Sciences, Foundation for Research and Technology-Hellas (FORTH/ICE-HT), 26504 Patras, Greece
* Correspondence: geoba@chemeng.upatras.gr (G.B.); karanikolos@chemeng.upatras.gr (G.N.K.)

1. Introduction

The increasing concentration of carbon dioxide (CO_2) in the atmosphere is widely recognized as one of the main drivers of climate change [1,2]. Rapid industrial development, growing global energy demand and continued reliance on fossil fuels have contributed to unprecedented greenhouse gas emissions [3,4]. In response to these challenges, carbon capture, utilization and storage (CCUS) technologies have emerged as essential components of global decarbonization strategies. These technologies aim not only to reduce CO_2 emissions from industrial sources but also to enable the transition toward more sustainable energy and chemical production systems [5,6].

Carbon capture and conversion processes encompass a wide range of approaches, including chemical absorption and adsorption, membrane separation, chemical looping combustion, and direct air capture (DAC) [7]. Once captured, CO_2 can either be stored in geological formations or converted into valuable chemicals and fuels. These pathways have the potential to transform CO_2 from an environmental liability into a useful resource, thus contributing to the development of circular carbon systems [8].

Despite considerable advances over recent decades, several technical and economic challenges remain. High energy requirements for solvent regeneration, limitations in catalyst performance for CO_2 conversion, uncertainties related to long-term geological storage, and infrastructure constraints continue to hinder large-scale deployment of CCUS technologies [9,10]. Addressing these issues requires interdisciplinary research efforts combining chemical engineering, catalysis, materials science, chemistry, and geosciences, as well as process systems engineering.

The Special Issue entitled "CO_2 Capture and Conversion Processes: Recent Trends and Future Perspectives" brings together recent research contributions addressing key aspects of CO_2 capture, storage, utilization and system-level evaluation. The eight papers included in this collection cover a wide range of topics, from solvent-based capture technologies and chemical looping combustion to geological storage studies, catalytic conversion pathways and carbon footprint assessment methods. Together, these contributions provide valuable insights into the current state of research and highlight potential directions for future development.

2. Overview of Contributions

The contributions published in this Special Issue cover a wide variety of topics related to CO_2 capture and conversion processes, highlighting the multidisciplinary nature of this rapidly evolving research field. They can be grouped into four thematic areas: advanced CO_2 capture technologies, geological storage and adsorption mechanisms, CO_2

 https://doi.org/10.3390/pr14091332

utilization and catalytic conversion pathways, and system-level analyses related to carbon management and emerging energy systems (Figure 1).

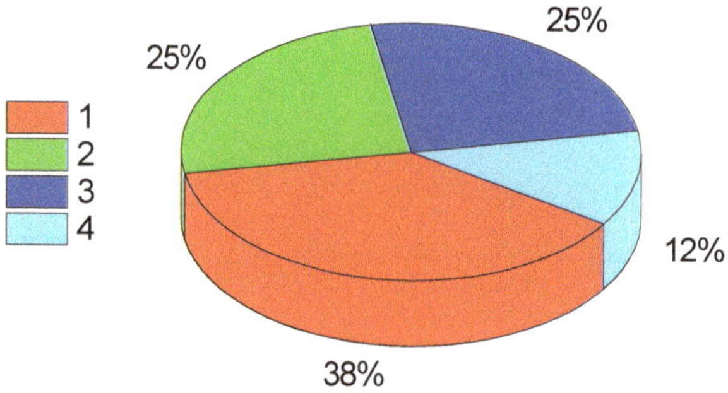

Figure 1. Publication distribution (%) based on subject: advanced CO_2 capture technologies (1, red), geological storage and adsorption (2, green), technology reviews and system-level assessment (3, blue), and CO_2 utilization and catalytic conversion (4, cyan).

A significant portion of the contributions focus on advances in CO_2 capture technologies, particularly those aimed at improving the efficiency, monitoring and operational performance of capture systems. Han et al. [11] investigate the relationship between electrical conductivity and CO_2 absorption capacity in aqueous di-methyl-ethanolamine (DMEA) solutions. In their study, they demonstrate that electrical conductivity measurements can be used to estimate CO_2 loading during absorption, leading to the development of an empirical equation capable of predicting absorption capacity in real time. Such monitoring approaches could offer a practical and low-cost tool for improving operational control in solvent-based carbon capture systems.

Another study addressing solvent-based capture technologies is presented by Wang et al. [12], in which the authors examine the performance of phase-change absorbents designed for CO_2 capture from blast furnace gas, focusing on solvent systems based on monoethanolamine (MEA) and 1-(2-aminoethyl)piperazine (AEP). Their results indicate that certain solvent mixtures, particularly those involving AEP combined with n-propanol, exhibit promising absorption capacities along with relatively low regeneration energy requirements. These findings highlight the potential of phase-change absorbents as an energy-efficient alternative for industrial CO_2 capture processes.

Complementing solvent-based approaches, Tao et al. [13] explore an alternative capture concept. Chemical looping combustion (CLC) is considered a promising technology because it inherently separates CO_2 during combustion by transferring oxygen from the air to fuel through a solid oxygen carrier. In this study, the authors evaluate the redox performance and reaction kinetics of a natural iron ore used as an oxygen carrier. The results suggest that natural iron ores may represent a cost-effective and environmentally friendly option for large-scale CLC systems, although operational parameters must be carefully controlled to avoid sintering and performance degradation.

Beyond capture technologies, this collection also includes papers related to geological storage and adsorption mechanisms, which play a critical role in the long-term sequestration of captured CO_2. AlNajdi et al. [14] investigate the mineralogical composition and pore structure characteristics of the Lista Shale (a specific geological rock formation), which acts as a caprock above a proposed CO_2 storage reservoir in the North Sea. Through detailed mineralogical and pore-system analyses, the authors assess the ability of the formation to function as a reliable top seal. Their findings indicate that the clay-rich matrix and restricted pore throat sizes contribute to complex fluid pathways and low permeability, thus supporting the potential of the formation to contain injected CO_2 in the long term.

https://doi.org/10.3390/pr14091332

Additional insights into subsurface CO_2 behavior are presented by Ortiz Cancino and Santos Santos [15]. The authors experimentally investigate the adsorption behavior of methane and carbon dioxide on shale samples with exceptionally high total organic carbon contents. Their results reveal that CO_2 consistently exhibits higher adsorption capacities than CH_4 under the tested conditions, demonstrating the strong affinity of shale formations for CO_2. These findings contribute to a better understanding of gas–rock interactions in shale systems and support the potential use of such formations for both CO_2 storage and enhanced gas recovery.

Another important research direction represented in this Special Issue concerns CO_2 utilization and catalytic conversion pathways, which aim to transform captured CO_2 into valuable chemicals and fuels. The review article by Sajnani et al. [16] provides a comprehensive overview of the catalytic hydrogenation of CO_2 to methanol. Combining bibliometric analysis with a detailed examination of catalytic systems and reaction mechanisms, the authors highlight the importance of Cu-based catalysts, which offer a balance between catalytic activity, selectivity, and economic feasibility. Their analysis identifies key research trends and highlights the potential of methanol synthesis as a pathway for CO_2 valorization and sustainable carbon management.

In addition to studies focusing on specific technologies, this collection also includes contributions that examine broader technological and system-level perspectives related to carbon management. Thiedemann and Wark [17] present a comprehensive overview of current carbon capture technologies. The authors summarize the main carbon capture approaches, namely post-combustion capture, pre-combustion capture, oxy-fuel combustion, chemical looping and DAC, while also discussing storage options such as geological sequestration and mineralization. Furthermore, they highlight pathways for utilizing captured CO_2 in the production of chemicals and fuels, thus offering a concise overview of the technological landscape and associated challenges.

Finally, Chan et al. [18] address the broader environmental performance of emerging low-carbon technologies by proposing a simplified methodology for evaluating the carbon footprint of chemical production systems during the conceptual design stage. By introducing the concepts of product carbon intensity and economic carbon intensity, the study provides a framework for comparing alternative process configurations. Applying the methodology to different blue hydrogen production scenarios, the authors demonstrate how early-stage carbon footprint analysis can help identify sustainable process designs.

3. Challenges and Future Perspectives

The studies included in this Special Issue illustrate the diverse research efforts currently underway to address the challenges associated with CO_2 capture and conversion processes. Advances in solvent development, catalytic conversion, geological storage evaluation, and process assessment methodologies highlight the progress that has been made in recent years.

However, several challenges remain before these technologies can be widely deployed at an industrial scale. One of the most significant challenges in CO_2 capture technologies is the high energy requirement associated with solvent regeneration. Although new absorbent systems and alternative capture technologies such as membrane and adsorption technologies and chemical looping show promising results, further improvements are needed to reduce energy consumption and operational costs while enhancing scalability.

From the perspective of geological storage, ensuring the long-term integrity of storage sites remains a critical issue. Detailed studies of caprock properties, pore structures, and geochemical interactions, such as those presented in this collection, are essential for evaluating the reliability of potential storage formations. Continued research in this area

https://doi.org/10.3390/pr14091332

will help reduce uncertainties associated with CO_2 sequestration and support the safe deployment of large-scale storage projects.

The utilization of captured CO_2 as a feedstock for chemical production also represents an important pathway toward sustainable carbon management. Processes such as CO_2 hydrogenation to methanol have attracted significant research interest, but large-scale implementation will depend on further advances in catalyst design and integration with renewable hydrogen production technologies.

Finally, system-level evaluation tools will play an increasingly important role in guiding the development of low-carbon technologies. Simplified assessment methods, such as the carbon intensity metrics proposed in this Special Issue, can support decision-making during early-stage process design and help identify more sustainable technological solutions.

Overall, the contributions presented in this collection provide valuable insights into the current state of research and development for CO_2 capture and conversion processes. Continued collaboration between researchers, industry stakeholders, and policymakers will be essential to translate these scientific advances into practical technologies at scale, capable of contributing to global decarbonization efforts.

Funding: This research received no external funding.

Acknowledgments: We are thankful to all the authors who submitted their impressive work to this Special Issue and to the reviewers for their time and effort in reviewing the manuscripts.

Conflicts of Interest: The authors declare no conflicts of interest.

References

1. Varghese, A.M.; Karanikolos, G.N. CO_2 capture adsorbents functionalized by amine–bearing polymers: A review. *Int. J. Greenh. Gas Control* **2020**, *96*, 103005. [CrossRef]
2. Lacis, A.A.; Schmidt, G.A.; Rind, D.; Ruedy, R.A. Atmospheric CO_2: Principal Control Knob Governing Earth's Temperature. *Science* **2010**, *330*, 356–359. [CrossRef] [PubMed]
3. Le Quéré, C.; Peters, G.P.; Friedlingstein, P.; Andrew, R.M.; Canadell, J.G.; Davis, S.J.; Jackson, R.B.; Jones, M.W. Fossil CO_2 emissions in the post-COVID-19 era. *Nat. Clim. Change* **2021**, *11*, 197–199. [CrossRef]
4. Zhang, Z.; Pan, S.-Y.; Li, H.; Cai, J.; Olabi, A.G.; Anthony, E.J.; Manovic, V. Recent advances in carbon dioxide utilization. *Renew. Sustain. Energy Rev.* **2020**, *125*, 109799. [CrossRef]
5. Hepburn, C.; Adlen, E.; Beddington, J.; Carter, E.A.; Fuss, S.; Mac Dowell, N.; Minx, J.C.; Smith, P.; Williams, C.K. The technological and economic prospects for CO_2 utilization and removal. *Nature* **2019**, *575*, 87–97. [CrossRef] [PubMed]
6. Boot-Handford, M.E.; Abanades, J.C.; Anthony, E.J.; Blunt, M.J.; Brandani, S.; Mac Dowell, N.; Fernández, J.R.; Ferrari, M.-C.; Gross, R.; Hallett, J.P.; et al. Carbon capture and storage update. *Energy Environ. Sci.* **2014**, *7*, 130–189. [CrossRef]
7. D'Alessandro, D.M.; Smit, B.; Long, J.R. Carbon Dioxide Capture: Prospects for New Materials. *Angew. Chem. Int. Ed.* **2010**, *49*, 6058–6082. [CrossRef] [PubMed]
8. Artz, J.; Müller, T.E.; Thenert, K.; Kleinekorte, J.; Meys, R.; Sternberg, A.; Bardow, A.; Leitner, W. Sustainable Conversion of Carbon Dioxide: An Integrated Review of Catalysis and Life Cycle Assessment. *Chem. Rev.* **2018**, *118*, 434–504. [CrossRef] [PubMed]
9. Bui, M.; Adjiman, C.S.; Bardow, A.; Anthony, E.J.; Boston, A.; Brown, S.; Fennell, P.S.; Fuss, S.; Galindo, A.; Hackett, L.A.; et al. Carbon capture and storage (CCS): The way forward. *Energy Environ. Sci.* **2018**, *11*, 1062–1176. [CrossRef]
10. McLaughlin, H.; Littlefield, A.A.; Menefee, M.; Kinzer, A.; Hull, T.; Sovacool, B.K.; Bazilian, M.D.; Kim, J.; Griffiths, S. Carbon capture utilization and storage in review: Sociotechnical implications for a carbon reliant world. *Renew. Sustain. Energy Rev.* **2023**, *177*, 113215. [CrossRef]
11. Han, S.-J.; Han, J.Y.; Wee, J.-H. Real-Time Estimation of CO_2 Absorption Capacity Using Ionic Conductivity of Protonated Di-Methyl-Ethanolamine (DMEA) and Electrical Conductivity in Low-Concentration DMEA Aqueous Solutions. *Processes* **2024**, *12*, 2495. [CrossRef]
12. Wang, Y.; Cheng, F.; Li, J.; Liu, Y.; Wang, H.; Li, Z.; Yang, X. Development of MEA-Based and AEP-Based CO_2 Phase Change Absorbent. *Processes* **2025**, *13*, 92. [CrossRef]
13. Tao, C.; Liang, Q.; Li, Q.; He, M.; Shen, X.; Wang, H.; Wang, M.; Wang, X. Performance Evaluation and Kinetic Analysis of an Iron Ore as Oxygen Carrier in Chemical Looping Combustion. *Processes* **2025**, *13*, 2949. [CrossRef]

https://doi.org/10.3390/pr14091332

14. AlNajdi, N.; Worden, R.H.; Utley, J.E.P. The Palaeocene Lista Shale: A Planned Carbon Capture and Storage Top Seal for the East Mey CO_2 Storage Site. *Processes* **2024**, *12*, 2773. [CrossRef]
15. Ortiz Cancino, O.P.; Santos Santos, N. Experimental Adsorption Study of Pure CH_4 and CO_2 on Organic-Rich Shales from the Cesar-Ranchería Basin, Colombia. *Processes* **2025**, *13*, 2108. [CrossRef]
16. Sajnani, S.; Memon, M.A.; Memon, S.A.; Kumar, A.; Mehvish, D.; Ameen, S.; Mukarama; Zhou, W.; Liu, Y. CO_2 to Methanol Conversion: A Bibliometric Analysis with Insights into Reaction Mechanisms, and Recent Advances in Catalytic Conversion. *Processes* **2025**, *13*, 314. [CrossRef]
17. Thiedemann, T.M.; Wark, M. A Compact Review of Current Technologies for Carbon Capture as Well as Storing and Utilizing the Captured CO_2. *Processes* **2025**, *13*, 283. [CrossRef]
18. Chan, E.Y.H.; Adi Putra, Z.; Tan, R.R.; Wan, Y.K.; Foo, D.C.Y. Carbon Footprint Analysis of Chemical Production: A Case Study of Blue Hydrogen Production. *Processes* **2025**, *13*, 1254. [CrossRef]

https://doi.org/10.3390/pr14091332

processes

MDPI

Article

Performance Evaluation and Kinetic Analysis of an Iron Ore as Oxygen Carrier in Chemical Looping Combustion

Congxi Tao [1], Qian Liang [1,*], Qingmei Li [1], Minghai He [1], Xuhui Shen [1], Hao Wang [1], Ming Wang [1] and Xudong Wang [2,*]

1 China Resources Building Materials Technology Holdings Co., Ltd., Shenzhen 518000, China
2 Jiangsu Key Laboratory of Process Enhancement and Energy Equipment Technology, School of Mechanical and Power Engineering, Nanjing Tech University, Nanjing 211816, China
* Correspondence: liangqian11@cr-bmt.com (Q.L.); xdwang_seu@seu.edu.cn (X.W.)

Abstract

Chemical looping combustion (CLC) provides an inherently cost-effective method for carbon capture by employing a solid oxygen carrier (OC) to transfer lattice oxygen from air to fuel. The search for low-cost, high-performance natural OCs is crucial for the large-scale deployment of this technology. A natural iron ore containing 41.34% Fe_2O_3 was systematically evaluated as OC for the CLC of CO. Its redox performance was quantified in a fixed-bed reactor between 750 °C and 900 °C with CO concentrations of 10–20%. Multi-cycle tests were conducted to assess stability. Kinetic analysis of the initial cycles was performed using an integral model fitting method. Multi-cycle tests revealed that the fresh ore achieved peak conversions of 48.9% at 750 °C and 77.2% at 900 °C. However, severe sintering occurred beyond 850 °C after the first cycle, causing approximately a 50% drop in OC conversion. Interestingly, once sintered, a self-activation phenomenon was observed during subsequent cycles; the OC conversion slowly recovered from 32% to 37% from the second to the fifteenth cycle under the aggressive conditions (900 °C, 20% CO). Kinetic analysis of the initial cycles (before sintering) revealed low apparent activation energies, ranging from 15.93 to 19.13 kJ mol^{-1}, which are significantly lower than the typical literature values for iron-based ores. This work underscores the potential of natural iron ores as economical and sustainable OCs for CO-rich fuels. The observed self-activation ability of the sintered OC is a promising finding for long-term operation. The results also highlight the critical importance of operating conditions to avoid deep reduction and sintering, necessitating a high solids inventory and a moderate oxygen-to-fuel ratio in practical CLC systems.

Keywords: chemical looping combustion; iron ore; oxygen carrier; performance evaluation; reaction kinetics; self-activation ability

1. Introduction

The issue of global warming caused by CO_2 emissions has been the focus of research for many years, especially after 2015 when the Paris Agreement was signed [1–4]. However, in early 2025, official reports indicated that the Earth's average temperature had increased by more than 1.5 °C compared to pre-industrial levels [5]. The data reiterated the urgent need to curb carbon emissions, which reached an unprecedented peak in 2024 [6]. In the energy sector, fossil fuels play an important role as the cornerstone for energy safety. As the imperative for decarbonization grows, there is heightened focus on carbon capture and storage technologies within this domain [7–9].

There have been different ways for carbon capture in fossil fuel utilization, such as integrated gasification combined cycle, CO_2 sorption in flue gas, oxy-fuel combustion, and so on [10–13]. Of these, chemical looping combustion (CLC) has garnered significant attention, owing to its inherent CO_2 separation during combustion, low-NO_x formation, and high exergy efficiency, etc. [13–16]. In CLC, fuel and air are processed in different reactors, namely the fuel reactor (FR) and air reactor (AR). Oxygen from the air is transferred to the fuel via an oxygen carrier (OC), typically a metal oxide. That avoids the direct contact between the fuel and air. Consequently, combustion products in the FR are not diluted by the nitrogen in air. After condensation and purification of the flue gas at the FR side, it yields a high-concentration CO_2 stream for capture. The principle of CLC is shown in Figure 1.

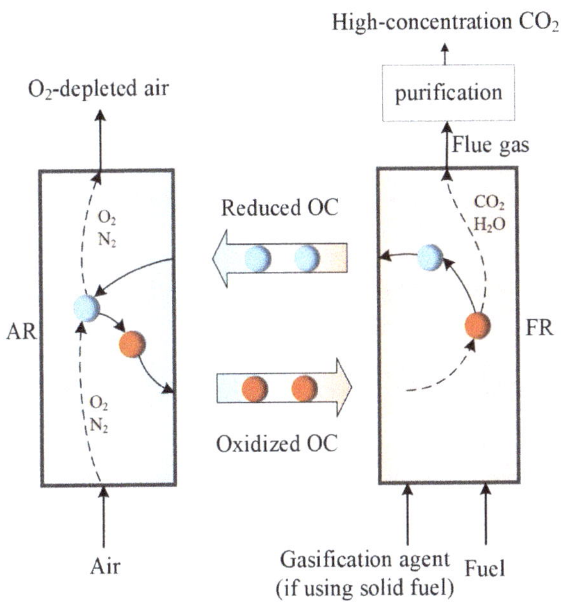

Figure 1. Principle of CLC.

OC circulates between two reactors, participating alternately in redox reactions, which is crucial to CLC [15,17]. Its properties directly determine the combustion efficiency, carbon capture performance, long-term stability, and operational cost. When selecting an oxygen carrier, it is necessary to consider its performance from multiple aspects, such as the reaction kinetics, oxygen carrying capacity, and environmental performance [17]. Among various OCs, Fe-based OC exhibits favorable properties in terms of reactivity and environmental impact [15,18,19]. The preparation of Fe-based OCs typically involves using Fe_2O_3 as the active component with Al_2O_3, ZrO_2, TiO_2, SiO_2, and other inert substances [20–22]. By adjusting ratios and processes, controllable oxygen carriers can be synthesized. However, the scalability and cost-effectiveness of the synthetic process remain significant challenges [23]. Researchers recognize the economic viability of using natural ore as an oxygen carrier and consider it an inevitable choice for large-scale CLC systems [24–27]. The selection of appropriate OC necessitates a comprehensive investigation into its reaction performances.

The prevalent iron ores encompass ilmenite and hematite, among others. To examine the combustion performance of such ores as OC, the intermediate gasification products like CO, H_2, and CH_4 from the solid fuel combustion process are usually employed as fuels to study their reaction characteristics with iron ore [28,29]. Leion et al. conducted an investigation into the reaction performance of various ilmenite minerals in CLC using a laboratory-scale fluidized bed. Their findings affirmed the viability of using natural ores as oxygen carriers and established that these OCs possess commendable strength and fluidization characteristics [27]. Adánez et al. employed CO, CH_4, and H_2 as gaseous fuels in a thermogravimetric analyzer (TGA) to study the reaction performance of ilmenite in CLC. They observed an enhancement in the reaction performance of ilmenite during the initial few redox cycles, which is referred to as the activation process. The duration of this activation process was found to be contingent upon the composition of the gas fuel [30]. For performance evaluation, experiments are usually conducted to derive kinetic parameters such as the apparent activation energy for the reaction between OCs and fuel. Significant progress has been made in the research on the reaction characteristics of natural iron ore, and some of the main relevant literature is summarized in Table 1.

Table 1. Performance evaluation using Fe-based iron ore as OC.

OC Types	Active Compositions	Test Fuel	Reactor	Apparent Activation Energy	Reaction Order	Reference
Kryvbas iron ore	84.86% Fe_2O_3	CH_4	TGA	42.0 kJ mol^{-1}	1.98	[31]
Canadian ilmenite	55.8% Fe_2O_3 32.6% TiO_2	CO	TGA	115 kJ mol^{-1}	0.67	[32]
Norwegian ilmenite	11.2% Fe_2O_3 54.6% Fe_2TiO_5	H_2 CO CH_4	TGA	109.2 kJ mol^{-1} 113.3 kJ mol^{-1} 165.2 kJ mol^{-1}	1	[33]
Chinese hematite	66% Fe_2O_3	CO	TGA	110.75 kJ mol^{-1}	1.5	[34]
Chinese hematite	66% Fe_2O_3	CO	Fluidized bed reactor	74.48 kJ mol^{-1}	1	[35]
Ilmenite	46.4% Fe_2O_3	CH_4	TGA	62.4 kJ mol^{-1}	0.52	[36]
Canadian ilmenite	10.4% Fe_2O_3 + 30% Fe_2TiO_5	CH_4	TGA	106.7 kJ mol^{-1}	0.7	[37]
Spanish iron ore	76.5% Fe_2O_3	H_2 CO CH_4	TGA	81.1 kJ mol^{-1} 76.1 kJ mol^{-1} 257 kJ mol^{-1}	1	[38]
Lean iron ore	35.21% Fe_2O_3 10.0% $CaSO_4$	CH_4 CO	TGA	62 kJ mol^{-1} 56 kJ mol^{-1}	0.5	[39]
Chinese iron ore	44.16% Fe_2O_3	CH_4	TGA	157.5kJ mol^{-1} 126.9 kJ mol^{-1}	1 2	[40]

The activation energy among varying iron ores exhibits significant diversity, attributable to the different distribution locations and active component concentrations inherent in natural iron ores, as well as a variety of microstructure changes (such as pore characteristics, surface structure, etc.). In recent years, numerous pilot-scale CLC units have been constructed and set into operation [14,15,24,26,41]. There is a substantial demand for low-cost and efficient OC. Though many works have focused on the performance evaluation and selection for OCs, only a few types of OCs have been tested on pilot-scale or sub-industrial CLC units. Previously, our group has designed and established a pilot-scale CLC circulating fluidized bed reactor [13,42]. A kind of iron ore OC has been used for the CLC of coal. Satisfactory CLC performances were obtained. While the feasibility of natural iron ores as OCs is established, this work focuses on an ore with a comparatively low Fe_2O_3 content (41.34%) but high inherent SiO_2 and Al_2O_3, evaluating its performance under aggressive conditions. Testing ores with lower active content but high mechanical strength is highly relevant for cost-cutting. The primary focus of this study was to evaluate the macroscopic redox performance and reaction kinetics of a naturally sourced iron ore under typical CLC conditions.

2. Experimental Section

2.1. Material

As previously noted, both the OC circulation and inventory are much larger in a pilot-scale CLC unit. The expense associated with synthetic OC is prohibitively high, and it always exhibits relatively low mechanical strength. A kind of iron ore produced in Hebei, China, is chosen due to its properties of low cost and proper Fe content in prior selections.

The selected iron ore was crushed and sieved to obtain a particle size fraction from 350 to 830 μm. Then, the ore particle is calcined at 1000 °C for 2 h in the atmosphere of air to guarantee sufficient oxidation. For the precalcination step, it is not merely for oxidation but is a critical pre-treatment to achieve a stable initial microstructure and maximum oxidation state before the redox cycles begin. This pre-treatment at high temperature is also a common method in the study of OC performance in the field of CLC. The calcined particle is used for the OC in following experiment. Its compositions are analyzed using X-ray Fluorescence (XRF) Spectrometer, typed Rigaku ZSX Primus III+ (Rigaku Corporation, Tokyo, Japan). The results are shown in Table 2.

Table 2. Main components of the iron ore OC.

Components	Fe_2O_3	SiO_2	Al_2O_3	CaO	MgO	Others
Weight	41.34%	32.17%	11.85%	3.77%	1.47%	9.40%

It is seen that this kind of OC has relatively high contents of SiO_2 and Al_2O_3, which are always used as inert supports in synthetic OC. This natural ore has a relatively high mechanical strength due to this reason. While the factors such as particle size, fuel type, and reactor design can indeed influence the observed reaction rates and apparent kinetics, a detailed comparative analysis of these parameters falls beyond the intended scope.

2.2. Experimental Setup and Procedures

A fixed-bed reactor system is employed to test the CLC performance of this iron ore OC, which is shown in Figure 2. The experimental setup consists of a fixed-bed reactor, gas supply and flow rate controller, temperature controller, and a gas processing and analyzer.

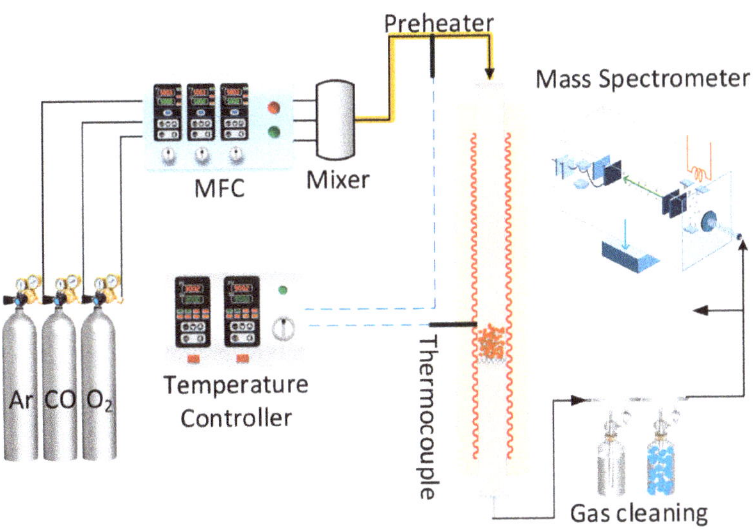

Figure 2. Fixed-bed reactor setup for OC evaluation.

The primary reactor body is made of stainless steel, which has an inner diameter of 8 mm with a height of 600 mm. Within the reactor, a porous sintered plate is located in the flat-temperature zone to place the OC particles. Electrical heating is employed to maintain the reaction temperature, which is measured using a thermocouple. The various gases are quantified using MFC, then mixed in a mixer. At the outlet of the mixer, the gas line is preheated. A temperature controller, equipped with two separated instruments, is used to control the reactor and preheater temperatures through fuzzy PID logic. As the flue gas exits from the reactor, it is cleaned and dried in pure water and silica gel. Subsequently, the cleaned gas is analyzed by a mass spectrometer (MS) to obtain the gas concentrations. To ensure the highest accuracy, the absolute calibration of the MS was not solely reliant on the MFCs. Before each experiment, the MFCs and MS were calibrated using a primary standard: a soap-film flow meter. This calibration was performed at the reactor outlet, after all conditioning elements, thus directly correlating the MS signal to the volumetric flow rate of the gas mixture at ambient temperature and pressure. This procedure directly validates the entire chain of flow and concentration measurements against a primary standard and accounts for any minor deviations in the MFC calibrations or the physical state of the gas.

Prior to the OC evaluation experiment, 15.000 g of OC is weighed and placed into the reactor. Then, the Ar is introduced to purge the reaction zone, while the reactor is simultaneously heated to the setting temperature. While the reactor temperature reaches the operating temperature, a stream of CO gas, at a specific concentration and diluted by Ar, is introduced into the reactor for the reduction of OC. This process continues for 10 min, after which the pure Ar gas is reintroduced for an additional 5 min to sweep the reactor. Following this, a stream of 21% O_2, diluted by Ar, is introduced to regenerate the OC for 10 min. The cycle concludes with the reintroduction of pure Ar to sweep the reactor in preparation for the next redox cycle. When all redox cycles are finished, the electrical heater is turned off and the Ar is introduced for cooling the system. Throughout the various stages of the redox cycles, the total gas volume flow rate is maintained at 800 mL min^{-1}. Fixed reaction times ensure that each cycle and each experiment is performed identically, allowing for a direct and fair comparison of the final conversion and deactivation trends. Also, this setting of the reaction time is long enough to capture the reaction profile and reach a pseudo-steady state in gas composition. In addition, the oxidation of reduced iron phases to Fe_2O_3 is a highly exothermic and typically much faster process than reduction, especially in the presence of air. The 10 min period with air is exceedingly long to ensure complete re-oxidation of OC before the next cycle begins.

2.3. Experimental Data Processing

The gas compositions at the reactor outlet are analyzed to determine the gaseous productions. Typically, the reactivity of the OC is gauged by its reduction, given that the oxidation process is quite fast. During the reduction process, the iron ore OC may be reduced to different phases, as shown in Equation (1):

$$Fe_2O_3 \rightarrow Fe_3O_4 \rightarrow FeO \rightarrow Fe, \tag{1}$$

The theoretical weight loss of the OC in a reduced state is different depending on the final phases represented in Equation (1). In order to quantify the conversion of the OC utilized in this work, the final state of the OC is considered to be Fe. During the experiment, the conversion of the OC X is defined by the following equation:

$$X = \frac{m_{ox} - m_t}{m_{ox} - m_{red}}, \tag{2}$$

where the m_{ox} is the weight of the iron ore OC in fully oxidized state, namely the weight fed into the reactor; m_{red} is the OC weight in theoretical reduced state; m_t is the OC weight at time t during reduction process, which should be calculated by the lattice oxygen offered by the OC.

During the reduction process, the CO_2 is generated due to the participation of lattice oxygen provided by the OC. The molar flow of the CO_2 n_{CO_2} can be determined as follows:

$$n_{CO_2} = \int_0^t n_{red} x_{CO_2} dt, \tag{3}$$

where n_{red} is the total gas molar flow rate during reduction period, which can be calculated by the volume flow rate of gaseous products F_{red}; x_{CO_2} represents the gas concentration as measured by a mass spectrometer.

The inert gas does not participate in the reduction of OC. Therefore, the mass balance of the inert gas, Ar, can be used to determine the total flow rate of the flue gas. The use of an inert gas as an internal standard for mass balance calculations is indeed the universal and fundamental method in the CLC field for determining absolute gas flow rates and solid conversion. This method is well-established in the literature. Based on that, the F_{red} can be calculated using the balance of Ar in the reaction system, namely:

$$F_{red} = \frac{F_{Ar}}{1 - x_{CO} - x_{CO_2}}, \tag{4}$$

where F_{Ar} is the flow rate of the carrier gas Ar in the CO gas; x_{CO} is the concentration of CO in the gaseous products.

The increase in mass from CO to CO_2 is equivalent to the decrease in mass of OC during the reduction period. This can be quantified by the following calculation:

$$m_{ox} - m_t = \frac{1000 n_{CO_2}}{22.4} \times \frac{16}{44} = \frac{2000 n_{CO_2}}{125.4}, \tag{5}$$

3. CLC Performance of the Iron Ore as Oxygen Carrier

In a practical CLC process, numerous factors influence the OC reactivity, including reaction temperature, fuel characteristics, reaction time, etc. In this part, the effects of two key factors, namely temperature and the concentration of gaseous fuel, are investigated. Blank experiments were conducted under identical conditions (flow rate, temperature, particle size) using a bed of inert quartz sand with the same size and volume as the OC. This directly measures the system's intrinsic response time, including the mixing in the preheater, reactor void volume, and the MS response. The response time has been treated when the OC is used as bed material for the CLC experiment.

3.1. *Effects of Reaction Temperature*

In a common CLC unit, the reaction temperature is a key factor affecting the reactivity of the OC. When the CLC reaction temperature is changed from 750 °C to 900 °C, the inlet CO concentration is kept at 10%. Each redox experiment is conducted for five cycles. The gaseous products during the OC reduction period are shown in Figure 3.

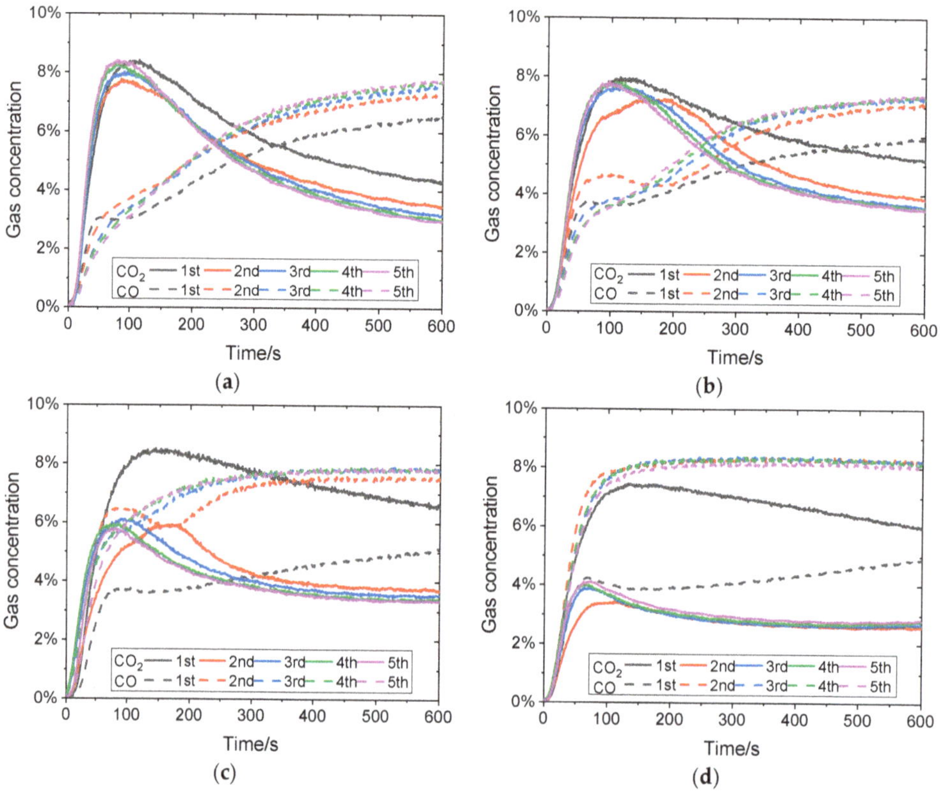

Figure 3. Gas concentrations at different reaction temperatures: (**a**) 750 °C; (**b**) 800 °C; (**c**) 850 °C; and (**d**) 900 °C.

Once the CO is introduced into the reactor, it reacts with the OC, resulting in the production of CO_2. However, due to the short residence time of CO, it cannot be totally converted, thus both CO and CO_2 are measured at the outlet of the reactor. As illustrated in Figure 3a, the CO_2 concentration reaches its maximum of 8.35% in the first cycle at 118 s, then it decreases over time. The peak CO_2 concentration in the second cycle is 7.59%, which is lower than that in the first cycle. Additionally, the time to reach it is also slightly earlier. In the subsequent three cycles, the peak CO_2 concentration exceeds that of the second cycle, and the time to reach these peaks shortens. The fifth cycle records the highest CO_2 concentration, reaching 8.37% at approximately 89 s. When the reduction period ends, the CO_2 concentrations in these cycles decrease to 4.31%, 3.46%, 3.17%, 3.01%, and 2.98%, respectively. Notably, the rate of decrease in the first cycle is significantly slower, indicating a higher conversion of CO compared to the subsequent cycles, where the rates are similar. The CO concentrations increase over the reaction time within these cycles. A higher concentration of CO_2 in the first cycle suggests greater CO conversion. Its concentration is noticeably lower than that in other cycles.

At 800 °C, the CO_2 and CO concentrations exhibit a pattern similar to that at 750 °C. However, there are notable differences: the time required to reach the maximum is longer, and the subsequent decline is more gradual, resulting in a broader peak of CO_2 concentration as illustrated in Figure 3b. At the final seconds of the reduction periods, the CO_2 concentrations in these cycles decrease to 5.19%, 3.85%, 3.59%, 3.49%, and 3.50%, respectively, which are higher than those at 750 °C. This suggests an increased formation of CO_2

while CO is being consumed. Moreover, it is found that the final CO_2 concentrations in these cycles become relatively stable with the increase in cycle number.

The gas concentrations at 850 °C markedly differ from those observed previously. In the initial redox cycle, CO_2 achieves its maximum of 8.51% at approximately 144 s. Subsequently, it decreases to 6.53% at 600 s with a moderate decline rate. Notably, the CO_2 concentration remains elevated throughout the reduction period, demonstrating a high conversion of CO and OC. The CO concentration first increases and then decreases to a low value of 3.58% almost at the same time as the maximum CO_2 concentration. As the reaction progresses, the CO concentration experiences a slight uptick. The CO_2 concentrations in the following cycles are much lower than those in the first cycle. The maximum values of them are 5.92%, 6.10%, 5.92%, and 5.77%, respectively. With the exception of the second cycle, the durations to reach these peaks are quite short, near 80~90 s. In these cycles, the CO concentration keeps a relatively similar variation trend and level, demonstrating a similar conversion of CO. However, discernible deviations from the first cycle suggest potential sintering of the OC.

At 900 °C, the CO_2 concentration variation is similar to that at 850 °C during the initial cycle. However, its maximum value is lower than that at 850 °C. In the following cycles, the CO_2 concentrations remain relatively low, with maximum values of 3.42%, 3.87%, 3.99%, and 4.11%, respectively. On the contrary, the CO concentrations are significantly higher. This phenomenon confirms the previous speculation about the potential sintering of oxygen carriers at elevated temperatures. In these cases, the OC conversion is calculated and shown in Figure 4.

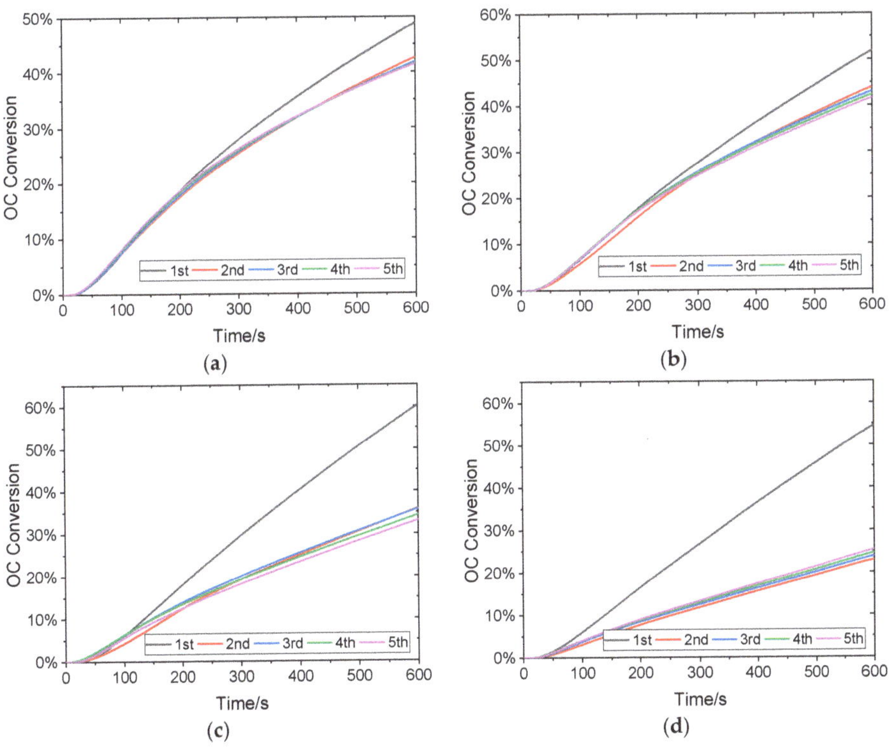

Figure 4. OC conversions at different reaction temperatures: (**a**) 750 °C; (**b**) 800 °C; (**c**) 850 °C; and (**d**) 900 °C.

The results indicate that the OC conversion during the first cycle surpasses that of subsequent cycles. The OC conversions in the first cycles reach 48.89%, 51.85%, 60.19%, and 54.55%, respectively, when the temperature increases from 750 to 900 °C. Clearly, a temperature increase is advantageous for OC conversion, bolstering the reduction reaction between the OC and CO. However, due to the potential of high-temperature sintering, the OC conversion decreases at 900 °C. When the Fe-containing phase in OC is totally reduced to Fe_3O_4 or FeO, the expected OC conversion should be 3.33% or 10%, respectively. However, under these conditions, the observed OC conversions are much higher, suggesting that the Fe-containing phase in the OC has been converted into the Fe phase from FeO.

At 750 °C, the OC conversion maintains a relatively high value, ranging from 41.47% to 42.63% in the following cycles. Although these values are lower than those in the first cycle, they persist within a stable range, demonstrating a consistent reactivity of the OC. When the reaction temperature is 800 °C, the OC conversions in the following cycles are not only higher than those at 750 °C but also approximately 8~10% lower than that of the first cycle. Though the OC conversion at 850 °C in the first cycle is the highest, it decreases to a low level near 35% in the following cycles. A similar trend is observed at a temperature of 900 °C. These findings align with previous results of the gas concentrations, revealing the potential severe deterioration of OC reactivity in CLC. This is due to the deep reduction of the Fe-based OC. Typically, the suitable reduced phase of OC is considered to be Fe_3O_4. Upon further reduction to a lower valence state, there would be a huge amount of reaction heat generated during the oxidation period, which consequently leads to the sintering of OC.

However, there is an interesting phenomenon that has been observed in the cases at 900 °C. Though the OC has potential to be sintered, the OC conversions from the second to fifth cycles are 22.95%, 23.83%, 24.58%, and 25.33%, respectively, indicating an upward trend—a pattern that is unique to this temperature, as the conversions decrease at other temperatures. That is suggestive of a self-activation behavior of the sintered OC. In typical pilot-scale CLC units, the large OC circulation rate prevents reduction to FeO or Fe phases. However, under certain extreme sintering conditions, a gradual improvement in the reactivity of the OC can be seen. The implications of this phenomenon warrant further investigation.

3.2. Effects of CO Concentration

With the reaction temperature maintained at 800 °C, the inlet CO concentration is changed to 10%, 15%, and 20%. The results under the condition of 10% CO concentration have been given in the previous part. Under other conditions, the gas concentrations are shown in Figure 5.

As illustrated in Figure 3b, the peak CO_2 concentration in the first redox cycle is approximately 7.98%. There is a noticeable trend of increasing peak CO_2 concentrations in subsequent cycles when the inlet CO concentration rises. Consistent with prior observations, the concentration in the first cycle is the highest. With an inlet CO concentration of 15%, the observed CO_2 concentration surpasses that with 10% inlet CO in the corresponding cycle. In this instance, the CO_2 concentration peaks at 11.66% at 110 s during the first cycle and then decreases to 6.46% at 600 s. This concentration remains higher than those in the following cycles. By the end of these cycles, the CO_2 concentration decreased to 4.29%, 4.18%, 4.13%, and 4.20%, respectively. Correspondingly, the CO concentration in the first cycle is the lowest.

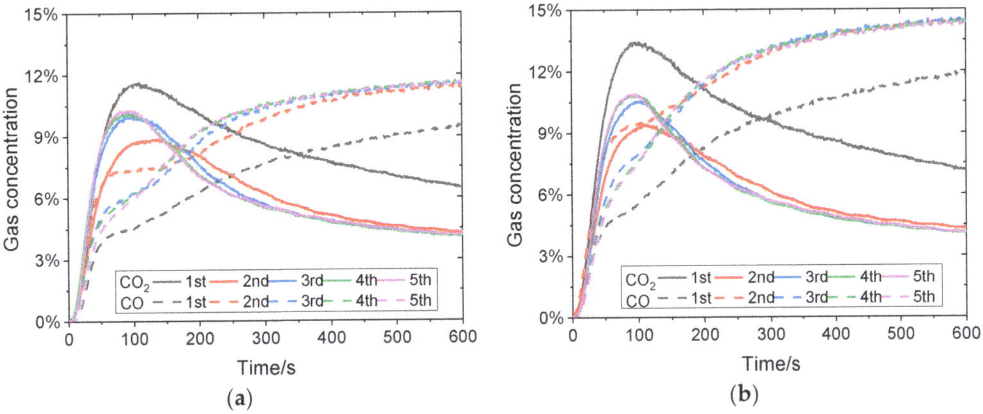

Figure 5. Gas concentrations with different inlet CO concentrations: (**a**) 15%; (**b**) 20%.

When the inlet CO concentration reaches 20%, the peak CO_2 concentration across five cycles is 13.44%, 9.45%, 10.58%, 10.89%, and 10.90%, respectively. The CO_2 concentration at the final second of the first cycle is 7.20%, which is higher than that in the previous condition. In these cases, the sintering of OC becomes unavoidable due to the high conversion in the first cycle.

Under these conditions, the OC conversions are shown in Figure 6.

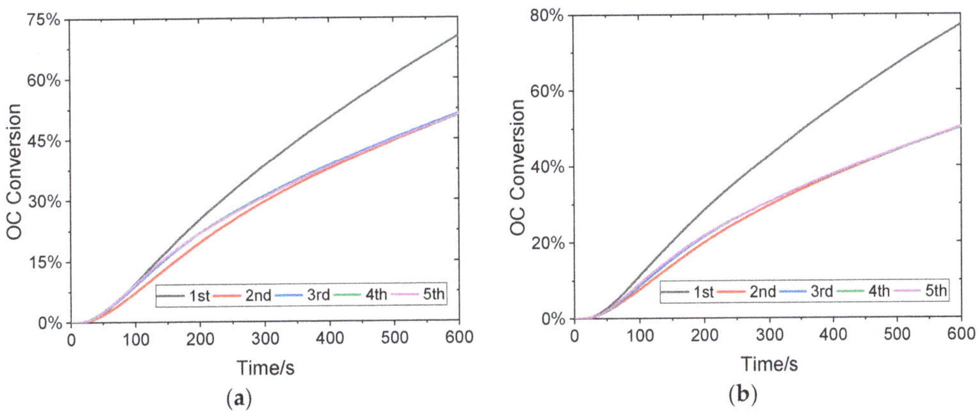

Figure 6. OC conversions with different inlet CO concentrations: (**a**) 15%; (**b**) 20%.

With a 15% inlet CO concentration, the OC conversion in the first cycle achieves 70.39%, significantly surpassing the 51.85% observed with an inlet CO concentration. This suggests that an increased CO concentration positively impacts its conversion, attributed to a higher partial pressure in the atmosphere. In the second cycle, the conversion rate of OC, which is directly proportional to the CO_2 concentration, is relatively slow, which leads to a low conversion of 50.95% at the end of the reduction period. From the third to fifth cycles, the OC conversions have similar variation trends, with no obvious difference among them. At approximately the initial 200 s, the conversion rate in these three cycles is higher than that in the second cycle, yielding a higher conversion. However, the conversion rate in these cycles becomes slower, and the OC conversions gradually cope with that in the second cycle. The final OC conversions in these cycles are similar, which are 51.35%, 50.93%, and 51.04%, respectively.

While 20% CO is employed as fuel, the OC conversion in the first cycle peaks at 77.22%, making it the highest performance under the test conditions in this work. This indicates that over half of the Fe-containing phase has been converted to Fe. During the regeneration period, the sintering of the OC would become severe, leading to the deterioration of the OC reactivity. In the following four cycles, the OC conversions reach 50.42%, 50.16%, 50.17%, and 50.36%, respectively, which are slightly lower than those under the condition with 15% CO as fuel.

3.3. Self-Activation of Sintered OC

As previously noted, a potential self-activation process of the sintered OC is noticed in multiple redox cycles. To further verify this capability from the perspective of OC conversion, an experiment under more rigorous conditions is conducted. The reaction temperature is set as 900 °C, and the inlet CO concentration is set as 20%. The CO_2 concentrations and the OC conversions in fifteen redox cycles are shown in Figure 7.

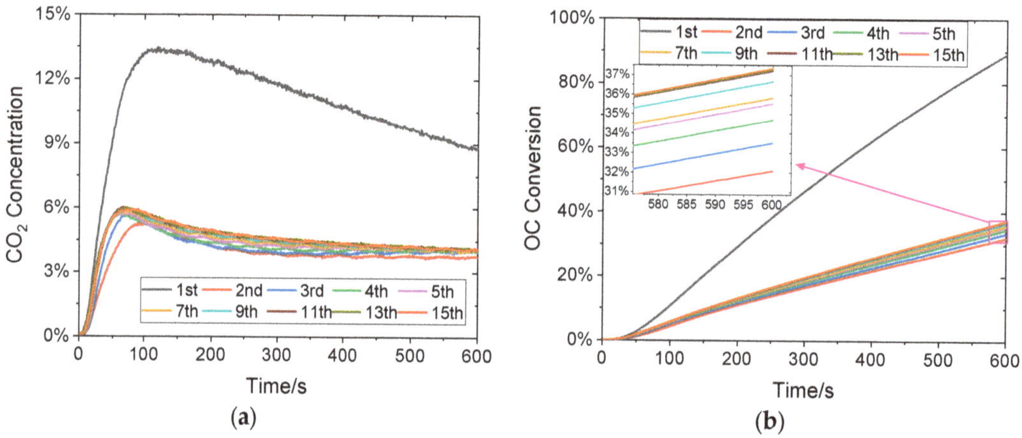

Figure 7. CO_2 concentrations and the OC conversions in 15 redox cycles: (**a**) CO_2 concentrations; (**b**) OC conversions.

The first reduction cycle is substantially different from the rest, having a much higher CO_2 concentration (maximum of 13.47% at 119 s) and a broader reaction profile. This suggests a rapid initial reduction step followed by fast kinetics towards completion. This distinct initial behavior is attributed to the activated phase of the fresh OC, potentially involving the initial reduction of surface oxides and structural changes upon its first exposure to the reducing atmosphere. Similar to the previous result, the sintering occurs after the first redox cycle. From the second cycle onwards, the CO_2 concentration follows a stable pattern, characterized by significantly lower concentrations and a more defined peak shape. Peak CO_2 concentrations for cycles 2–15 were generally in the range from 5.27% to 6.02%. This stabilization signifies the establishment of a consistent reduction mechanism and surface chemistry after the sintering in the first cycle. The consistent peak height and shape across cycles 2 through 15 suggest that the oxygen release capacity and reaction kinetics are relatively constant during this period. Sintering of the OC leads to several detrimental effects. It decreases the active surface area available for gas–solid reactions between CO and OC. In addition, high temperatures cause pore collapse or widening, restricting CO diffusion into the particle interior and reducing access to internal oxygen storage sites.

Similarly, a pronounced distinction in OC conversion is immediately evident between the first reduction cycle and all subsequent cycles. The initial reduction period exhibits markedly higher reaction rates and achieves substantially greater conversion at any given time point compared to later cycles. For instance, at 600 s, OC conversion reached 89.30% in the first cycle. In stark contrast, the second cycle achieves only 32.07% conversion. The conversion trajectories for cycles 2, 3, 4, 5, 7, 9, 11, 13, and 15 largely overlap or show only marginal improvements in later cycles compared to cycle 2, but with minor variations. It is seen that the final OC conversion gradually increased from 32.07% in the second cycle to 37.33% in the fifteenth cycle. The consistency in the conversion from the second to fifteenth cycles suggests that the most severe sintering occurs during the initial cycle, after which the particle morphology stabilizes, albeit in a state with inherently lower reactivity than the fresh material. The very minor, gradual increase in conversion observed in later cycles could potentially indicate limited crack formation or surface restructuring over extended cycling, slightly mitigating the effects of sintering. This agrees with the previous hypothesis about the self-activation ability of sintered OC, although it does not reverse the dominant effect of potential sintering. The high content of inert SiO_2 and Al_2O_3 (see Table 2) likely acts as a natural support matrix, providing mechanical integrity and potentially limiting the extent of sintering, which may contribute to this observed self-activation ability. The long-term cycling (50–100 cycles) under non-destructive conditions and mechanical attrition testing is the gold standard for applied OC development. Future work will focus on it to verify the long-term behavior of OC under moderate conditions.

It is noteworthy that the present study primarily focuses on the macroscopic redox performance and reaction kinetics of the iron ore OC. The primary aim was to evaluate the feasibility and bulk reactivity of this low-cost, natural ore under CLC conditions. The conclusions regarding sintering and self-activation are therefore inferred from the consistent and reproducible trends observed in the gas conversion and OC conversion data presented in Sections 3.1–3.3. This study focuses on CO as a representative fuel to probe the intrinsic reduction kinetics and stability of the ore; its performance with other fuels such as H_2 and CH_4 is a critical subject for future research.

4. Kinetic Analysis

Kinetic parameters are important for the OC performance evaluation. There are different methods to determine the reaction kinetics, such as isoconversional analysis, model fitting method, and so on. Isoconversional analysis is a powerful tool for revealing multi-mechanism processes. Isoconversional and model-fitting methods are complementary. The former is excellent for estimating activation energy without model assumptions, especially for complex processes, while the latter is valuable for directly testing specific mechanistic hypotheses. The primary goal of our kinetic analysis was to identify the most probable reaction mechanism that describes the reduction process of this specific ore. Also, the apparent activation energy was obtained to compare with the literature. Therefore, the kinetic parameters of this OC in CLC are calculated based on the model fitting method, which is widely used in this field.

4.1. Model Fitting Method

In the gas–solid reaction of OC reduction, the conversion is always related to the reaction temperature. As demonstrated earlier, the sintering occurs in the first redox cycle when the temperature exceeds 850 °C and 10% CO is used as the reduction gas. Therefore, the results in the first cycles at temperatures from 750 °C to 850 °C are employed to determine the kinetic parameters. Typically, the conversion rate can be expressed as a function of temperature and OC conversion during the reduction period, as described below:

$$\frac{\mathrm{d}X}{\mathrm{d}t} = k(T)f(X), \tag{6}$$

where $f(X)$ is the model function during the reduction period; $k(T)$ is the Arrhenius rate constant, which is given as:

$$k(T) = A\exp\left(-\frac{E}{RT}\right), \tag{7}$$

where A and E are two kinetic parameters, namely the pre-exponential factor and apparent activation energy; R denotes the gas constant.

The integral of the model function $f(X)$ is $g(X)$. While the reaction is conducted isothermally, the $g(X)$ is expressed as:

$$g(X) = \int_0^X \frac{\mathrm{d}X}{f(X)} = k(T)t, \tag{8}$$

The common kinetic models are given in Table 3.

Table 3. Some common kinetic models for $f(X)$ and $g(X)$ [40].

Kinetic Model	$f(X)$	$g(X)$
Kinetics-order models	$f_1 = 1 - X$ $f_2 = (1 - X)^2$ $f_3 = (1 - X)^3$	$g_1 = -\ln(1 - X)$ $g_2 = (1 - X)^{-1} - 1$ $g_3 = [(1-X)^{-2} - 1]/2$
Diffusion model	$f_4 = 1/(2X)$ $f_5 = 1/[-\ln(1 - X)]$ $f_6 = (3/2)(1 - X)^{2/3}[1 - (1 - X)^{1/3}]$	$g_4 = X^2$ $g_5 = (1 - X)\ln(1 - X) + X$ $g_6 = [1 - (1 - X)^{1/3}]^2$
Contraction model	$f_7 = 2(1 - X)^{1/2}$ $f_8 = 3(1 - X)^{2/3}$	$g_7 = 1 - (1 - X)^{1/2}$ $g_8 = 1 - (1 - X)^{1/3}$
Nucleation model	$f_9 = 2(1 - X)[-\ln(1 - X)]^{1/2}$ $f_{10} = 3(1 - X)[-\ln(1 - X)]^{2/3}$	$g_9 = [-\ln(1 - X)]^{1/2}$ $g_{10} = [-\ln(1 - X)]^{1/3}$

There are two different methods for fitting the models using these functions. While $f(X)$ is employed, it necessitates the calculation of $\mathrm{d}X/\mathrm{d}t$, which is highly sensitive to the noises and fluctuations. To avoid these effects, this work employs the integral function $g(X)$ to fit the conversion. It is seen from Equation (8) that when the model is suitable to describe the reduction of OC, the $g(X)$ should exhibit a linear relationship with reaction time.

When the functions in Table 3 are used, the conversion variation against time is used to obtain the $g(X)$. Then the g(X) is fitted using a linear function via the linear least squares method. The data samples for $g(X)$ and the linear fitting results are shown in Figure 8.

It is seen that certain kinetic models exhibit an approximate linear relationship with reaction time. This suggests that these models are suitable for describing the gas–solid reaction. For choosing the reliable models, the coefficient of determination R^2 for each fitting result is given in Table 4.

Table 4. R^2 of each linear fitting result.

T (°C)	g_1	g_2	g_3	g_4	g_5	g_6	g_7	g_8	g_9	g_{10}
750	0.9933	0.9714	0.9714	0.9071	0.8879	0.8655	0.9949	0.9951	0.8289	0.3878
800	0.9835	0.9461	0.9461	0.8776	0.8545	0.828	0.9924	0.9902	0.8843	0.5104
850	0.9580	0.8913	0.8913	0.8433	0.8129	0.7774	0.9785	0.9727	0.9453	0.6787

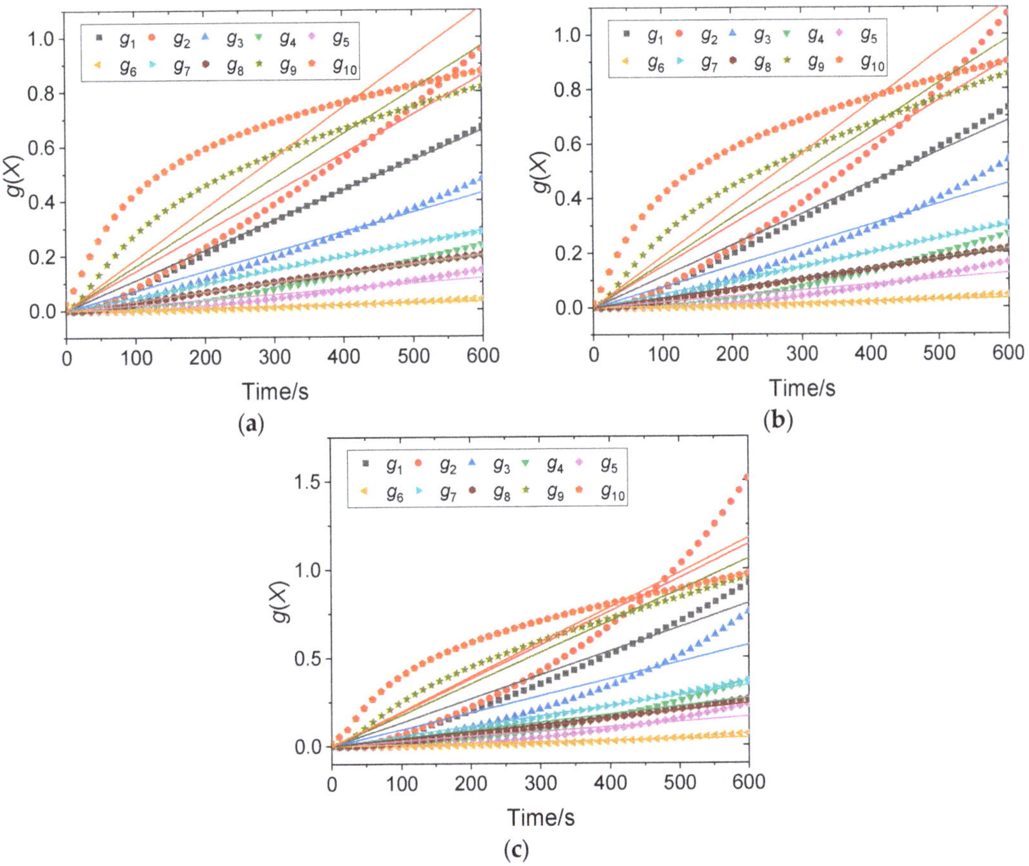

Figure 8. Examination and linear fitting of different models: (**a**) 750 °C; (**b**) 800 °C; and (**c**) 850 °C.

From these quantitative results, the g_1, g_7, and g_8 are regarded as possible kinetic models for the reduction of this iron ore OC due to a high R^2 over 95% at different temperatures.

4.2. Determination of Kinetic Parameters

Once the kinetic models have been determined, their kinetic parameters can be derived from the fitting results. The slope of the fitted line is directly related to the $k(T)$ as illustrated in Equation (8). By taking the logarithm of Equation (7), it can be obtained as follows:

$$\ln k(T) = \ln A - \frac{E}{R} \cdot \frac{1}{T} \quad (9)$$

Using the fitting results in Figure 8, both $k(T)$ and T are known. By substituting these values into Equation (9), a linear relationship between $k(T)$ and T can be obtained. The slope and intercept of this relationship can be employed to calculate the pre-exponential factor and apparent activation energy. Given three possible kinetic models, the relationship between ln$k(T)$ and $1/T$ is shown in Figure 9.

According to the relationship illustrated in Figure 9, the apparent activation energies of the OC reduction in these models can be obtained as listed in Table 5.

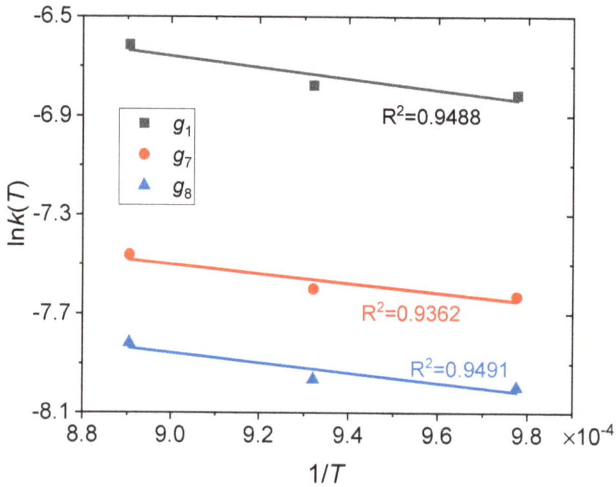

Figure 9. Relationship between ln$k(T)$ and $1/T$ for different models.

Table 5. Calculated apparent activation energy in these models.

Model	g_1	g_7	g_8
Apparent activation energy	19.13 kJ mol^{-1}	15.93 kJ mol^{-1}	16.94 kJ mol^{-1}

As shown in Table 5, the apparent activation energy of this OC for CLC varies in the range from 15.93 kJ mol^{-1} to 19.13 kJ mol^{-1}, which is much lower than previous results as listed in Table 1. This OC is suitable for CLC; however, in the practical CLC system, a large equivalence ratio of OC to the fuel is needed to avoid a deep reduction of the OC, which causes the sintering of OC. The low apparent activation energies determined for the reduction reaction, coupled with the best fit to the contracting volume model and purely surface–chemical controlled model, provide insight into the likely reaction mechanism. The values obtained here are remarkably low, suggesting the chemical reaction at the interface is very facile. The fit to the models g_7 and g_8 indicates that the overall reaction rate is significantly influenced by the diffusion of oxygen ions through the product layer (FeO/Fe) or the inward movement of the reaction interface. The large particle size used in this study (20–40 mesh) further promotes such diffusional limitations.

5. Conclusions

In contrast to the high-Fe_2O_3 ores prevalent in the literature, this study demonstrates that an iron ore with a lower active content, but high inert support, can exhibit favorable reactivity and a unique self-activation phenomenon post-sintering. The redox performance of a natural iron ore used as OC in CLC was evaluated under different temperatures and CO concentrations. The reaction kinetics were analyzed based on a model-fitting method. From the results, key conclusions are obtained as follows:

(1) Increasing the reaction temperature from 750 °C to 850 °C enhances the initial OC conversion, whereas operating at 900 °C instigates swift potential sintering, as indicated by a pronounced drop in CO_2 yield.
(2) High CO concentration not only expedites the initial conversion but also intensifies sintering. This reaffirms the imperative for meticulous control over reduction depth.
(3) A notable self-activation effect was noted in sintered OC during prolonged cycling at 900 °C with 20% CO, which led to a gradual increase in conversion from 32.07% to 37.33% over 15 cycles.
(4) Kinetic modeling confirms that the reduction of this OC adheres to a contracting volume model or first-order kinetic mechanism, characterized by an apparent activation energy of 15.93–19.13 kJ mol^{-1}.

This work provides a fundamental evaluation of a natural iron ore's performance under CO reduction, underscoring its potential as an economical OC for processes utilizing syngas or similar CO-rich fuels. To leverage these benefits in pilot-scale or full-scale CLC units, it is crucial to maintain a high solids inventory and a moderate oxygen-to-fuel ratio. This approach limits the OC reduction to Fe_3O_4/FeO, thereby preventing activity loss due to sintering. However, its performance with other fuels like CH_4 and H_2, crucial for broader industrial applications, remains to be investigated in future work.

Author Contributions: Conceptualization, C.T. and X.W.; methodology, Q.L. (Qingmei Li) and Q.L. (Qian Liang); software, Q.L. (Qian Liang); validation, Q.L. (Qingmei Li), M.H. and X.S.; formal analysis, Q.L. (Qian Liang) and M.H.; investigation, C.T., M.W. and X.W.; resources, C.T. and H.W.; data curation, M.W. and H.W.; writing—original draft preparation, C.T. and X.W.; writing—review and editing, Q.L. (Qian Liang); supervision, Q.L. (Qian Liang) and X.W.; funding acquisition, X.W. All authors have read and agreed to the published version of the manuscript.

Funding: This research was funded by the National Natural Science Foundation of China (52406141), the Natural Science Foundation of Jiangsu Province (BK20240547), and the Natural Science Foundation of the Jiangsu Higher Education Institutions of China (23KJB470016).

Data Availability Statement: The original contributions presented in this study are included in the article. Further inquiries can be directed to the corresponding author.

Conflicts of Interest: Authors Congxi Tao, Qian Liang, Qingmei Li, Minghai He, Xuhui Shen, Hao Wang, Ming Wang were employed by the company China Resources Building Materials Technology Holdings Co., Ltd. The remaining author declare that the research was conducted in the absence of any commercial or financial relationships that could be construed as a potential conflict of interest. The company had no role in the design of the study; in the collection, analyses, or interpretation of data; in the writing of the manuscript, or in the decision to publish the results.

References

1. Seneviratne, S.I.; Rogelj, J.; Séférian, R.; Wartenburger, R.; Allen, M.R.; Cain, M.; Millar, R.J.; Ebi, K.L.; Ellis, N.; Hoegh-Guldberg, O.; et al. The many possible climates from the Paris Agreement's aim of 1.5 °C warming. *Nature* **2018**, *558*, 41–49. [CrossRef] [PubMed]
2. Boon, J. Sorption-enhanced reactions as enablers for CO_2 capture and utilisation. *Curr. Opin. Chem. Eng.* **2023**, *40*, 100919. [CrossRef]
3. Chen, L.; Msigwa, G.; Yang, M.; Osman, A.I.; Fawzy, S.; Rooney, D.W.; Yap, P.S. Strategies to achieve a carbon neutral society: A review. *Environ. Chem. Lett.* **2022**, *20*, 2277–2310. [CrossRef]
4. Matthews, H.D.; Wynes, S. Current global efforts are insufficient to limit warming to 1.5 °C. *Science* **2022**, *376*, 1404–1409. [CrossRef]
5. Tollefson, J. Earth breaches 1.5 °C climate limit for the first time: What does it mean? *Nature* **2025**, *637*, 769–770. [CrossRef]
6. International Energy Agency. Global Energy Review 2025. Available online: www.iea.org/reports/global-energy-review-2025/co2-emissions (accessed on 8 August 2025).

7. Hong, F.; Qi, Y.; Yang, Z.; Yu, L.; Guang, X.; Diao, J.; Sun, B.; Liu, H. Recent advances of CO_2 hydrogenation to methanol. *DeCarbon* **2025**, *8*, 100111. [CrossRef]
8. Wang, F.; Guan, D.; Wu, C.; Zhang, X.; Wang, G. Numerical study on induction heating enhanced methanol steam reforming for hydrogen production. *DeCarbon* **2024**, *6*, 100075. [CrossRef]
9. Dou, B.; Song, Y.; Liu, Y.; Feng, C. High temperature CO_2 capture using calcium oxide sorbent in a fixed-bed reactor. *J. Hazard. Mater.* **2010**, *183*, 759–765. [CrossRef]
10. Dubey, A.; Arora, A. Advancements in carbon capture technologies: A review. *J. Clean. Prod.* **2022**, *373*, 133932. [CrossRef]
11. Wilberforce, T.; Olabi, A.; Sayed, E.T.; Elsaid, K.; Abdelkareem, M.A. Progress in carbon capture technologies. *Sci. Total Environ.* **2021**, *761*, 143203. [CrossRef]
12. Osman, A.I.; Hefny, M.; Abdel Maksoud, M.I.A.; Elgaragy, A.M.; Rooney, D.W. Recent advances in carbon capture storage and utilisation technologies: A review. *Environ. Chem. Lett.* **2021**, *19*, 797–849. [CrossRef]
13. Wang, X.; Wang, X.; Kong, Z.; Shao, Y.; Jin, B. Auto-thermal operation and optimization of coal-fueled separated gasification chemical looping combustion in a pilot-scale unit. *Chem. Eng. J.* **2020**, *383*, 123159. [CrossRef]
14. Adánez, J.; Abad, A. Chemical-looping combustion: Status and research needs. *Proc. Combust. Inst.* **2019**, *37*, 4303–4317. [CrossRef]
15. Adánez, J.; Abad, A.; Mendiara, T.; Gayán, P.; Diego, L.F.; García-Labiano, F. Chemical looping combustion of solid fuels. *Prog. Energy Combust. Sci.* **2018**, *65*, 6–66. [CrossRef]
16. Wang, X.; Shao, Y.; Jin, B. Spatiotemporal statistical characteristics of multiphase flow behaviors in fuel reactor for separated-gasification chemical looping combustion of solid fuel. *Chem. Eng. J.* **2021**, *412*, 128575. [CrossRef]
17. Lyngfelt, A. 11—Oxygen carriers for chemical-looping combustion. In *Calcium and Chemical Looping Technology for Power Generation and Carbon Dioxide (CO_2) Capture*; Woodhead Publishing: Cambridge, UK, 2015; pp. 221–254. [CrossRef]
18. Mendiara, T.; Pérez, R.; Abad, A.; Diego, L.F.; García-Labiano, F.; Gayán, P.; Adánez, J. Low-Cost Fe-Based Oxygen Carrier Materials for the iG-CLC Process with Coal. 1. *Ind. Eng. Chem. Res.* **2012**, *51*, 16216–16229. [CrossRef]
19. Guan, Y.; Liu, Y.; Lin, X.; Wang, B.; Lyu, Q. Research Progress and Perspectives of Solid Fuels Chemical Looping Reaction with Fe-Based Oxygen Carriers. *Energy Fuels* **2022**, *36*, 13956–13984. [CrossRef]
20. De Vos, Y.; Jacobs, M.; Van Der Voort, P.; Driessche, I.V.; Snijkers, F.; Verberckmoes, A. Optimization of spray dried attrition-resistant iron based oxygen carriers for chemical looping reforming. *Chem. Eng. J.* **2017**, *309*, 824–839. [CrossRef]
21. Abad, A.; Gayán, P.; García-Labiano, F.; Diego, L.F.; Izquierdo, M.T.; Mendiara, T.; Adánez, J. Relevance of oxygen carrier properties on the design of a chemical looping combustion unit with gaseous fuels. *Greenh. Gases Sci. Technol.* **2023**, *13*, 125–143. [CrossRef]
22. Cheng, D.; Yong, Q.; Zhao, Y.; Gong, B.; Zhang, J. Study on the Interaction of the Fe-Based Oxygen Carrier with Ashes. *Energy Fuels* **2020**, *34*, 9796–9809. [CrossRef]
23. Liu, X.; Bu, H.; Zou, G.; Wu, X.; Zheng, C.; Ma, J.; Zhao, H. Industrial-Scale Preparation of Biore Oxygen Carriers for Chemical Looping Combustion via a Hydroforming Method. *Energy Fuels* **2024**, *38*, 6156–6170. [CrossRef]
24. Li, Z.; Wang, Y.; Li, W.; Wei, G.; Liu, X.; Lin, S.; Li, J.; Li, D.; Meng, Q.; Nie, L.; et al. Demonstration of a 5-MW_{th} Chemical Looping Combustion Unit Fueled by Lignite. *Engineering* **2025**, *in press*. [CrossRef]
25. Wang, X.; Shao, Y.; Jin, B. Thermodynamic evaluation and modelling of an auto-thermal hybrid system of chemical looping combustion and air separation for power generation coupling with CO_2 cycles. *Energy* **2021**, *236*, 121431. [CrossRef]
26. Ströhle, J.; Orth, M.; Epple, B. Design and operation of a 1MW_{th} chemical looping plant. *Appl. Energy* **2014**, *113*, 1490–1495. [CrossRef]
27. Leion, H.; Mattisson, T.; Lyngfelt, A. Use of Ores and Industrial Products as Oxygen Carriers in Chemical-Looping Combustion. *Energy Fuels* **2009**, *23*, 2307–2315. [CrossRef]
28. De Vos, Y.; Jacobs, M.; Van Der Voort, P.; Driessche, I.V.; Snijkers, F.; Verberckmoes, A. Development of Stable Oxygen Carrier Materials for Chemical Looping Processes—A Review. *Catalysts* **2020**, *10*, 926. [CrossRef]
29. Liu, F.; Liu, J.; Yang, Y. Review on the Theoretical Understanding of Oxygen Carrier Development for Chemical-Looping Technologies. *Energy Fuels* **2022**, *36*, 9373–9384. [CrossRef]
30. Adánez, J.; Cuadrat, A.; Abad, A.; Gayán, P.; Diego, L.F.; García-Labiano, F. Ilmenite Activation during Consecutive Redox Cycles in Chemical-Looping Combustion. *Energy Fuels* **2010**, *24*, 1402–1413. [CrossRef]
31. Ksepko, E.; Babinski, P.; Evdou, A.; Nalbandian, L. Studies on the redox reaction kinetics of selected, naturally occurring oxygen carrier. *J. Therm. Anal. Calorim.* **2016**, *124*, 137–150. [CrossRef]
32. Lu, X.; Rahman, R.A.; Lu, D.Y.; Ridha, F.N.; Duchesne, M.A.; Tan, Y.; Hughes, R.W. Pressurized chemical looping combustion with CO: Reduction reactivity and oxygen-transport capacity of ilmenite ore. *Appl. Energy* **2016**, *184*, 132–139. [CrossRef]
33. Abad, A.; Adánez, J.; Cuadrat, A.; García-Labiano, F.; Gayán, P.; Diego, L.F. Kinetics of redox reactions of ilmenite for chemical-looping combustion. *Chem. Eng. Sci.* **2011**, *66*, 689–702. [CrossRef]

34. Su, M.; Ma, J.; Tian, X.; Zhao, H. Reduction kinetics of hematite as oxygen carrier in chemical looping combustion. *Fuel Process. Technol.* **2017**, *155*, 160–167. [CrossRef]
35. Su, M.; Zhao, H.; Tian, X.; Zhang, P.; Du, B.; Liu, Z. Intrinsic Reduction Kinetics Investigation on a Hematite Oxygen Carrier by CO in Chemical Looping Combustion. *Energy Fuels* **2017**, *31*, 3010–3018. [CrossRef]
36. Ksepko, E.; Babiński, P.; Nalbandian, L. The redox reaction kinetics of Sinai ore for chemical looping combustion applications. *Appl. Energy* **2017**, *190*, 1258–1274. [CrossRef]
37. Khakpoor, N.; Mostafavi, E.; Mahinpey, N.; Siegler, H.D.H. Oxygen transport capacity and kinetic study of ilmenite ores for methane chemical-looping combustion. *Energy* **2019**, *169*, 329–337. [CrossRef]
38. Mendiara, T.; Abad, A.; de Diego, L.; García-Labiano, F.; Gayán, P.; Adánez, J. Reduction and oxidation kinetics of Tierga iron ore for Chemical Looping Combustion with diverse fuels. *Chem. Eng. J.* **2019**, *359*, 37–46. [CrossRef]
39. Wang, X.; Liu, H.; Jin, B.; Zhao, J.; Sun, C.; Snape, C.E. Experimental Evaluation of a Chinese Sulfur-Containing Lean Iron Ore as the Oxygen Carrier for Chemical-Looping Combustion. *Ind. Eng. Chem. Res.* **2016**, *55*, 428–435. [CrossRef]
40. Wang, X.; Wang, X.; Shao, Y.; Jin, Z.; Jin, B. Reactivity of a Chinese lean iron ore as oxygen carrier: Kinetics and characterization. *Thermochim. Acta* **2018**, *670*, 114–122. [CrossRef]
41. Lyngfelt, A. Chemical Looping Combustion: Status and Development Challenges. *Energy Fuels* **2020**, *34*, 9077–9093. [CrossRef]
42. Wang, X.; Wang, X.; Zhang, S.; Kong, Z.; Jin, Z.; Shao, Y.; Jin, B. Test Operation of a Separated-Gasification Chemical Looping Combustion System for Coal. *Energy Fuels* **2018**, *32*, 11411–11420. [CrossRef]

processes

Article

Experimental Adsorption Study of Pure CH_4 and CO_2 on Organic-Rich Shales from the Cesar-Ranchería Basin, Colombia

Olga Patricia Ortiz Cancino * and Nicolas Santos Santos

Petroleum Engineering School, Universidad Industrial de Santander, Bucaramanga 680002, Santander, Colombia; nicolas@uis.edu.co
* Correspondence: oportizc@uis.edu.co

Abstract

This study presents a comprehensive experimental evaluation of pure methane (CH_4) and carbon dioxide (CO_2) adsorption on organic-rich shale samples from the Cesar-Ranchería Basin, Colombia. Adsorption isotherms were measured at two temperatures (50 °C and 80 °C) and up to 3 MPa using a manometric method. The data were fitted using the Langmuir model. The samples exhibit high total organic carbon (TOC) contents, ranging from 33.44% to 69.63%, but surprisingly low BET surface areas (1–7 m^2/g), an uncommon combination in shale systems. Despite these low surface areas, the samples showed notable adsorption capacities, particularly for CO_2, which consistently outperformed CH_4 across all conditions. Maximum CO_2 adsorption capacities reached up to 1.6 mol/kg, while CH_4 values peaked at 0.49 mol/kg. The Langmuir parameters reflect a stronger affinity and greater capacity for CO_2, supporting its potential role in enhanced gas recovery and CO_2 sequestration. These findings contribute to understanding gas–shale interactions in thermally immature and highly organic-rich formations and highlight the importance of mineralogy and organic matter characteristics beyond surface area alone. This work provides novel insights into the adsorption behavior of Colombian shales and serves as a valuable reference for future gas in-place estimations and shale reservoir evaluations in similar geological contexts.

Keywords: gas adsorption; adsorption isotherms; Cesar-Ranchería Basin; gas storage

1. Introduction

The increasing global energy demand, paired with the urgent need to reduce greenhouse gas emissions, has driven a significant shift toward cleaner and more sustainable energy sources. In this context, natural gas—mainly composed of methane (CH_4)—has gained prominence as a transitional fuel due to its lower carbon intensity compared to coal and oil [1]. Unconventional gas reservoirs, particularly organic-rich shales, have become a strategic target for energy development in numerous countries, including Colombia, where sedimentary basins such as Cesar-Ranchería offer promising exploration opportunities [2,3].

Shale gas is predominantly stored through adsorption mechanisms on organic matter (kerogen) and fine-grained mineral surfaces [4]. This mode of storage becomes increasingly relevant in formations characterized by high total organic carbon (TOC) content, nanoporous structures, and low permeability [5,6]. A critical aspect in evaluating these systems is understanding how various geochemical and mineralogical features—such as specific surface area, maturity, and mineral composition—influence gas adsorption behavior under reservoir-like conditions [7].

 https://doi.org/10.3390/pr13072108

Experimental studies have shown that both methane and carbon dioxide exhibit strong affinities for organic-rich matrices, with CO_2 typically displaying higher adsorption capacities than CH_4 due to its quadrupolar nature and greater polarizability [8,9]. This makes shale formations relevant not only for gas production but also as potential candidates for CO_2 geological storage [10,11]. However, adsorption performance can vary significantly depending on the textural and compositional heterogeneity of the shale, as well as the prevailing pressure and temperature conditions [12].

A growing body of research has addressed adsorption behavior in shales from North America and parts of Europe and Asia [13–15]. Nonetheless, detailed experimental data for South American shales—particularly those from Colombia—remain scarce. Previous investigations in the Cesar-Ranchería Basin have primarily focused on hydrocarbon potential, thermal maturity, and TOC distributions [16–18], with limited attention given to the adsorption behavior of greenhouse gases or the implications for both energy and climate strategies.

There are ongoing debates regarding the dominant adsorption sites in shales. While some authors argue that organic matter is the primary adsorption phase [19], others highlight the significant role of clay minerals, particularly illite and kaolinite, to gas uptake [20]. Additionally, the influence of thermal maturity on adsorption remains debated; higher maturity often correlates with increased microporosity but may also reduce adsorption due to changes in kerogen structure [21,22].

This study addresses these gaps by presenting an experimental investigation of pure CH_4 and CO_2 adsorption on shallow shale samples from the Cesar-Ranchería Basin. These samples are characterized by unusually high TOC values (33.44–69.63 wt%) combined with low specific surface areas (1–7 m^2/g), a combination rarely reported in the literature. By integrating adsorption isotherms with geochemical and mineralogical analyses, this work aims to clarify how these contrasting properties affect gas storage. The findings contribute to a deeper understanding of gas–shale interactions in tropical basins and provide valuable data for future gas in-place estimations, energy exploitation, and carbon management initiatives in the region.

2. Materials and Methods

2.1. Sample Characterization

In this study, four core samples were used for obtain original experimental data for CH_4 and CO_2 adsorption, from ANH-La Loma-1 and ANH-Carretalito-1 wells in the Cesar-Ranchería Basin (CRB). The samples analyzed in this study were obtained from shallow depths ranging between 45 and 150 m.

The CRB is a foreland sedimentary basin located in northeastern Colombia, extending across the southern part of the La Guajira department and the northeastern sector of the Cesar department. It is geographically bounded to the northwest by the pre-Cretaceous rocks of the Sierra Nevada de Santa Marta, to the north by the Oca Fault, to the southeast by the pre-Cretaceous outcrops of the Serranía de Perijá, and to the southwest by the Bucaramanga-Santa Marta Fault system. The basin covers an approximate area of 11,668 km^2 and is characterized by a complex tectonic history resulting from the interaction between the South American Plate and the Caribbean Plate [23,24]. The basin's structural configuration and stratigraphic succession make it a geologically attractive region for the exploration of both conventional and unconventional hydrocarbons [17,24].

In the laboratory, samples were crushed using a mechanical grinder to obtain a fine and homogeneous powder with an estimated particle size below 250 μm. All analyses were conducted on dried samples, prepared under controlled conditions to preserve their physicochemical properties. The samples are named S1 and S2 (ANH-La Loma-1) and S3 and S4 (ANH-Carretalito-1).

2.2. Mineralogical and Geochemical Characterization

The mineralogical composition of the samples was determined by X-ray diffraction (XRD) using a Bruker D8 Advance diffractometer with Cu-Kα radiation in the DRX Laboratory of Universidad Industrial de Santander.

The dominant mineral phases are summarized in Table 1. Quartz and kaolinite were the most abundant constituents in samples S1 and S2, while sample S3 exhibited high contents of whewellite and dickite. Several secondary minerals, including pyrite, natrojarosite, and lizardite, were also detected. No XRD results were available for sample S4.

Table 1. Mineralogical composition of samples from CRB (XRD results).

Mineral	% in S1	% in S2	% in S3
Quartz	45.07	43.84	19.35
Natrojarosite	14.80		
Lizardite-1T	13.89		
Pyrite	11.39	6.34	
Gypsum	1.72		
Barite	1.60		
Spinel	1.38		
Kaolinite	9.13	45.37	
Chamosite-1MIIb	1.02		
Jarosite		2.77	
Anatase		1.68	
Whewellite			57.1
Dickite			23.55

Geochemical properties of the shale samples were evaluated to determine their hydrocarbon generation potential, thermal maturity, and organic richness. Total organic carbon (TOC) content was measured to assess the quantity of organic matter present in the rock matrix, with values ranging from 33.44 wt% to 69.63 wt%, indicating exceptionally high organic richness. Pyrolysis parameters, including S_1 (free hydrocarbons), S_2 (hydrocarbons generated through pyrolysis), and S_3 (CO_2 from kerogen oxidation), were obtained using standard Rock-Eval techniques. Derived indices such as the hydrogen index (HI), oxygen index (OI), and production index (PI) were used to characterize the type and maturity of organic matter. The samples exhibit low Tmax values (420–425 °C), suggesting immature to early mature stages of thermal evolution. The low PI values are consistent with a limited level of hydrocarbon generation, while the HI and OI values indicate a predominance of Type II and mixed Type II/III kerogen. Vitrinite reflectance (Ro%) was also measured and corroborates the Rock-Eval Tmax data, supporting the interpretation of early maturity. These geochemical characteristics confirm that the analyzed shales possess a high organic content but have undergone limited thermal alteration, which may influence their adsorption behavior and gas storage potential. The most important measurements are summarized in Table 2.

Table 2. Geochemical parameters.

Sample ID	TOC (wt%)	Tmax (°C)	S_1 (mg HC/g Rock)	S_2 (mg HC/g Rock)	S_3 (mg CO_2/g Rock)	HI (mg HC/g TOC)	OI (mg CO_2/g TOC)	PI (S1/(S1 + S2))	Ro (%)
S1	33.44	420	2.45	110.10	1.70	329.22	5.08	0.02	0.36
S2	53.67	422	2.11	155.98	2.65	290.62	4.94	0.01	0.39
S3	66.58	425	1.45	143.70	10.78	215.84	16.19	0.01	0.51
S4	69.63	424	1.05	113.49	10.44	162.99	14.99	0.01	0.53

TOC: total organic carbon; Tmax: temperature at maximum pyrolysis yield; S_1: free hydrocarbons; S_2: hydrocarbons generated through pyrolysis; S_3: CO_2 released during pyrolysis; HI: hydrogen index; OI: oxygen index; PI: production index; Ro: vitrinite reflectance. These parameters indicate high organic richness and early thermal maturity across the analyzed samples.

2.3. *Gas Adsorption Experiments*

Adsorption experiments were performed using a home-made high-pressure (HP) manometric device. Before each measurement, samples were degassed at 110 °C under vacuum for at least 12 h to ensure the complete removal of pre-adsorbed gases and moisture. The adsorption isotherms were measured at two temperatures (50 °C and 80 °C) and up to 3 MPa to simulate typical subsurface reservoir conditions in shallow to intermediate-depth formations and to explore the temperature-dependent behavior of gas uptake. These temperature values have also been widely used in similar experimental studies for evaluating physisorption in shale systems [10,13,25]. All experiments were performed in duplicate to ensure reproducibility, and the results were corrected for system void volume. Equilibrium was assumed once pressure fluctuations were less than 0.001 over 45 min, for S2 it took about 90 min. The claimed uncertainty is better than 3% [26].

The amount of gas is measured by monitoring the pressure drop of a fixed, known volume containing the adsorbent sample. For the manometric method, the measuring device consists of a dosing cell (31.5 cm^3), and an adsorption cell (16.78 cm^3) with calibrated volumes, equipped with a high-precision pressure transducer. The entire system must be maintained under constant temperature conditions. Three two-way valves allow the separation of the dosing cell from the adsorption cell. Isothermal conditions are ensured by a PID regulator (Eurotherm 3208, Eurotherm Automation, Dardilly, France), monitored through the use of two thermocouples placed on each of the cells. This equipment can operate over broad ranges of pressure (0–3.0 MPa) and temperature (303.15 K–423.15 K).

The adsorption isotherms were fitted using the Langmuir adsorption model, widely applied to describe CH_4 and CO_2 adsorption on organic-rich shales due to its simplicity and ability to provide reliable estimates of monolayer adsorption capacity under subsurface conditions. The mathematical formulation followed the three-parameter Langmuir model as described by Gensterblum et al. [27].

Parameter estimation was carried out using the Solver add-in in Microsoft Excel, applying the generalized reduced gradient (GRG) nonlinear algorithm to minimize the sum of squared residuals (SSE) between experimental and modeled adsorption values. This approach ensures robust optimization for nonlinear regression with continuous variables. To assess the uncertainty of the fitted parameters, confidence intervals were determined by individually varying either n_L or p_L while keeping the other fixed at its optimum value, until a 5% increase in SSE was achieved compared to the minimum SSE of the fit. This method provides a quantitative measure of parameter sensitivity and reliability.

2.4. *Surface Area and Pore Structure Characterization by N_2 Adsorption–Desorption*

Textural properties of the shale samples were evaluated by N_2 adsorption–desorption at 77 K using a Micromeritics 3FLEX™ instrument. Before analysis, samples were degassed at 110 °C under vacuum for 12 h using a VacPrep 061 (Micromeritics, Micromeritics, Miami, FL, USA) to remove moisture and volatiles. The specific surface area was determined using the Brunauer–Emmett–Teller (BET) method. Pore size distributions were derived from the desorption branches of the isotherms using the Barrett–Joyner–Halenda (BJH) method, as recommended for slit-shaped pore analysis. N_2 adsorption–desorption isotherms were measured over a relative pressure (P/P_0) range of 0.01–0.99.

3. Results

3.1. Mineralogical Composition and Its Role in Adsorption

X-ray diffraction (XRD) analysis revealed diverse mineralogical compositions among the samples. S1 and S2 were rich in quartz (45.07% and 43.84%, respectively) and kaolinite, with minor phases such as pyrite, natrojarosite, and lizardite. S3 showed a dominant presence of whewellite (57.1%) and dickite (23.55%), while mineralogical data were not available for S4. Despite repeated attempts, including additional XRD and exploratory XRF analyses, no reliable data could be obtained for S4, likely due to sample alteration or contamination during handling or storage. No additional material from S4 is available for further analysis.

Clays such as kaolinite and dickite can enhance adsorption through increased surface area and the development of active sites [28]. However, the low BET values observed—even in samples with substantial clay content—suggest that mineralogy alone does not account for the differences in gas uptake. This indicates a complex interplay between mineral phases, thermal maturity, and the structure of the organic matrix [29,30]. The notable adsorption capacity of S4, despite the lack of mineralogical data, may be partly explained by a mineralogical composition similar to that of S3, as both samples originate from the same well. This possibility, along with potential structural features favoring gas uptake, highlights the need for further study.

3.2. Geochemical Properties and Thermal Maturity

Total organic carbon (TOC) contents ranged from 33.44 to 69.63 wt%, with particularly high values in samples S3 and S4. Nevertheless, the specific surface areas (BET) were relatively low (1–7 m^2/g), suggesting a predominance of non-porous or poorly connected organic domains, likely due to the immature thermal state. This has been reported in other studies where high TOC did not correlate with higher surface area or adsorption when the pore network within the kerogen was poorly developed [31,32].

Tmax values ranged from 420 to 425 °C and vitrinite reflectance (Ro) values from 0.36 to 0.53%, placing all samples in the immature to early stage. According to the extended HI classification, S1 to S3 are associated with Type II/III kerogen and S4 with Type III kerogen. Figure 1 shows a plot of hydrogen index (HI) versus Tmax for samples S1 to S4. The dashed horizontal lines indicate the boundaries between kerogen types according to the extended HI classification: Type II/III (200–400 mg HC/g TOC), Type III (50–200 mg HC/g TOC), and Type IV (<50 mg HC/g TOC) [33]. Vertical dashed lines indicate thermal maturity stages at 430 °C (onset of maturity) and 450 °C (transition to postmature stage). The samples are primarily in the immature to early mature window, with kerogen types ranging from Type II/III to Type III.

These results are consistent with previous studies showing that high TOC in immature shales does not necessarily imply higher adsorption capacity due to the lack of developed microporosity in the organic matrix [31,32]. This geochemical profile provides essential context for interpreting the adsorption behavior presented in the following sections.

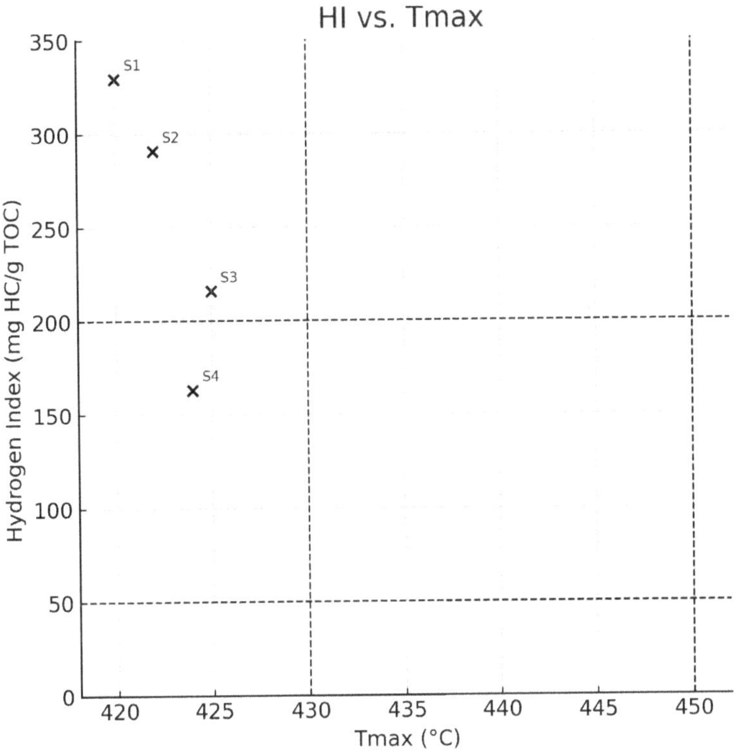

Figure 1. Tmax–HI graph. Source: own elaboration.

3.3. Adsorption Behavior of CH_4 and CO_2

The adsorption isotherms of pure methane (CH_4) and carbon dioxide (CO_2) were measured at 50 and 80 °C for all samples. In every case, CO_2 displayed higher adsorption capacity than CH_4 across the entire pressure range (0–3 MPa), consistent with its higher molecular polarizability and stronger affinity for organic matter [25,34].

The highest overall adsorption capacity was observed in sample S4, followed by S3, S2, and S1. This ranking only partially correlates with BET surface area, which ranged from 1 m^2/g (S2) to 7 m^2/g (S4). The fact that this trend does not strictly follow TOC values indicates that additional factors—such as pore accessibility and mineral structure—also play a key role in controlling gas uptake [5,29,31].

The complete set of adsorption data is presented in Figures 2–5. Each figure displays the experimental data points together with the fitted Langmuir model curves, allowing for a clear visual assessment of fit quality. The corresponding Langmuir parameters and coefficients of determination (R^2) are summarized in Tables 3–6, confirming the strong agreement between experimental data and model predictions and supporting the applicability of the Langmuir model to these organic-rich shale samples. The data point symbols are coded as follows: filled triangles represent CO_2 adsorption at 50 °C, filled squares represent CO_2 adsorption at 80 °C, open triangles represent CH_4 adsorption at 50 °C, open squares represent CH_4 adsorption at 80 °C, and solid lines represents Langmuir model fit. This consistent coding applies across Figures 2–5 and facilitates the direct comparison of gas behavior across temperatures and samples.

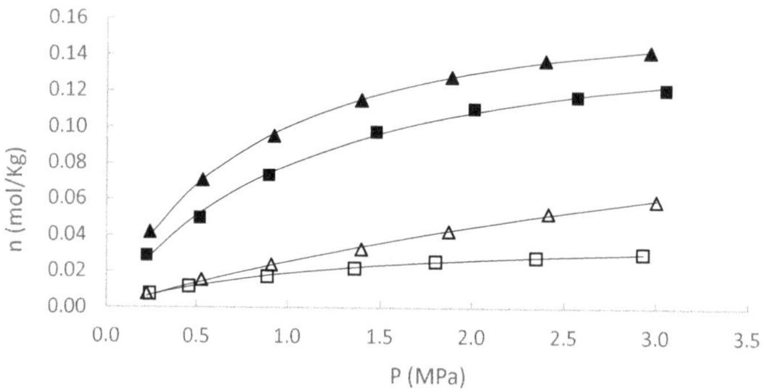

Figure 2. Adsorption isotherms and Langmuir fit for S1, ▲ CO_2 adsorption at 50 °C; ■ CO_2 adsorption at 80 °C; △ CH_4 adsorption at 50 °C; □ CH_4 adsorption at 80 °C; solid lines: Langmuir model fit.

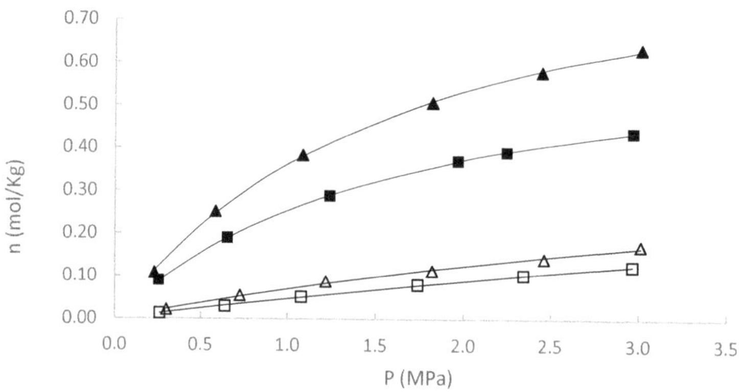

Figure 3. Adsorption isotherms and Langmuir fit for S2, ▲ CO_2 adsorption at 50 °C; ■ CO_2 adsorption at 80 °C; △ CH_4 adsorption at 50 °C; □ CH_4 adsorption at 80 °C; solid lines: Langmuir model fit.

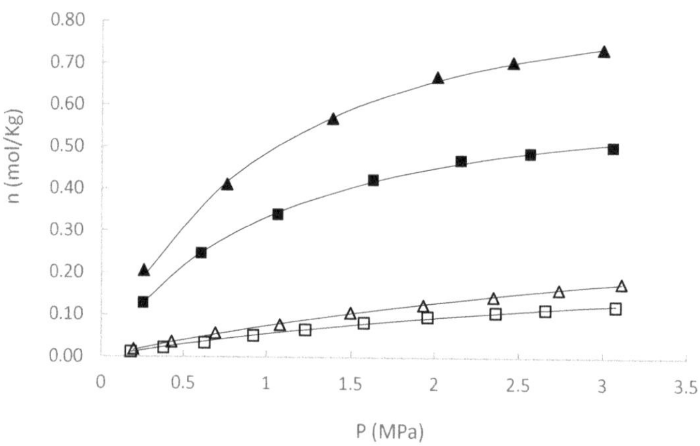

Figure 4. Adsorption isotherms and Langmuir fit for S3, ▲ CO_2 adsorption at 50 °C; ■ CO_2 adsorption at 80 °C; △ CH_4 adsorption at 50 °C; □ CH_4 adsorption at 80 °C; solid lines: Langmuir model fit.

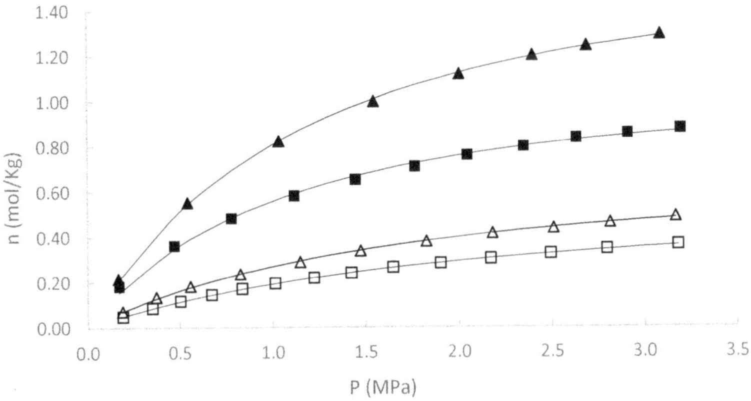

Figure 5. Adsorption isotherms and Langmuir fit for S4, ▲ CO_2 adsorption at 50 °C; ■ CO_2 adsorption at 80 °C; △ CH_4 adsorption at 50 °C; □ CH_4 adsorption at 80 °C; solid lines: Langmuir model fit.

Table 3. Langmuir model fitting parameters and 95% confidence intervals for S1.

Experiment	n_L (mol/kg) [95% CI]	p_L (MPa) [95% CI]	Δn	R^2
CH_4 at 50 °C	0.2134 [0.2149–0.2118]	7.2346 [7.3029–7.1672]	0.0005	0.9970
CH_4 at 80 °C	0.0456 [0.0454–0.0458]	1.3625 [1.3492–1.3760]	0.0002	0.9951
CO_2 at 50 °C	0.1986 [0.1990–0.1982]	0.9486 [0.9533–0.9438]	0.0004	0.9928
CO_2 at 80 °C	0.1837 [0.1845–0.1830]	1.2885 [1.3015–1.2756]	0.0007	0.9976

Table 4. Langmuir model fitting parameters and 95% confidence intervals for S2.

Experiment	n_L (mol/kg) [95% CI]	p_L (MPa) [95% CI]	Δn	R^2
CH_4 at 50 °C	0.6390 [0.6436–0.6345]	7.9081 [7.9806–7.8366]	0.0014	0.9945
CH_4 at 80 °C	0.9517 [0.9491–0.9544]	18.6652 [18.6195–18.7132]	0.0004	0.9997
CO_2 at 50 °C	1.1090 [1.1112–1.1068]	2.0109 [2.0109–2.0030]	0.0016	0.9974
CO_2 at 80 °C	0.7515 [0.7521–0.7510]	1.9051 [1.9077–1.9026]	0.0004	0.9972

Table 5. Langmuir model fitting parameters and 95% confidence intervals for S3.

Experiment	n_L (mol/kg) [95% CI]	p_L (MPa) [95% CI]	Δn	R^2
CH_4 at 50 °C	0.5756 [0.5776–0.5736]	6.5794 [6.6096–6.5493]	0.0006	0.9991
CH_4 at 80 °C	0.3553 [0.3575–0.3531]	5.3011 [5.3473–5.2560]	0.0007	0.9989
CO_2 at 50 °C	1.0938 [1.0976–1.0899]	1.2038 [1.2145–1.1932]	0.0037	0.9970
CO_2 at 80 °C	0.7467 [0.7486–0.7447]	1.2004 [1.2083–1.1927]	0.0017	0.9954

These results provide a robust basis for the detailed discussion of adsorption mechanisms and controlling factors presented in Section 4.

Table 6. Langmuir model fitting parameters and 95% confidence intervals for S4.

Experiment	n_L (mol/kg) [95% CI]	P_L (MPa) [95% CI]	Δn	R^2
CH_4 at 50 °C	0.8306 [0.8326–0.8287]	2.0418 [2.0535–2.0349]	0.0011	0.9962
CH_4 at 80 °C	0.6608 [0.6619–0.6598]	2.4018 [2.4087–2.3950]	0.0005	0.9968
CO_2 at 50 °C	1.9964 [1.9991–1.9937]	1.4314 [1.4360–1.4267]	0.0022	0.9929
CO_2 at 80 °C	1.2662 [1.2718–1.2607]	1.2371 [1.2508–1.2235]	0.0040	0.9923

Experimental data were correlated using the three-parameter Langmuir model described by Gensterblum et al. [27] and applied by Gasparik et al. [35], which is given as follows:

$$n_{ads}^{excess} = n_L \frac{p}{p + p_L}\left(1 - \frac{\rho_g(p,T)}{\rho_{ads}}\right) = n_{ads}^{absolute}\left(1 - \frac{\rho_g(p,T)}{\rho_{ads}}\right) \quad (1)$$

where n_{ads}^{excess} is the adsorbed amount of gas (mol/kg) at pressure p (MPa), p_L is the Langmuir pressure (the pressure at which half of the Langmuir volume is adsorbed), n_L is the amount adsorbed (mol/kg) when the monolayer is completely filled (Langmuir maximum capacity), ρ_g is the gas density (kg/m^3) to a p and T, and ρ_{ads} is the adsorbed phase density, which was assumed as a fixed value of 421 kg/m^3 for CH_4 [36] and 1027 kg/m^3 por CO_2 [37].

The standard deviation was calculated according to Pozo et al. [38] as follows:

$$\Delta n = \frac{1}{N}\sqrt{\sum_{1}^{N}\left(n_{exp} - n_{fit}\right)^2} \quad (2)$$

where N is the number of data points; n_{exp} is the experimental adsorption value; and n_{fit} is the calculated value at each adsorption pressure.

The parameters of the Langmuir model obtained from nonlinear regression are presented in Tables 3–6 for each sample, gas, and temperature condition. The fitting results include the maximum adsorption capacity (n_L), Langmuir pressure (P_L), and the fitting parameter (Δn). The values of Δn indicate that the Langmuir model provides a good fit to the experimental data, supporting its applicability to describe CH_4 and CO_2 adsorption on these organic-rich shales. Differences in n_L and P_L across the samples reflect variations in adsorption affinity and capacity, which are influenced by geochemical maturity, mineralogy, and surface area.

The parameter estimation was carried out using the Solver add-in in Microsoft Excel, applying the generalized reduced gradient (GRG) nonlinear algorithm to minimize the sum of squared residuals between experimental and modeled values. This method is widely used for nonlinear regression problems involving continuous variables and constraints. The consistency of the results and the low Δn values across all fits confirm the robustness of the approach.

To assess the uncertainty in the Langmuir parameters, confidence intervals were estimated by varying each parameter individually while keeping the other fixed at its optimum value. Solver (Microsoft Excel version 1808 Build 10417.20020, GRG Nonlinear) was used to identify the range of n_L and P_L values that resulted in an increase of 5% in the sum of squared errors (SSE) compared to the minimum SSE obtained in the fit. The resulting upper and lower bounds reflect the sensitivity of the parameters to the model fit and provide a measure of the reliability of the estimated values.

In addition to the standard deviation (Δn), the goodness of fit was quantitatively assessed by calculating the sum of squared errors (SSE) for each isotherm. The SSE values were found to be consistently low across all samples and conditions, supporting the adequacy of the Langmuir model in describing the adsorption behavior. Although advanced statistical tests such as the Fisher test were not applied in this study, the robustness of the results is demonstrated by the combination of low SSE values, high coefficients of determination ($R^2 > 0.99$ in most cases), and narrow confidence intervals. Together, these metrics provide strong evidence for the reliability of the fitted parameters. Values in brackets in Tables 3–6 represent the 95% confidence intervals, estimated by varying one parameter at a time while holding the others fixed, until the SSE increased by 5% relative to the minimum SSE of the best fit.

3.4. Textural Analysis by N_2 Adsorption–Desorption

N_2 adsorption–desorption analyses provided additional insight into the pore structure of the shale samples. The isotherms (Figure 6) display Type IV behavior with H3–H4 hysteresis loops, characteristic of mesoporous materials with slit-shaped or layered pore structures.

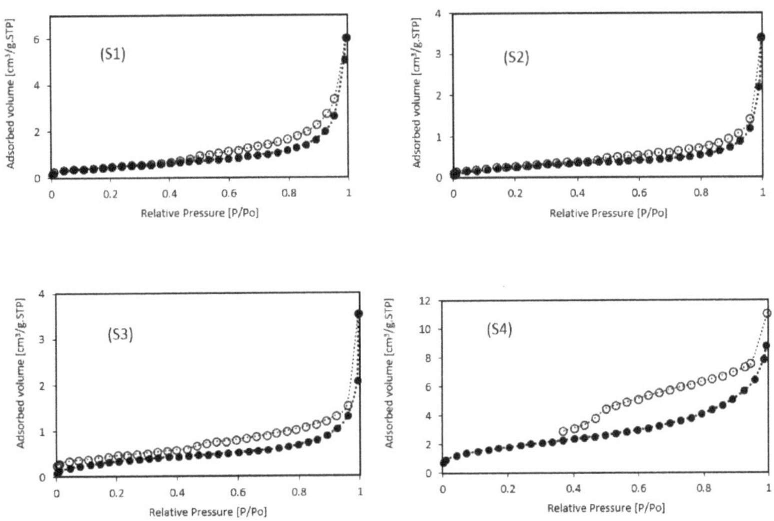

Figure 6. N_2 adsorption–desorption isotherms at 77 K for shale samples S1, S2, S3, and S4. Filled symbols represent adsorption branches, and open symbols represent desorption branches.

Among the samples, S4 exhibited the highest N_2 uptake, consistent with its larger BET surface area (7 m^2/g), while S1 and S2 showed the lowest adsorbed volumes, reflecting their more limited pore development. S3 presented intermediate characteristics, with modest N_2 adsorption compared to S4.

The pore size distribution curves (Figure 7), derived from the desorption branches using the Barrett–Joyner–Halenda (BJH) method, revealed that the samples are dominated by mesopores (2–50 nm). S4 displayed a broader mesopore distribution and greater cumulative pore volume, which may partially explain its superior CH_4 and CO_2 adsorption performance. In contrast, S1 and S2 showed narrower distributions concentrated in the 3–10 nm range, while S3 featured a more uniform distribution across the mesopore region.

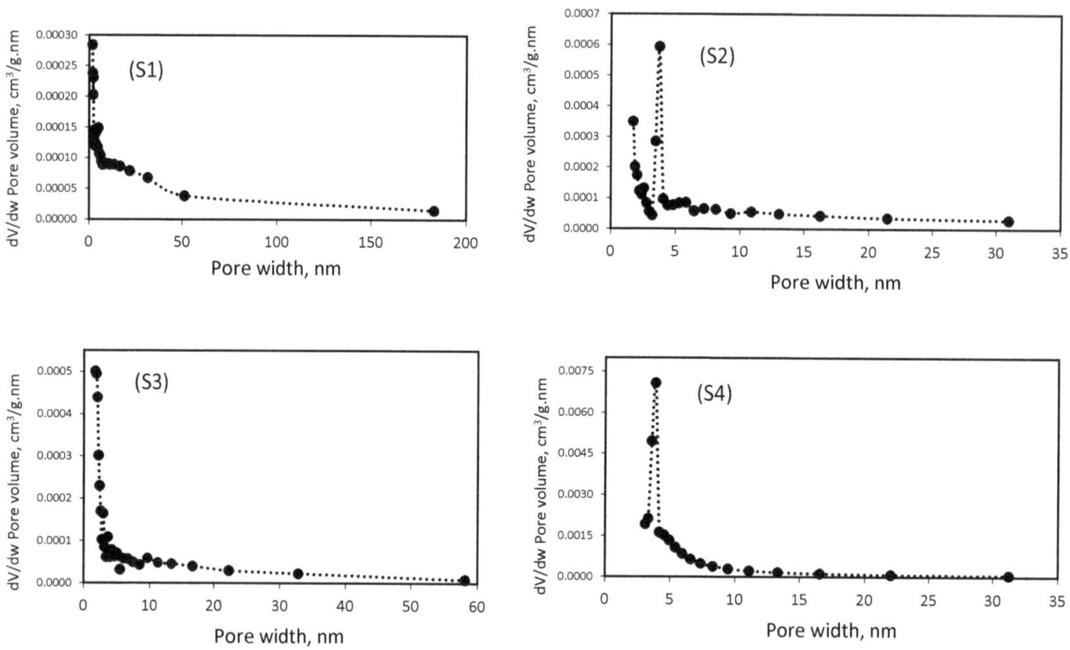

Figure 7. Pore size distribution curves of samples (S1–S4) derived from N_2 desorption using the BJH method. dV/dw differential pore volume plotted against pore width.

These textural features, combined with geochemical and mineralogical data, highlight that gas storage capacity in these thermally immature, organic-rich shales is not solely determined by organic content, but is strongly influenced by pore structure, connectivity, and mesopore development.

4. Discussion

4.1. Influence of Organic Matter and Thermal Maturity

The organic-rich nature of the studied shales, with TOC values ranging from 33.44 to 69.63 wt%, suggests high hydrocarbon generation potential. However, the low BET surface areas observed (1–7 m^2/g) confirm that organic content alone does not account for the observed adsorption capacities. This decoupling between TOC and adsorption potential has been reported in other studies, where immature or poorly porous organic matter contributes little to gas storage due to limited internal pore development [29,31].

The limited pore accessibility in these thermally immature samples (Tmax 420–425 °C; Ro 0.36–0.53%) likely arises not only from underdeveloped kerogen porosity but also from the structural characteristics of the macromolecular organic matrix. As recent studies have shown, immature shales often exhibit poorly connected pore systems and low internal surface areas despite high TOC levels [39,40]. Furthermore, adsorption may be modulated by the type and morphology of kerogen, which influence the distribution and nature of sorption sites [41,42]. Interactions between organic matter and mineral phases—such as kaolinite, dickite, and whewellite identified in our samples—may also alter surface chemistry and pore accessibility [43].

The weak correlation observed between TOC and adsorption capacity in our data, along with the mesopore distributions revealed by N_2 adsorption–desorption analyses, highlights that gas uptake in these shales reflects a complex interplay between organic structure, maturity, pore network development, and matrix composition.

4.2. Role of Mineralogy in Gas Storage

Mineralogical composition played a significant role in differentiating adsorption behavior among the samples. S1 and S2 exhibited high quartz and kaolinite contents, while S3 was dominated by whewellite and dickite. Although clays such as kaolinite and dickite are often linked to increased surface area and sorption sites, in this study their presence did not correspond to elevated BET values or superior gas uptake. This finding reinforces previous reports that mineral composition alone does not control adsorption unless accompanied by suitable textural development and pore network connectivity [30].

The presence of secondary minerals such as natrojarosite and lizardite in S1 may partially obstruct pore accessibility or modify gas interactions through their polar surfaces. Conversely, S3's mineralogy, combined with its more uniform mesopore distribution, likely contributed to its intermediate adsorption performance.

For S4, no XRD data could be obtained despite repeated measurements, including supplementary XRF analyses, likely due to sample alteration or contamination during handling or storage. No additional material was available for further analysis. Nevertheless, the superior adsorption performance of S4 and its broad mesopore distribution observed in N_2 adsorption–desorption analyses suggest that its pore structure compensated for the lack of specific mineralogical data. This highlights the importance of textural properties in gas storage, beyond mineralogy alone. Future work will focus on obtaining new core material to complete its mineralogical characterization.

4.3. Textural Properties and Pore Structure

The N_2 adsorption–desorption isotherms and pore size distribution analyses provided further insights into the textural characteristics of the studied samples. All isotherms exhibited Type IV-like behavior with H3 hysteresis loops, indicative of slit-shaped pores, limited pore connectivity, and layered structures, which are typical of immature or poorly developed organic matrices. The pore size distributions, derived using the BJH method, revealed that the dominant pores fell within the mesopore range (2–50 nm), with incremental pore volumes peaking between approximately 4 and 20 nm depending on the sample.

Among the samples, S4 displayed the highest cumulative pore volume and BET surface area, which aligns with its superior adsorption performance for both CH_4 and CO_2. S3 exhibited an intermediate cumulative pore volume with a relatively uniform mesopore distribution, contributing to its noteworthy adsorption despite its low BET value. In contrast, S1 and S2 presented the lowest cumulative pore volumes and narrower pore size distributions, consistent with their lower gas uptake.

These results emphasize that pore accessibility and connectivity—rather than TOC or mineralogy alone—are key determinants of adsorption behavior in these shales. The relatively low microporosity observed across all samples reflects their immature thermal state, as significant micropore development is typically associated with advanced kerogen transformation [33].

Overall, the textural evidence supports the interpretation that gas storage capacity in these organic-rich shales is governed by a complex interplay of organic matter content, mineralogy, and pore structure, with the latter exerting a decisive influence. These findings underscore the relevance of detailed pore structure characterization in evaluating shale gas potential and CO_2 sequestration capacity.

4.4. Performance and Langmuir Interpretation

The experimental isotherms and Langmuir modeling confirm that CO_2 exhibits higher adsorption capacity and affinity than CH_4 across all samples and temperatures, which is consistent with its higher polarizability and quadrupole moment [25,27]. The Langmuir parameters show that CO_2 not only achieves greater maximum adsorption (n_L) but also requires lower equilibrium pressure (P_L) to reach saturation, indicating more efficient sorption dynamics and stronger interactions with the shale matrix.

The increase in adsorption capacity at the lower temperature (50 °C) compared to 80 °C reinforces the dominance of physisorption mechanisms [13]. Sample S4 consistently exhibited the highest adsorption capacities for both gases, in agreement with its larger BET surface area, higher cumulative pore volume, and elevated TOC. In contrast, S3 showed remarkable adsorption performance despite its modest BET value, likely due to the favorable contribution of its unique mineralogy (e.g., whewellite and dickite), relatively uniform mesopore distribution, and intermediate maturity level. This reinforces the view that adsorption capacity is governed by the complex interplay of textural, mineralogical, and geochemical properties rather than any single parameter alone [32,33].

Although several isotherm models—including Sips, Toth, Freundlich, and Dubinin–Radushkevich—could be applied to describe gas adsorption on shale, the Langmuir model was selected in this study due to its simplicity, theoretical clarity, and well-documented applicability to organic-rich formations. Previous studies have shown that the Langmuir model provides sufficiently accurate fits for CH_4 and CO_2 adsorption in shale systems, particularly under the moderate pressure conditions (0–3 MPa) typical of early-stage reservoir evaluations [10,13,27,30]. The Langmuir parameters—maximum capacity (n_L) and Langmuir pressure (P_L)—are physically meaningful and facilitate inter-sample and inter-study comparisons, adding practical value for resource assessment.

The consistently low fitting residuals (Δn), narrow confidence intervals, and high coefficients of determination ($R^2 > 0.99$ in most cases) across all samples support the robustness of the Langmuir fits. While alternative models might better capture surface heterogeneity, the Langmuir model proved adequate for the objectives of this study. Future research could explore multi-model fitting and advanced statistical analyses to provide deeper insights into the heterogeneous adsorption behavior of these complex shale matrices.

4.5. *Implications for Shale Gas Potential*

The findings of this study highlight that high TOC alone is not a reliable predictor of adsorption capacity in organic-rich shales. Instead, shale gas potential should be assessed through a holistic understanding of organic maturity, mineralogical framework, and pore structure. The results show that even samples with exceptionally high TOC can exhibit limited adsorption if the pore network is poorly developed, as seen in the thermally immature samples of this study. Samples S1 to S3, dominated by Type II/III kerogen, may achieve enhanced adsorption performance with further maturation, as thermal evolution is known to promote the development of microporosity and improve pore connectivity within the organic matrix. In contrast, S4, despite its immature state, demonstrates significant adsorption capacity, likely reflecting a combination of favorable pore structure and mineralogical characteristics.

These observations are consistent with recent studies emphasizing the dynamic interplay between geochemical, mineralogical, and textural factors in controlling gas sorption behavior in shale formations [25,34,35]. The pore size distributions, mesoporosity, and cumulative pore volumes measured in this study further illustrate the critical role of textural properties in determining adsorption potential—underscoring that gas storage capacity results from the combined effects of composition and structure rather than individual parameters.

It is important to note that this study is based on four samples from shallow depths within two wells and therefore may not fully capture the heterogeneity of the Cesar-Ranchería Basin. Future investigations incorporating a broader spatial, stratigraphic, and maturity range will be essential to validate these findings and support more comprehensive resource assessments for this basin. Such studies should also explore the evolution of pore systems with increasing thermal maturity and diagenetic alteration to better predict shale gas potential over geological timescales.

5. Conclusions

This study presents a comprehensive experimental assessment of CH_4 and CO_2 adsorption on thermally immature, organic-rich shale samples from the Cesar-Ranchería Basin in Colombia. Based on the integrated analysis of adsorption behavior, geochemical properties, mineralogy, and pore structure, the following conclusions can be drawn.

CO_2 consistently exhibited higher adsorption capacity and stronger affinity than CH_4 under all tested conditions, due to its greater polarizability and quadrupole moment. This confirms the potential of these shales for CO_2 geological storage in addition to gas production.

The adsorption capacity varied among the samples, with S4 showing the highest performance. This result is partly explained by its larger BET surface area and cumulative pore volume. However, the correlation with TOC was weak, underscoring that gas uptake is governed by a complex interplay of organic content, thermal maturity, mineralogy, and especially pore structure.

Despite their exceptionally high TOC contents (33.44–69.63 wt%), the samples exhibited low BET surface areas (1–7 m^2/g) and limited microporosity, consistent with their immature to early mature thermal state (Tmax: 420–425 °C; Ro: 0.36–0.53%). This highlights that TOC alone is not a reliable predictor of adsorption capacity in thermally immature shales.

Pore structure, as revealed by N_2 adsorption–desorption and BJH analysis, played a decisive role in adsorption performance. S4 showed the broadest mesopore distribution and highest cumulative pore volume, aligning with its superior gas uptake. In contrast, S1 and S2 exhibited narrower mesopore ranges and lower pore volumes, which limited their adsorption.

The Langmuir model provided an excellent fit to the experimental data, as demonstrated by low residuals, narrow confidence intervals for fitted parameters, and high coefficients of determination ($R^2 > 0.99$ in most cases). This supports its continued use for early-stage evaluations, although future studies could benefit from multi-model comparisons to account for surface heterogeneity.

Overall, these findings emphasize the need for integrated evaluation of shale gas potential, considering not just TOC or mineral content but also textural characteristics and thermal maturity. Such holistic assessments are essential for accurately estimating gas in-place and designing CO_2 sequestration strategies in the Cesar-Ranchería Basin and similar geological settings.

Author Contributions: Conceptualization, O.P.O.C. and N.S.S.; data curation, O.P.O.C.; formal analysis, O.P.O.C.; investigation, O.P.O.C.; methodology, O.P.O.C.; project administration, N.S.S.; resources, N.S.S.; supervision, O.P.O.C.; validation, O.P.O.C. and N.S.S.; visualization, O.P.O.C.; writing—original draft, O.P.O.C.; writing—review and editing, O.P.O.C. and N.S.S. All authors have read and agreed to the published version of the manuscript.

Funding: This research was funded under Contract No. FP44842-454-2017 (UIS-MINCIENCIAS), derived from Agreement 730/327 of 2016 (ANH-MINCIENCIAS).

Data Availability Statement: The data presented in this study are available on request from the corresponding author; the data are not publicly available due to privacy restrictions.

Acknowledgments: Authors acknowledge the GIT of Universidad Industrial de Santander for their support of this work.

Conflicts of Interest: The authors declare no conflicts of interest.

References

1. U.S. Energy Information Administration. *International Energy Outlook 2023*; U.S. EIA: Washington, DC, USA, 2023.
2. Aguilera, R.F. Assessing shale gas potential globally: Energy return on investment (EROI) and other challenges. *Energy Policy* **2014**, *64*, 701–710.
3. Agencia Nacional de Hidrocarburos (ANH). *Potencial de Recursos en Yacimientos No Convencionales de Colombia*; ANH: Bogotá, Colombia, 2020.
4. Chalmers, G.R.L.; Bustin, R.M. The organic matter distribution and gas adsorption capacity of the Lower Cretaceous shales of northeastern British Columbia, Canada. *Int. J. Coal Geol.* **2007**, *70*, 223–239. [CrossRef]
5. Ross, D.J.K.; Bustin, R.M. Characterizing the shale gas resource potential of Devonian–Mississippian shales in western Canada. *AAPG Bull.* **2008**, *92*, 87–125. [CrossRef]
6. Curtis, M.E.; Sondergeld, C.H.; Rai, C.S. Structural characterization of gas shales on the nanoscale. *Microsc. Microanal.* **2012**, *18*, 694–695.
7. Wang, C.; Liu, H.H. A review on gas adsorption in shale nanopores: Modeling advances and challenges. *J. Nat. Gas Sci. Eng.* **2021**, *94*, 104127.
8. Gasparik, M.; Ghanizadeh, A.; Gensterblum, Y.; Krooss, B.M.; Littke, R. CO_2 and CH_4 adsorption capacity of European shales. *Energy Procedia* **2013**, *37*, 5243–5252.
9. Busch, A.; Alles, S.; Gensterblum, Y.; Prinz, D.; Dewhurst, D.N.; Raven, M.D.; Stanjek, H.; Krooss, B.M. Carbon dioxide storage potential of shales. *Int. J. Greenh. Gas Control* **2008**, *2*, 297–308. [CrossRef]
10. Heller, R.; Vermylen, J.; Zoback, M. Experimental investigation of gas adsorption in shales. *Fuel* **2014**, *115*, 378–393.
11. Rexer, T.F.T.; Benham, M.J.; Aplin, A.C.; Thomas, K.M. Methane adsorption on shale under simulated geological temperature and pressure conditions. *Energy Fuels* **2013**, *27*, 3099–3109. [CrossRef]
12. Zhang, T.; Ellis, G.S.; Ruppel, S.C.; Milliken, K.; Yang, R. Effect of organic matter properties on methane sorption in shales. *Int. J. Coal Geol.* **2012**, *103*, 89–98.
13. Pan, Z.; Connell, L.D. Comparative adsorption of CH_4, CO_2 and N_2 on organic–inorganic shales. *J. Nat. Gas Sci. Eng.* **2015**, *27*, 82–95.
14. Liu, X.; Zhou, Y.; Hu, Q.; Zhang, Y. Pore structure and CH_4/CO_2 adsorption capacity of high-TOC shales. *Fuel* **2018**, *226*, 125–135.
15. Wang, S.; Liu, H.; Yang, R.; Zhang, L.; Li, Y. Adsorption of CO_2 and CH_4 in gas shale nanopores: Insights from molecular simulations. *Langmuir* **2014**, *30*, 13960–13968.
16. Ríos-Reyes, C.A.; Buitrago, H.; Pulido, J.C. Thermal maturity and organic geochemistry of shales from the Cesar-Ranchería Basin. *J. South Am. Earth Sci.* **2017**, *76*, 283–297.
17. Buitrago, H.; Arango, M.; De La Parra, M. Evaluation of unconventional resource potential in northern Colombia. In Proceedings of the AAPG International Conference and Exhibition, Cartagena, Colombia, 4–17 September 2014.
18. Agencia Nacional de Hidrocarburos (ANH). *Atlas Geológico de Colombia, Volumen IV: Cuenca Cesar-Ranchería*; ANH: Bogotá, Colombia, 2015.
19. Tang, X.; Ripepi, N.; Liu, B.; Liu, Z. Methane adsorption in shales: Influence of organic matter vs. clay content. *Fuel* **2016**, *174*, 560–570.
20. Clarkson, C.R.; Bustin, R.M. Variation in permeability and storage capacity of shale gas reservoirs with mineralogy. *Int. J. Coal Geol.* **2000**, *38*, 33–45.
21. Bustin, R.M.; Chalmers, G.R. Gas adsorption capacity and chemical composition of thermally mature shales. *Int. J. Coal Geol.* **2010**, *82*, 235–248.
22. Guo, X.; Bustin, R.M.; Ning, Z. Impact of maturity and pore structure on methane adsorption in organic-rich shales. *Mar. Pet. Geol.* **2020**, *117*, 104399.
23. Agencia Nacional de Hidrocarburos (ANH). Ronda Colombia 2021: Cuenca Cesar-Ranchería. 2021. Available online: https://www.anh.gov.co/es/ (accessed on 20 April 2025).
24. Silva, C.; Toro, J.; Martínez, J.D. Integrated structural and basinal analysis of the Cesar–Ranchería Basin, Colombia: Implications for its tectonic history and petroleum systems. In *Petroleum Basins and Hydrocarbon Potential of the Andes of Peru and Bolivia*; AAPG Memoir 117; American Association of Petroleum Geologists: Tulsa, OK, USA, 2020; pp. 349–374.
25. Busch, A.; Gensterblum, Y.; Krooss, B.M.; Littke, R. Methane and carbon dioxide adsorption–diffusion in organic-rich shales. *Int. J. Coal Geol.* **2008**, *76*, 89–101. [CrossRef]

26. Pino, D.; Plantier, F.; Bessieres, D. Experimental determination of the adsorption isotherms in gas mixtures under extended pressure and temperature range. *J. Therm. Anal. Calorim.* **2014**, *117*, 1469–1477. [CrossRef]
27. Gensterblum, Y.; Bode, M.; de Weireld, G.; Prinz, D.; Littke, R.; Busch, A.; Krooss, B.M. European inter-laboratory comparison of high-pressure CO_2 sorption isotherms. I: Activated carbon. *Carbon* **2009**, *47*, 2958–2969. [CrossRef]
28. Curtis, M.E.; Ambrose, R.J.; Sondergeld, C.H.; Rai, C.S. Structural characterization of gas shales on the micro- and nano-scales. *J. Nat. Gas Sci. Eng.* **2012**, *8*, 28–39. [CrossRef]
29. Heller, R.; Zoback, M. Adsorption of methane and carbon dioxide on gas shale and pure mineral samples. *J. Unconv. Oil Gas Resour.* **2014**, *8*, 14–24. [CrossRef]
30. Guo, X.; Bustin, R.M. Pore structure evolution and fluid flow in thermally matured organic-rich shales. *Mar. Pet. Geol.* **2020**, *117*, 104399. [CrossRef]
31. Tang, X.; Ripepi, N.; Liu, B.; Liu, Z. Quantitative evaluation of the relationship between surface area and methane adsorption in organic-rich shales. *Fuel* **2016**, *182*, 107–114. [CrossRef]
32. Liu, X.; Zhou, Y.; Hu, Q.; Zhang, Y. Relationship between TOC and methane adsorption in shales: Experimental evidence from different formations. *Fuel* **2018**, *226*, 125–135. [CrossRef]
33. Peters, K.E.; Walters, C.C.; Moldowan, J.M. *The Biomarker Guide*, 2nd ed.; Cambridge University Press: Cambridge, UK, 2005; Volume 1.
34. Gasparik, M.; Ghanizadeh, A.; Gensterblum, Y.; Krooss, B.M.; Littke, R. Geological controls on the methane storage capacity of European shales. *Energy Procedia* **2013**, *40*, 511–520. [CrossRef]
35. Gasparik, M.; Ghanizadeh, A.; Bertier, P.; Gensterblum, Y.; Bouw, S.; Krooss, B.M. High-pressure methane sorption isotherms of black shales from the Netherlands. *Energy Fuels* **2012**, *26*, 4995–5004. [CrossRef]
36. Weniger, P.; Kalkreuth, W.; Busch, A.; Krooss, B.M. High pressure methane and carbon dioxide sorption on coal and shale samples from the Paraná Basin, Brazil. *Int. J. Coal Geol.* **2010**, *84*, 190–205. [CrossRef]
37. Gensterblum, Y.; Merkel, A.; Busch, A.; Krooss, B.M. High-pressure CH_4 and CO_2 sorption isotherms as a function of coal maturity and the influence of moisture. *Int. J. Coal Geol.* **2013**, *118*, 45–57. [CrossRef]
38. Pozo, M.; Pino, D.; Bessieres, D. Effect of thermal events on maturation and methane adsorption of Silurian black shales (Checa, Spain). *Appl. Clay Sci.* **2017**, *136*, 208–218. [CrossRef]
39. Wei, L.; Liu, D.; Pan, Z.; Chen, J.; Liu, H. Pore structure evolution of organic matter in immature shales and its effect on methane adsorption. *Fuel* **2023**, *344*, 128180. [CrossRef]
40. Zhang, H.; Song, Y.; Han, J.; Zhao, S. Thermal maturity controls on nanopore development and gas storage in lacustrine shales. *Int. J. Coal Geol.* **2022**, *253*, 104123.
41. Wang, Y.; Tang, X.; Ripepi, N. Influence of kerogen type and morphology on methane adsorption in immature shales: An experimental and modeling approach. *Energy Fuels* **2023**, *37*, 10567–10580. [CrossRef]
42. Li, Y.; Sun, Z.; Zhou, Y.; Bai, Y.; He, H. Kerogen-hosted nanopores and their control on gas adsorption in early-mature shale: Insights from FIB-SEM and gas adsorption techniques. *J. Nat. Gas Sci. Eng.* **2024**, *119*, 105571.
43. Zhou, W.; Hu, Q.; Liu, X.; Lu, S. Organic–inorganic interactions and their role in controlling pore accessibility and gas sorption in immature marine shales. *Mar. Pet. Geol.* **2022**, *143*, 105821. [CrossRef]

processes

MDPI

Article

Carbon Footprint Analysis of Chemical Production: A Case Study of Blue Hydrogen Production

Eric Y. H. Chan [1], Zulfan Adi Putra [2], Raymond R. Tan [3], Yoke Kin Wan [1] and Dominic C. Y. Foo [1,*]

[1] Department of Chemical and Environmental Engineering, University of Nottingham Malaysia, Broga Road, Semenyih 43500, Malaysia; ericchan9801@gmail.com (E.Y.H.C.); yokekin.wan@nottingham.edu.my (Y.K.W.)
[2] Exiron Baja Pratama, Jalan Raya Majalaya, Bandung 40394, Indonesia; zadiputra123@gmail.com
[3] Center for Engineering and Sustainable Development Research, De La Salle University, 2401 Taft Avenue, Manila 0922, Philippines; raymond.tan@dlsu.edu.ph
* Correspondence: dominic.foo@nottingham.edu.my

Abstract: Interest in hydrogen has grown as a means to decarbonize future energy systems. To maximize hydrogen's potential as the main energy carrier, the infrastructure for hydrogen production, distribution, and storage needs to be designed and developed at a global scale. Carbon footprint analysis is an important metric for ensuring that the environmental impact of the developed plant is kept at a minimum. However, application of conventional methods during the conceptual design stage is challenging due to lack of detailed process data coupled with the large number of potential designs to be vetted. As a result, there is a need to develop rapid screening techniques that can be used during the conceptual design stage to gauge potential carbon footprints. To address this issue, a simplified carbon footprint analysis method is proposed in this work. Two indices are introduced, i.e., "product carbon intensity" and "economic carbon intensity", to allow comprehensive analysis of the performance of design alternatives. By limiting the scope and basic economic analysis, the simplified carbon footprint analysis requires less data, and hence expedite the analysis process. The methodology is demonstrated through analysis of four design scenarios for blue hydrogen production. Among the scenarios, hydrogen production with both carbon capture and pre-reforming yielded better results based on product carbon intensity (2.43 kg CO_2/kg H_2), while design with only carbon capture performed better based on economic carbon intensity (11.25 kg CO_2/USD). Thus, high potential design scenarios were successfully identified based on the newly introduced indices.

Keywords: hydrogen economy; carbon intensity; techno-economy analysis; CO_2 reduction

1. Introduction

Energy is crucial for a high quality of life and global welfare, making the energy sector a massive market that is valued at 1.8 trillion USD in 2022 and expected to reach 3.9 trillion USD by 2032, with a compound annual growth rate (CAGR) of 8.04% [1]. As electricity demand rises due to increased population size, the supply of energy is also increasing, reaching 614 EJ in 2021 [2]. However, CO_2 emissions are also increasing, hitting 37.15 billion tons in 2022 [3]. To address climate change, the Paris Agreement aims to reduce emissions by 45% by 2030 and achieve net zero by 2050 [4]. The 2023 Conference of the Parties of the UNFCCC (COP28) highlighted slow progress and urged a faster transition away from fossil fuels [5]. A record number of 11 countries have committed to the net zero goal by increasing their production of renewable energy as a climate change mitigation strategy [6]. This commitment is reflected in the global production of renewable energy (excluding hydroelectric), which increased by 14% in 2022 to reach 40.9 EJ. Additionally, the production capacity of solar and wind energy showed an increasing trend, reaching 266 GW, with solar contributing 72% of that increase [7]. Despite these advancements, the variability of renewable energy sources, such as solar and wind power, poses challenges for grids, particularly in integrating these resources into existing infrastructure [8].

Hydrogen is a promising alternative to reduce the use of fossil fuels and plays a crucial role in addressing issues caused by climate change and the energy crisis [9]. It is an ideal option for storing surplus renewable energy [10]. As of the end of year 2021, approximately 47% of hydrogen is produced from natural gas, 27% from coal, 22% from oil as a by-product, and only around 4% from electrolysis [11]. Hydrogen fuels are categorized by color and are separated based on feed stock and production route. Grey hydrogen is produced using fossil fuels, such as oil and coal, which release greenhouse gases during production [12]. Green hydrogen is produced through electrolysis utilizing renewable energy sources and has zero carbon emissions [13,14]. Being entirely renewable, green hydrogen is a promising option for tackling climate change and achieving the energy transition. Many countries and regions have included the development of green hydrogen in their long-term energy plans, as it is a clean energy source that can be stored and transported [15,16]. The primary uses of hydrogen in future energy systems are supplying high-temperature process heat in the iron and steel industry [17]; serving as a fuel for long-range freight transport via internal combustion engines or fuel cells [18,19]; supplying domestic heating for decarbonization and commercial heating via boilers [20,21]; fueling stationary gas turbines [22], etc. However, the roundtrip efficiency of storing and utilizing energy via green hydrogen generally ranges between 30% and 40%; this efficiency reflects the energy losses that occur during the processes of electrolysis, hydrogen storage, and fuel cell conversion [23]. The roundtrip efficiency of a gaseous hydrogen energy storage system with a fuel cell as the hydrogen-to-power unit is around 42%, considering typical efficiency values of 60% for the fuel cell and 70% for the electrolyser, while neglecting the penalty for hydrogen conditioning [24]. Although this efficiency is lower than direct electricity storage solutions like batteries, green hydrogen offers unique advantages, such as long-term energy storage and flexibility in various applications.

Blue hydrogen plays a critical role in reducing emissions and transitioning to green hydrogen by providing a cleaner alternative to grey hydrogen. It involves integrating carbon capture and storage (CCS) technologies into the hydrogen production process to capture and store the CO_2 emissions generated from hydrogen production using fossil fuels [25]. Blue hydrogen can be as competitive as green hydrogen when it comes to their impact on the environment, as long as two conditions are met. The first is that the natural gas source used to produce blue hydrogen must have low associated greenhouse gas emissions. The second is that technologies with consistently high CO_2 capture rates must be employed. Current technological achievements suggests that a capture rate of 90% or above is consistently possible [26]. As of now, green hydrogen is approximately two times more expensive than blue hydrogen and four times more expensive than grey hydrogen. Up to 70% of the cost associated with green hydrogen production lies in the cost of its renewable energy input. The decrease in renewable energy cost will eventually result in green technology being an economically and environmentally attractive option. Green hydrogen is expected to break even with grey hydrogen in terms of production costs by year 2030 for locations where production of renewable energy is optimal and by year 2035 for regions with middling availability of renewable energy [27]. Black and brown hydrogen are the most environmentally damaging options, as they are produced from black coal or lignite (brown coal). It is estimated that their carbon intensity is 20–23 kg CO_2/kg H_2, which is approximately 10 times worse than that of green hydrogen; the latter is estimated to have a carbon intensity of 1–2 kg CO_2/kg H_2 [28].

Knowing the carbon footprint of the hydrogen product is vital. The carbon footprint is defined as the greenhouse gas emissions of a system as computed on a life cycle basis [29,30]. It can generally be broken down into Scope 1, 2, and 3 components consisting of direct emissions, emissions from purchased energy, and other value chain emissions, respectively. This established technique even led to the formulation of a series of standards by the International Organization for Standardization (ISO) to standardize the implementation of LCA internationally [31]. However, existing methodologies are not perfect. They are generally data-intensive, making computation of the total carbon footprint challenging. Scope 3 emissions are especially problematic, since their estimation relies on data sharing within the supply chains. Also, the standard methodology solely analyzes predefined system configurations; industrial decarbonization options need to be generated and analyzed separately. Efforts to further improve these methodologies have been made. One such method is *carbon emission pinch analysis* (CEPA), which was originally introduced for the analysis of *carbon-constrained energy planning* [32]. An extension of the technique was later developed to systematically evaluate carbon footprint reduction options [33]. Although simplified, the methodology's flexibility was demonstrated in multi-product plants [34]. Another work combining CEPA with input–output analysis was also developed and resulted in a more effective analysis method [35]. This hybrid approach was later used for the comparison of different scenarios of carbon footprint reduction in edible oil production [36].

The demand for better and more effective methods coincides with a 2022 survey where over half of the correspondents, who were from the scientific research community, considered the EIA procedure to be ineffective [37]. Heavy reliance on extensive data, often contain gaps that affect the overall quality of said data, remains a tall hurdle when conducting carbon footprint analysis. While difficulty surrounding the availability of specific original data can be circumvented through the use of secondary data, the credibility of the final analysis results is reduced as a trade-off. Additionally, complete collection of the required data in a timely fashion is uncommon due to the time-intensive nature of said process, which may lead to compounding delays to the delivery of the overall project [31]. Furthermore, boundary setting or scoping is also another significant hurdle faced when conducting a carbon footprint analysis. As more importance is placed on minimizing the environmental impact of human activities, some countries have incorporated the environmental impact assessment process into legislation, which was the impetus and origin of EIA. Thus, the ultimate test of scoping is approval from the planning committees of the project studied. An extensive scope may lead to additional expenses and delays, to a level where it may be rejected by the planning committees, while a limited scope renders the assessment too insignificant in terms of scientific and conservation value [38]. Such conflicting and subjective priorities of all the involved parties result in a difficult balancing act for those actually conducting the assessment, leading to the observed disconnect between the best practice presented in research and literature and those practices applied on the ground [39].

To contribute to the ongoing efforts to improve carbon footprint analysis, this work proposes a simple methodology for preliminary carbon footprint analysis that can be applied during the conceptual design stage. During the conceptual design stage, many details are not yet set in stone, as a large number of potential ideas are studied to determine their feasibility. It is hoped that the potential solutions may be shortlisted in order to proceed to a detailed design. A lack of detailed information results in the presence of many unknown facets of the overall project. This further exacerbates the difficulty of performing detailed carbon footprint analysis, such as through the use of conventional methods such as LCA. The lack of details leads to low-quality results, while the heavy workload required to explore potential solutions results in overall low work efficiency during the conceptual design stage. Therefore, by focusing on selected key indices, it is possible to simplify and optimize the carbon footprint analysis process, so as to provide results that are sufficient for use. In particular, any corporation has good information related to the production and financial aspects of their products and processes. Hence, analysis based on the carbon footprint of a unit product and/or unit investment would be appropriate. These results can then be used as reference to make appropriate design-related decisions during the early stages of design, where impact on the overall project schedule and economics is still minor. Doing so helps to achieve the goal of improving overall work efficiency during the conceptual design stage by assessing the many possible solutions and retaining only high potential designs.

A hydrogen production process is used as an example to demonstrate the methodology. The base case design and several design alternatives for carbon reduction are analyzed to evaluate their performance from a carbon emission perspective, along with considerations for production and economics.

The rest of this paper is organized as follows. The next section outlines the problem statement of this work, followed by a case study of steam methane reformation (SMR)-based hydrogen production. This is followed by a summary section where observations from the case study are compared and key takeaways of this work are presented. Any notable issues or observations regarding the data are also discussed in this section. Finally, the paper concludes with a section summarizing the key achievements and outlining directions for future work.

2. Problem Statement

The problem addressed in this paper can be described as follows:

- Given a state-of-the-art process for grey hydrogen production. Its carbon footprints are contributed through its energy use, in the form of steam and electricity.
- Alternative production units may be added in order to reduce the CO_2 footprint of the process; such units may include CCS (which leads to blue hydrogen production) or other units where hydrogen production may be enhanced. The newly added production units may incur additional capital and operating cost to the overall process.

The problem is to identify an economic-viable process pathway that has the lowest carbon footprint.

3. Methodology

A flowchart illustrating the proposed methodology is shown in Figure 1. For a given chemical process, a base case design is first developed. Next, a process simulation model is constructed using commercial software, where the mass and energy balances are obtained.

Heat integration is next carried out for the process, based on the established process integration principles [40]. Heat integration helps to reduce the energy consumption of the process, and hence leads to better economic performance and lower CO_2 emission. The carbon footprint and economic performance of the design alternatives are next evaluated. These steps are repeated for all design alternatives before a comparison may be made.

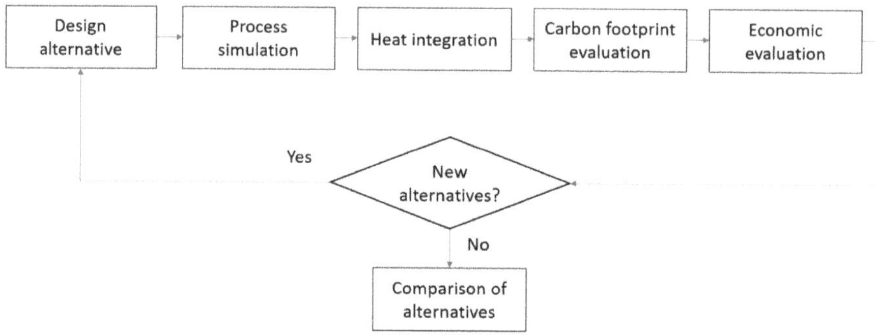

Figure 1. Flowchart for the proposed methodology.

In analyzing the carbon footprint, three types of carbon emission must be taken into consideration. Scope 1 and 2 emissions that are generated directly by the process and through its energy consumption can be determined from the simulation model (see Table 1). Similarly, the amount of CO_2 captured or reduced in each scenario can also be derived from the process simulation model. As the boundary of this study is limited to the production stage, scope 3 emissions that are not part of company value chain are excluded from the analysis [41].

Table 1. Definitions of the three types of emissions [42].

Type	Description
Scope 1	Direct greenhouse (GHG) emissions that occur from sources that are controlled or owned by an organization.
Scope 2	Indirect GHG emissions associated with the purchase of electricity, steam, heat, or cooling. Although the emissions physically occur at the facility where they are generated, they are accounted for in an organization's GHG inventory because they are a result of the organization's energy use.
Scope 3	Result from the activities of assets not owned or controlled by the organization in question. All emissions not included in an organization's scope 1 and 2 emissions are categorized as scope 3 emissions.

The carbon intensity of each design scenario was analyzed using two different indices, i.e., *product carbon intensity*, which is based on the amount of product made (expressed as kg CO_2/kg H_2), and *economic carbon intensity*, which is financial-based (expressed as kg CO_2/USD). Scope 1 emissions, E_1 (kg CO_2/h), can be derived through Equation (1):

$$E_1 = \sum_{i=1}^{n} C_{Gi} \tag{1}$$

where C_G is the amount of CO_2 generated by each process unit, which can be determined through process simulation. The summation is the total scope 1 emissions from the process.

For scope 2 emissions, E_2 (kg CO_2/h), all pieces of equipment that utilize electrical energy are taken into account through Equation (2):

$$E_2 = P_C \times C_E \tag{2}$$

where P_c is the total power consumption by the equipment (in kWh) within the process boundaries and C_E is the carbon intensity of the electricity purchased externally (in kg CO_2/kWh).

It is expected that carbon capture, utilization, and storage (CCUS) will be used for carbon footprint reduction in the near future. Through carbon capture (CC), CO_2 gas is isolated and removed from the system, which results in net reduction of the carbon footprint of the process. Therefore, the net carbon footprint of the process, C_N (kg CO_2/h), and product carbon intensity, CI_P, are given as:

$$C_N = E_1 + E_2 - E_{CCUS} \tag{3}$$

$$CI_P = \frac{C_N}{H} \tag{4}$$

where E_{CCUS} (kg CO_2/h) refers to the amount of CO_2 emissions captured, utilized, or reduced within the system, and H (kg/h) refers to the amount of hydrogen product.

From an economic standpoint, the operating cost, OP, of the process is the sum of the utility and feed costs required to maintain plant operation. Water and fuel consumption can be determined through simulation and calculation to determine water cost, C_w, and fuel cost, C_{Fuel}. They can be added to the feed cost, C_{Feed}, to yield the materials cost, C_m. The electric cost is calculated from the net power consumption, P_N (kWh), derived from equipment contributing scope 2 emissions and equipment such as turbines that are able to generate electricity while adjusting the pressure of gas streams. Equations (5)–(7) were used to calculate the operating cost of the process:

$$C_m = C_{Feed} + C_{Fuel} + C_w \tag{5}$$

$$P_N = P_C - P_G \tag{6}$$

$$OP = (P_N \times T) + C_m \tag{7}$$

where P_G (kWh) refers to any electrical power that is generated by the system and T refers to the tariff rate of the electricity to be purchased.

Additionally, the increase to both capital and operating costs need to be accounted for. The capital cost of each alternative design is calculated, while the extra capital cost (relative to the base case) for the newly added equipment is annualized through the annual worth method, based on an assumed minimum acceptable rate of return (MARR) and total design life. Doing so identifies the annual amount of money that needs to be invested into the plant on top of the annual operating cost for each design scenarios, AW. The sum of both AW and OP is the annualized cost, C_A, as shown in Equation (8), which can be used in Equation (9) to determine the economic carbon intensity, CI_E:

$$C_A = OP + AW \tag{8}$$

$$CI_E = \frac{C_N}{C_A} \tag{9}$$

Several design alternatives are being investigated, making process yield an important factor in comparing the design scenarios. The process yield provides a better understanding of the effect of the changes introduced in each design alternative compared to the base case. The process yield can be derived through Equation (10) by evaluating the theoretical amount of hydrogen recoverable, P_T (mol/h), and the actual production rate, P_A (mol/h). P_T can be derived from the feed molar flowrate and reaction formulas, both of which are pieces of information that can be acquired easily.

$$Yield(\%) = \frac{P_A}{P_T} \times 100 \tag{10}$$

Product carbon intensity represents the performance of the design from a production standpoint. A lower value is desirable to ensure low carbon emission per unit product made by the corporation. On the other hand, economic carbon intensity refers to the performance of the design from an economic perspective, where a lower value indicates lower emissions (than other alternatives) with the same amount of monetary investment. Alternatively, it also allows higher investment into other aspects of the process (such as process efficiency) while achieving the same CO_2 emissions.

4. Case Study

A case study of hydrogen production is used for illustration, which is based on the SMR process.

Figure 2 shows the simplified process flow diagram of the base case (see full PFD in Figure S1 in the Supplementary File. The process feed is natural gas received from the natural gas grid, whose composition was retrieved from published data, as shown in Table 2. As natural gas is the only raw material utilized in this process, the feed cost (C_{Feed}) in Equation (5) only accounts for the cost of the natural gas.

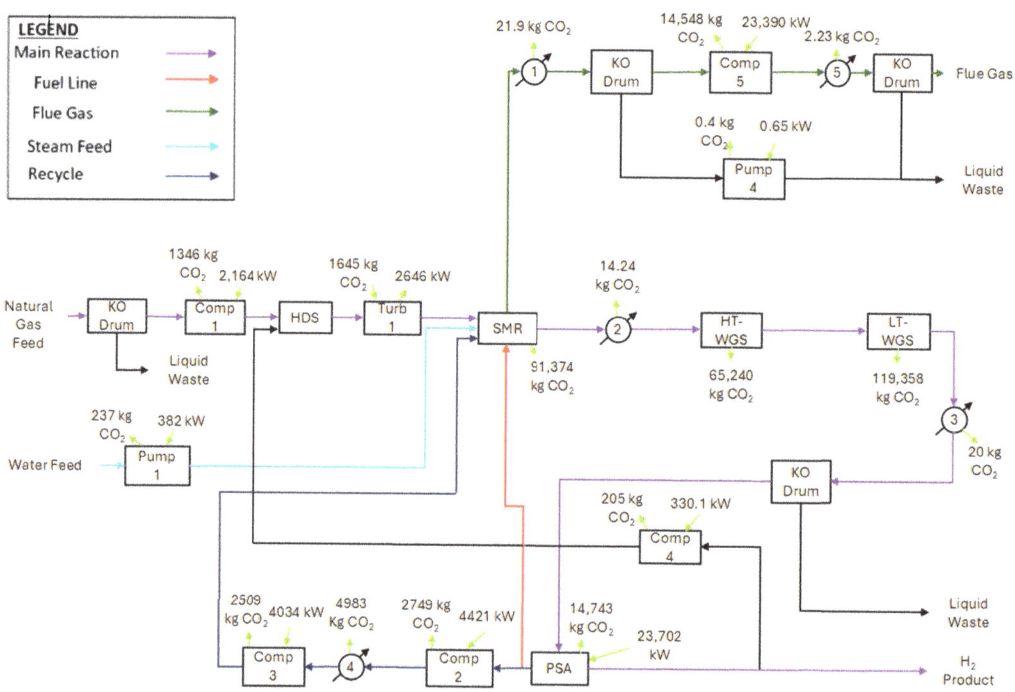

Figure 2. Simplified PFD for the base case.

As shown in Figure 2, the natural gas feed is first passed through a hydrodesulphurization (HDS) unit containing two packing materials, a hydrogenation catalyst and ZnO. Hydrogen is introduced to convert heavy sulfur-containing compounds (represented by mercaptan) to H_2S, which is removed when it passes through the ZnO packing. The H_2S gas reacts with the ZnO packing material to form zinc sulfide, which is regularly removed from the HDS unit and replaced with fresh ZnO.

Table 2. Natural gas feed composition [43].

Component	Molar Fraction
CH_4	0.7500
C2+	0.1500
CO_2	0.0300
N_2	0.0300
H_2S	0.0007
Mercaptan	0.0193
H_2O	0.0193

The resulted sulfur-free natural gas is sent to the SMR unit to undergo the reactions represented in Equations (11) and (12), before being sent to the isothermal water gas shift (WGS) reactors, The latter convert the CO formed by the SMR reaction to CO_2 and hydrogen; this completes the conversion of methane to a hydrogen product. The stoichiometry of the reaction is shown as Equations (11) and (12):

$$CH_4 + H_2O \rightarrow 3H_2 + CO \quad (11)$$

$$CO + H_2O \rightarrow CO_2 + H_2 \quad (12)$$

Upon completion of the WGS process, purification of the product is required to be compliant with standards. As a global standard for hydrogen fuel is yet to be established, its purity requirement varies accordingly. In this work, the targeted purity of hydrogen fuel was assumed to be 99.97% [44]. This was achieved through the use of pressure swing adsorption (PSA). The resultant tail gas, which is richer in CO_2, may be sent to the CC unit (see scenarios in latter section). To improve the final purity of the CO_2 product, unreacted hydrocarbons must first be removed.

High temperature is required for the reaction; thus, combustion of gaseous fuel was selected. Usage of tail gas as fuel gas is prioritized over recycling to reduce reliance on fuel that would increase the carbon footprint. The unreacted hydrocarbons in the form of methane and ethane, alongside unextracted hydrogen, provide some fuel value to the tail gas. Note that the operating temperature of the SMR needs to be controlled at about 900°, so to avoid deactivation of the catalyst due to coking from excess heat. Excess tail gas is recycled to the SMR, as the tail gas still contains unreacted methane which can help to improve yield of the hydrogen product. For every mole of methane, four moles of hydrogen can be recovered theoretically, as shown by Equations (11) and (12), respectively. Taking the upcoming largest blue hydrogen production hub that has recently been green lit as reference, the targeted production capacity is taken as 1 GWt [45]. To achieve the target, the required natural gas feed rate is approximately 5.9 Mmol/h, which equates to a theoretical yield of 24 Mmol H_2/h. The base case design, however, can only achieve a hydrogen production rate of 15 Mmol/h. Through Equation (10), the yield for the base case design is calculated as 63% (=15/24%), which acts as the baseline for further comparison.

The base case model was constructed using Aspen HYSYS v14 (www.aspentech.com), with simulation results given in Figure S2 in the Supplementary File. Upon completion of process simulation, heat integration and economic analysis were carried out using Aspen Energy Analyzer and Aspen Process Economic Analyzer (www.aspentech.com, accessed on 16 November 2024). Through heat integration, the optimal utility and operating cost can be determined for each design scenario.

From the simulation model, scope 1 emissions can be easily calculated based on the amount of CO_2 generated as a by-product (see reaction stoichiometry in Equation (12)) and on steam consumption. In order to quantify scope 2 emissions, it is assumed that the plant is located in Malaysia. Thus, its electric tariffs (in USD) and carbon intensity are used to quantify scope 2 emissions, along with power consumption information identified from the simulation model. These include the cooling water loop, cooling jackets of reactors, and pressure adjustment equipment (e.g., compressors and pumps). The calculation of the carbon intensity (C_E) of the Malaysian power mix is shown in Table 3. As shown, the carbon intensity was calculated based on the mix contribution [46]. On the other hand, the electric tariff of 0.072 USD for medium voltage general industrial use was assumed for economic evaluation [47].

Table 3. Carbon intensity of Malaysia's power mix.

Energy Source	Carbon Intensity (kg CO_2/kWh) [46]	Mix Contribution (%) [48]	Overall Carbon Intensity (kg CO_2/kWh)
Coal	1.001	50.9	0.510
Natural Gas	0.469	30.8	0.144
Hydropower	0.004	15.9	0.001
Bioenergy	0.018	1.9	0.000
Diesel	0.84	0.4	0.003
Overall Mix	-	100	0.658

The CO_2 emissions of each piece of equipment in the base case model were determined and are documented in Tables S1 (Scope 1 emissions) and S2 (Scope 2 emissions). These data were used to calculate the product carbon intensity (CI_P) using Equations (1)–(4). Equations (5)–(9) were then utilized to determine the economic carbon intensity (CI_E) of the base case model. For this base case model, the total Scope 1 and 2 emissions are reported as 279,471 and 38,835 kg/h, respectively (see Tables S1 and S2). Its turbine (power generation) and SMR process (consuming CO_2) contribute to a carbon footprint reduction of 5250 kg/h. Hence, its total carbon footprint is calculated using Equation (3) as 313,046 kg/h (=279,471 + 38,835 − 5250 kg/h), or 2479 kt/y, assuming an annual operating time (AOT) of 7920 h.

Figure 3 shows the product and economic carbon intensity values for the base case. The blue solid lines represent the total carbon emissions with respect to annual production and annualized cost. The process consists of four individual sections. The cumulative carbon emissions of these sections (dotted lines) were added to determine the total carbon emissions of the overall process, as shown in Figure 3. Hence, the latter can be viewed as the composite curve which represents the economic carbon intensity of the process (represented by the gradient of the composite curve).

On the other hand, the second solid line in green in Figure 3 represents the relationship between the process carbon emissions and hydrogen production. The gradient of the segment represents the product carbon intensity of the hydrogen production process. As the base case has an annual production of 238 kt of high purity hydrogen, its product carbon intensity is hence calculated as 10.4 kg CO_2/kg H_2 (=2479 kt CO_2/238 kt H_2), close to a previously reported value [49]. The derived annual operating cost is approximately 32 million USD, and the final economic carbon intensity of the base case is calculated as 77.30 kg CO_2/USD (=2479 kt CO_2/32 million USD). Note that the line representing the economic carbon intensity of the HDS section is not visible in Figure 3 due to its low impact on the overall process. These products and economic carbon intensities will serve as reference values where comparison can be made with other scenarios.

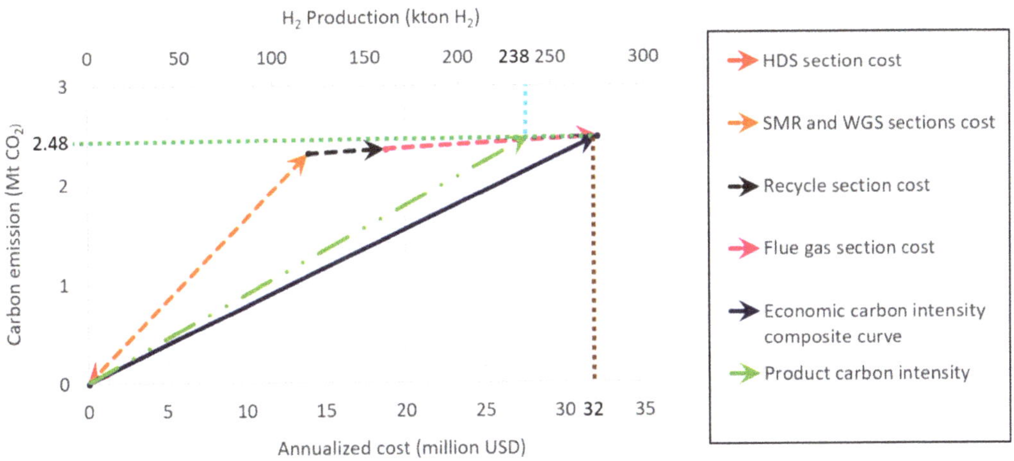

Figure 3. Product carbon intensity and economic carbon intensity diagram for the base case.

Once the base case is established, three proposed scenarios were introduced; they were simulated and analyzed for their economic and carbon footprint performances. In Scenario 1, a CC unit was added where CO_2 from the flue gas is captured for storage; this makes the blue hydrogen product. In scenario 2, a pre-reformation section was added, where additional process units are introduced, hoping to reduce the carbon footprint of the overall process. Adding the pre-reformation section was expected to improve the process yield. In Scenario 3, methanation was introduced after the pre-reformation section, so that the tail gas of the pressure swing absorber (PSA) could be recycled as both reactant and fuel for the SMR reactor. Methanation of tail gas improves the methane concentration of the SMR reactant feed, and hence was expected to enhance the yield of hydrogen in the SMR. Similar to the base case model, all scenarios were simulated in Aspen HYSYS v14, followed by heat integration and economic analysis with Aspen Energy Analyzer and Aspen Process Economic Analyzer (www.aspentech.com, accessed on 16 November 2024). In all scenarios, the capital cost (in USD) was annualized with MARR of 10% and 10 years of design life.

4.1. Scenario 1

Figure 4 shows the simplified PFD for scenario 1 (full PFD is found in Figure S3 in Supplementary Information file). As shown, a CC system based on monoethanolamine (MEA) absorber was utilized to capture CO_2 from the flue gas of reformer. The flue gas was rich in CO_2, water, and nitrogen and contained trace amounts of oxygen. By extracting CO_2 from the flue gas for carbon sequestration purposes, it was expected that the overall process carbon footprint would be reduced. Amine absorption was selected, as it is a matured technology commonly used for gas sweetening and CO_2 removal. The solvent used was a 20 wt% MEA in water mixture. MEA was selected among other amine options, as it is relatively lower cost, with lower focus on H_2S removal (sulfur removal is performed by the HDS unit) [50]. As the newly added MEA absorber system only affects the flue gas, the design alternative shown in scenario 1 possessed identical hydrogen production capacity as the base case (i.e., 238 kt).

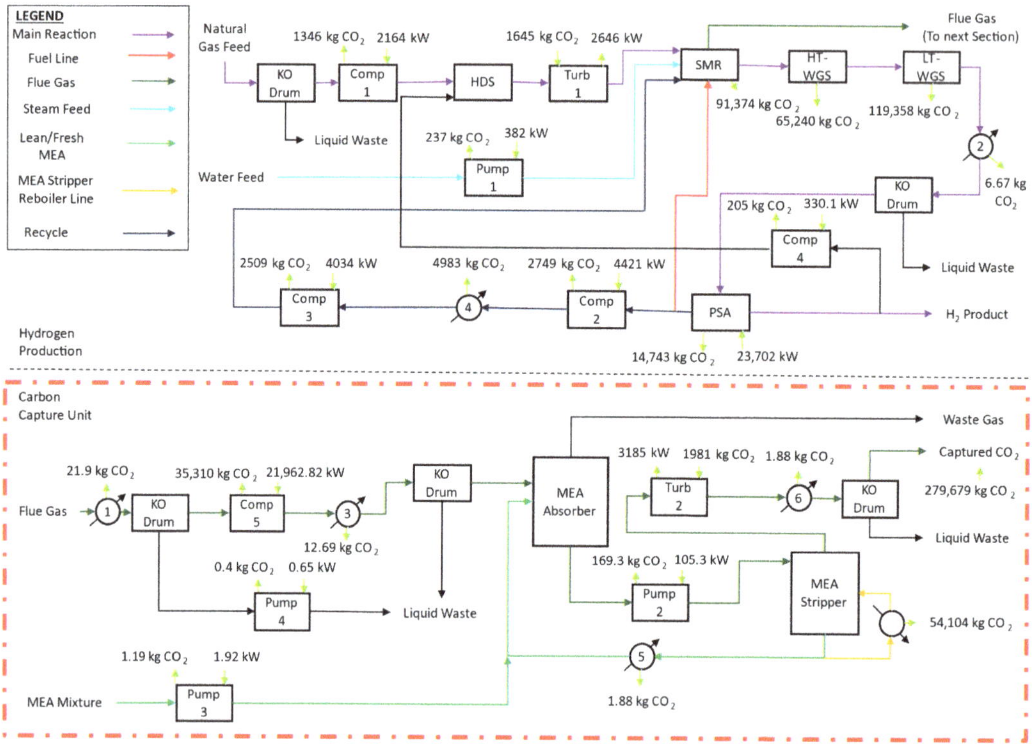

Figure 4. Simplified PFD for scenario 1 (the dotted line indicates the newly added CC unit).

However, there were some drawbacks to Scenario 1. The proposed change in Scenario 1 led to significant increase in the process equipment needed. This results in extra capital and operating costs, along with additional issues such as space constraints. Note that macro-level effects of the increasing design complexity, such as space constraints, additional workforce requirements, were not considered in this study. The effect of increased operating and capital costs is further discussed in Section 5.

The simulation model for this scenario is shown in Figure S4. The CO_2 emissions of all units are shown in Tables S1 (Scope 1) and S2 (Scope 2), while the carbon footprint reduction initiative with CC is reported in Table S3. Hence, the net carbon emissions of this design alternative were calculated using Equation (3) to be 93,286 kg/h, or 740 kt CO_2/y (with an AOT of 7920 h). This is a significant improvement, as it is a 70% reduction in carbon emission compared to the base case (2479 kt CO_2/y).

The annual operating cost for scenario 1 is 64.14 million USD and requires an additional 2.4 million USD in annual expenditures to install the new units. The product and economic carbon intensity values were calculated as 3.11 kg CO_2/kg H_2 (=740 kt CO_2/238 kt H_2) and 11.13 kg CO_2/USD (=740 kt CO_2/66.5 million USD), respectively. The cumulative carbon emission, hydrogen production, and annual cost information are plotted in Figure 5.

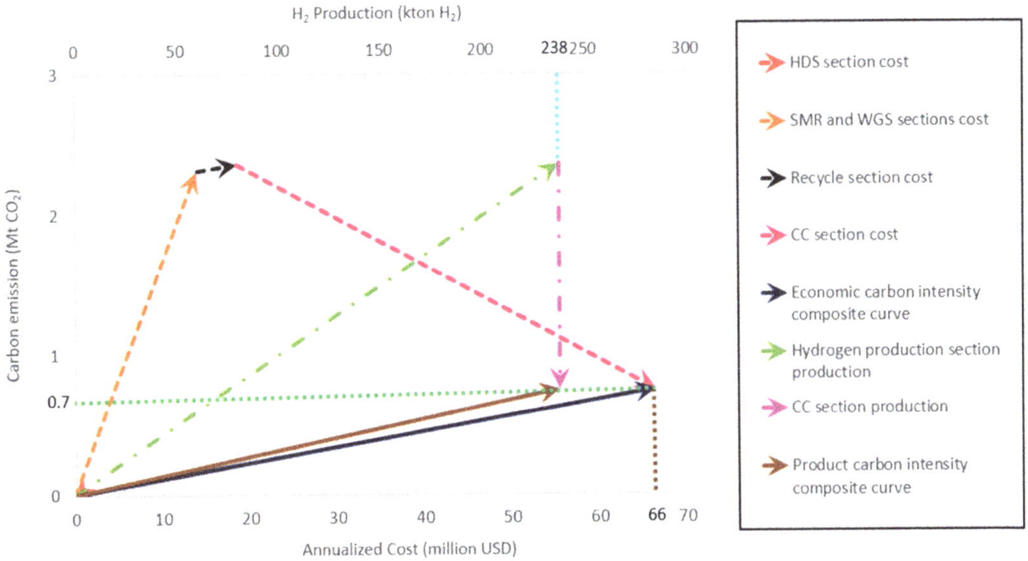

Figure 5. Product carbon intensity and economic carbon intensity diagram for scenario 1.

4.2. Scenario 2

Scenario 2 was built on the basis of Scenario 1, where a pre-reformation unit added between the HDS and SMR units, with simplified PFD in Figure 6 (see full PFD in Figure S5 in the Supplementary File). The newly added pre-reformation unit was operated adiabatically at 500 °C and 20 bar (lower compared to the main reformer, at 700 °C and 30 bar). Through the pre-reformation process, heavier hydrocarbons that are present within the feed (mainly ethane) can be broken down into methane. The reactions involved are given in Equations (13) and (14):

$$C_nH_{2n+2} + 2nH_2O \rightarrow (3n+1)H_2 + nCO_2 \quad (13)$$

$$CO_2 + 4H_2 \rightarrow CH_4 + 2H_2O \quad (14)$$

This additional step allowed the ethane component that was previously unutilized to be converted into the main reactant for the SMR process. In other words, pre-reformation increases the concentration of methane fed to the SMR reactor without changing the initial feed flowrate/composition. Hence, pre-reformation increases both the theoretical and actual yield of the overall process; the latter was calculated using Equation (10). Scenario 2 would produce hydrogen at a rate of 18 Mmol/h, resulting in a 75% yield (=18/24%), while the yields for the base case and scenario 1 are 63%. Similar to scenario 1, scenario 2 possesses similar limitations, such as increased costs, along with additional space and workforce requirements. The effect of the trade-off on the cost of improved production is further discussed in Section 5. The simulation model for scenario 2 is shown in Figure S6 in the Supplementary File.

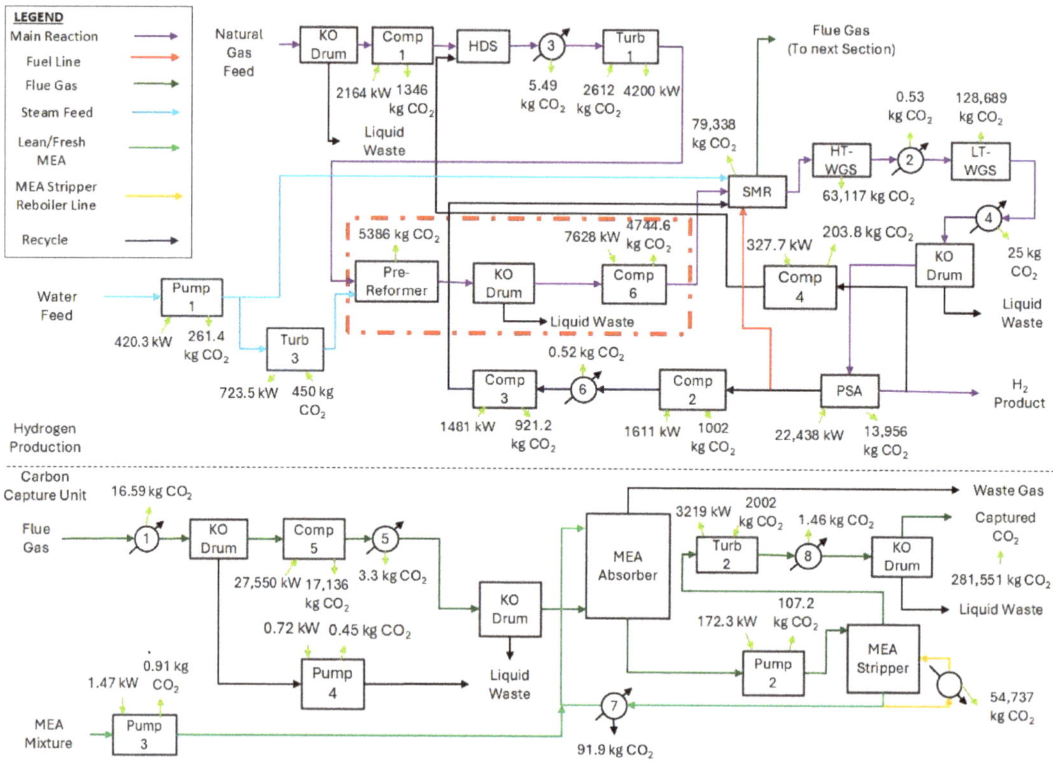

Figure 6. Simplified PFD for scenario 2 (the dotted line indicates the newly added pre-reformation unit).

With the addition of the pre-reformation process, the product carbon and economic intensity values were both affected. Scope 1 (331,250 kg CO_2/h; Table S1) and scope 2 (42,504 kg CO_2/h; Table S2) emissions, as well as the carbon footprint reduction initiative (286,908 kg CO_2/h; Table S3), were then used to calculate the net carbon emissions of this alternative using Equation (3) as 86,846 kg CO_2/h (=331,250 + 42,504 − 286,908 kg CO_2/h), or 688 kt CO_2/y with AOT of 7920 h. It is evidence that the scenario 2 carbon emissions were 72% lower than those in the base case (2479 kt CO_2/t), and this success is attributed to both the enhancement of the process yield and the presence of the CC unit.

Scenario 2 would incur 59.35 million USD in operational costs and requires an additional 3 million USD for the installation of the additional units (relative to the base case). The product carbon intensity was hence calculated as 2.43 kg CO_2/kg H_2 (=688 kt CO_2/283 kt CO_2). On the other hand, the economic carbon intensity was determined to be 11.02 kg CO_2/USD (=688 kt CO_2/62.4 million USD). The relationship between carbon emissions, hydrogen production, and the annualized cost of the process is shown in Figure 7.

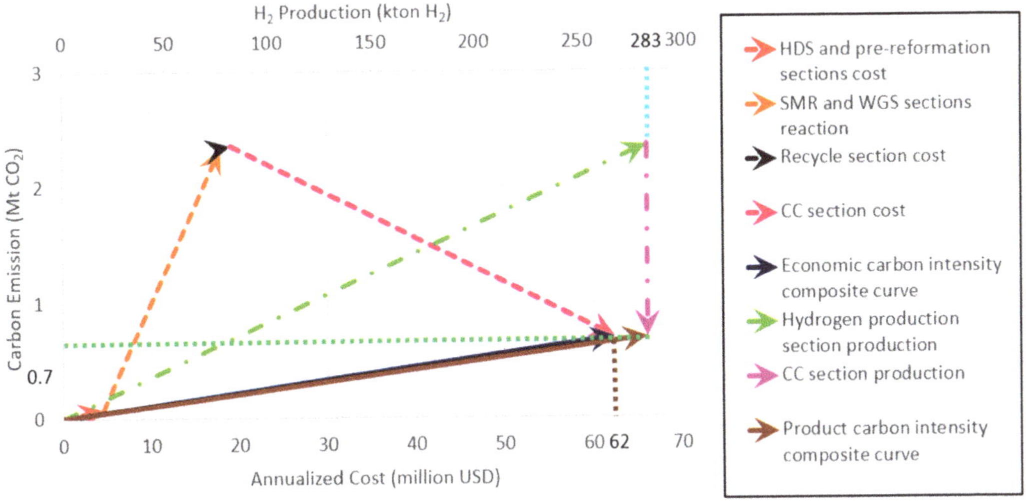

Figure 7. Product carbon intensity and economic carbon intensity diagram for scenario 2.

4.3. Scenario 3

The simplified PFD for scenario 3, which was built on the basis of scenario 2, is shown in Figure 8 (see full PFD in Figure S7 in the Supplementary File). The key highlight of scenario 3 is the addition of the methanation unit. The methanation unit is located after the PSA unit, before the tail gas (consists mainly CO_2) is recycled to the SMR reactor. Note that the presence of CO_2 does not degrade the performance of the various catalysts employed within the system. Excessive concentrations of CO_2, however, will have an adverse effect on the reactions taking place within the reactors. As nearly all reactions are reversible, the presence of CO_2 products (formed through the SMR and WGS reactions) results in the promotion of a reverse reaction, which led to reduced hydrogen yield. This phenomenon was observed in the simulation conducted for base case and scenario 1, where the side reaction operated in reverse when CO_2 composition reached 10 mol%.

The methanation process converts the captured CO_2 into methane by reacting it with hydrogen. Hence, methanation effectively turns the captured CO_2 into the main reactant. As the methanation process requires hydrogen as feed stock, which is supplied from the end product stream, this additional system acts as both a boost and limitation on the overall process performance. The simulation model indicated an overall decrease in hydrogen yield of 3% compared to that in scenario 2. As the pre-reformation process introduced in scenario 2 was retained, hydrogen was produced in excess; this allowed for the targeted production rate of hydrogen gas fuel to be achieved without any additional feed. The simulation model of scenario 3 is shown in Figure S8.

As before, adding the carbon footprints from the scope 1 and 2 emissions (Tables S1 and S2) and subtracting it from the reduction initiative (Table S3), this design alternative exhibited carbon emissions of 86,466 kg CO_2/h (=358,963 + 42,393 − 315,890 kg CO_2/h) or 677 kt CO_2/y, resulting in a 73% reduction in carbon emissions (compared to the base case).

The annual operating cost for the design alternative in scenario 3 was determined to be 58.16 million USD, while an additional 3.9 million USD was required annually for all the new equipment relative to the base case. The product carbon intensity of the new design was hence calculated as 2.50 kg CO_2/kg H_2 (=677 kt CO_2/270 kt H_2). On the other hand, the economic carbon intensity was calculated as 10.91 kg CO_2/USD

(=677 kt CO_2/62.1 million USD). The carbon emissions, hydrogen production, and annual costs are plotted in Figure 9.

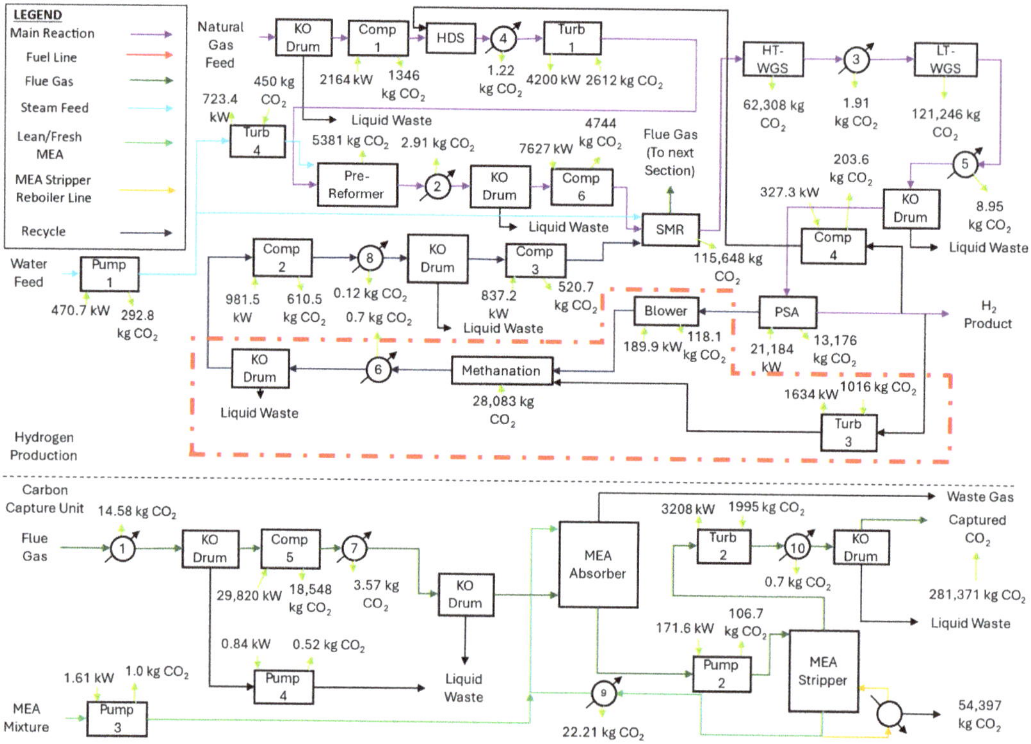

Figure 8. Simplified PFD for scenario 3 (the dotted line indicates the newly added methanation unit).

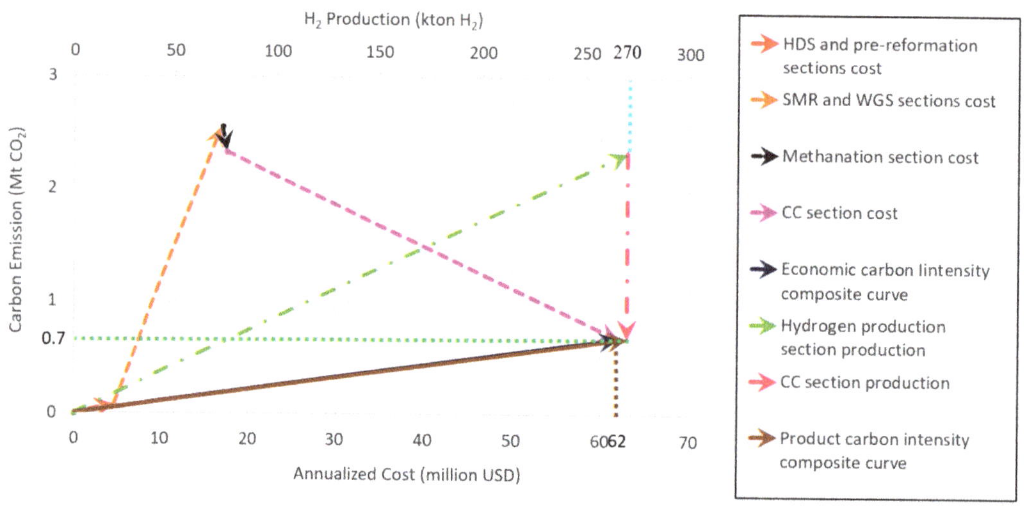

Figure 9. Product carbon intensity and economic carbon intensity diagram for scenario 3.

5. Summary

The performance of each design alternative is shown in Table 4.

Table 4. Performance of each case from a production perspective.

	Base Case	Scenario 1	Scenario 2	Scenario 3
Product CI ($kg\ CO_2/kg\ H_2$)	10.41	3.11	2.43	2.50
Yield	63%	63%	75%	72%
Capital cost (million USD)	3019	3034	3038	3043
Operating cost (million USD/y)	32.07	64.14	59.35	58.16
Annualized cost (million USD/y)	32.07 *	66.50	62.39	62.07
Economic CI ($kg\ CO_2/USD$)	77.30	11.13	11.02	10.91
Net profit (million USD) (net profit after accounting for feed cost and utility cost)	1781	1694	3332	2890
Payback period (Year)	1.69	1.79	0.91	1.05

* Based on Equation (8), the annualized cost (C_A) is the sum of the operating costs and the annualized capital cost difference between the design alternative and that of the base case. As no additional capital costs were incurred for the base case, its operating costs were taken as its annualized costs.

As shown in Table 4, all scenarios achieved the primary objective of reducing the overall carbon footprint of the process. Scenario 2 has the best performance, while scenario 3 is the close second.

The base case design does not feature any CC capabilities. The hydrogen produced through such designs is classified as grey hydrogen, with a product carbon intensity of 10.41 $kg\ CO_2/kg\ H_2$, which aligns with the results reported by Yan et al. [50]. Through the introduction of amine absorption as the CC method, hydrogen produced in scenario 1 is now categorized as blue hydrogen. As the CC unit is introduced at the end of base case design, it does not affect the production rate. With the added facilities to capture CO_2 in the flue gas, the product carbon intensity was reduced by 70% to 3.11 $kg\ CO_2/kg\ H_2$. This value is close to the value of 3.2 $kg\ CO_2/kg\ H_2$ reported by Yan et al. [49]. Through the introduction of pre-reformation, scenario 2 was able to achieve the lowest product carbon intensity, at 2.43 $kg\ CO_2/kg\ H_2$. The pre-reformation unit increases the production rate of hydrogen fuel gas by 19% to 283 kt/y, while minimizing its carbon footprint. Scenario 3 achieved the second best results, with a product carbon intensity of 2.50 $kg\ CO_2/kg\ H_2$. The difference in performance between scenarios 3 and 2 was mainly due to the reduction in yield caused by the need to feed part of the produced hydrogen into the methanation unit. Even though methanization achieved a higher reduction in product carbon intensity (73% compared to 72% in scenario 2), it was not significant enough to offset the negative effects of the yield reduction.

With an increased yield, the changes introduced in scenarios 2 and 3 also influence the total profits. The economic analysis for each design alternative is also included in Table 4, with detailed capital cost break down found in Table S4 (in the Supporting Information). As of 2023, the largest H_2 fuel supplier has set the price for H_2 fuel to 36 USD/kg [51]. The same value was used as the selling price of H_2 fuel for the economic analysis. The feed cost was instead set as 7.5 USD/mmBTU [52].

From an economic carbon intensity standpoint, the introduction of CC in scenario 1 had the largest impact. As economic carbon intensity is a function of process carbon emissions and annualized cost, scenario 3 exhibits better performance than scenario 2, which was considered best from a production perspective. Reduction in yield is not taken into account when considering economic carbon intensity, so the ability of methanation to reduce carbon emissions through CO_2 utilization is highlighted. Both scenarios 2 and 3 require more equipment than scenario 1, which resulted in higher capital costs. Note, however, that they have lower operating costs, due to their lower energy consumption; this is shown in Figure 10.

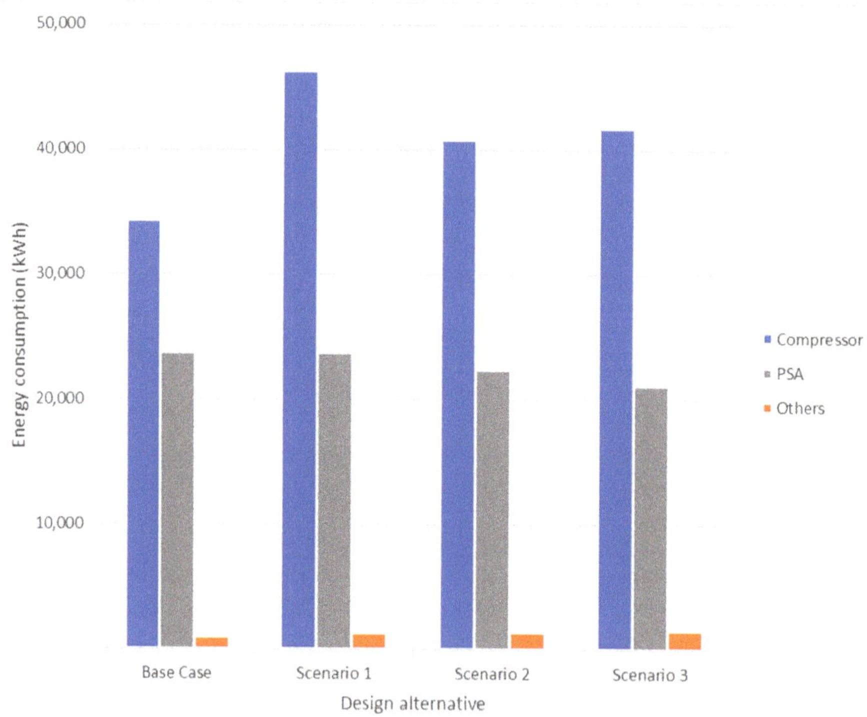

Figure 10. Energy consumption distribution of each design.

From the energy consumption distribution shown in Figure 10, only the compressors and PSA unit have a significant effect on the energy consumption of the process. Note that the energy consumption of the PSA unit is similar in all scenarios, while significant changes were observed for that of the compressors. Scenario 1 is observed to have a much higher energy consumption compared to the other scenarios. This is due to reduced flowrate of the two-stage compression stream. In Scenario 2, more tail gas was recycled as burner fuel, which does not require compression, whereas the methanation product stream in scenario 3 has a lower flowrate compared to tail gas feed.

6. Sensitivity Analysis

A sensitivity analysis was conducted with Scenario 3 as the basis to determine the impact of variations in various inputs on product carbon intensity and economic carbon intensity, with the results shown in Figures 11 and 12. For the former, the effect of fluctuations in process yield and CC efficiency (MEA system and methanation unit) on product carbon intensity was studied. From Figure 11, it is observed that the performance of the MEA systems had a significant effect on the overall product carbon intensity of Scenario 3, which was then followed by process yield and efficiency of the methanation unit. This is because the MEA system acts as the main CC technology responsible for bulk removal of CO_2 from the process streams, while the methanation unit acts as a supplementary system to boost the overall CC rate. The sensitivity analysis results suggest that the MEA system should be maintained with high efficiency, so as to lower the carbon footprint of the hydrogen product in the long term. Even though yield fluctuations had a less pronounced effect on product carbon intensity, Figure 11 shows that the product carbon intensity exponentially increased and decreased as the yield decreased and increased.

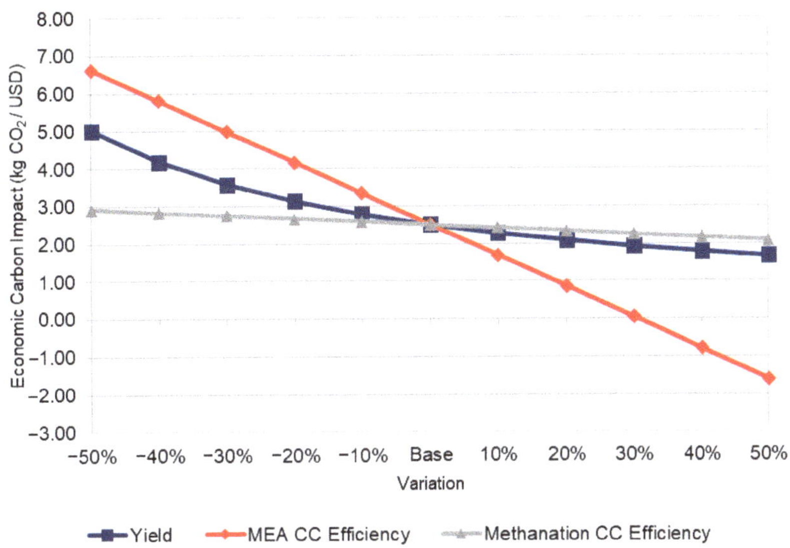

Figure 11. Sensitivity analysis of process yield and CC efficiency on product carbon intensity.

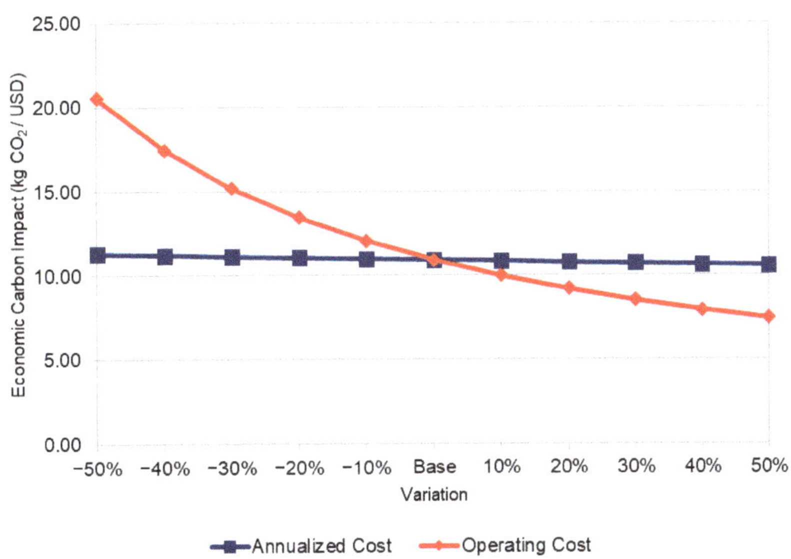

Figure 12. Sensitivity results of annualized capital and operating costs on economic carbon intensity.

As shown in Equation (8), economic carbon intensity is a function of both annualized capital costs and operating costs. A sensitivity analysis was carried out to determine the effect of these costs on the economic carbon intensity of Scenario 3. Figure 12 shows that increased operating costs resulted in an exponential decrease in economic carbon intensity, while annualized capital costs had little influence. As the annualized capital costs are based on the additional costs required compared to the base case, this amount (3.91 million USD) was significantly lower than the operating costs (58.16 million USD). Even though Scenario 3 had the highest capital costs due to the greatest number of pieces of equipment, its operating costs were the lowest (see Table 4). The reason for this was deviation in flowrates for the two-stage compressors used for recycling of the PSA tail gas.

It should be noted that the design alternatives with lower economic carbon intensity required higher monetary investment to achieve the same rate of CO_2 emissions. However, the increased investment may also result in improved economic performance (e.g., lower operating costs) in the long run.

7. Conclusions

In this work, a graphical technique for rapid screening of conceptual design alternatives based on carbon footprint is proposed. A case study consisting of a typical SMR-based blue hydrogen production process was analyzed for its produced carbon and economic carbon intensities. Three scenarios with different technologies (CC, pre-reformation, and methanation) were considered. Heat integration was performed for all scenarios to ensure the energy consumption of the designs was minimized before comparisons were made. All scenarios achieved varying levels of improvement over the base case, while achieving the production target. Scenarios 2 and 3 demonstrated better performance from product carbon intensity and economic carbon intensity perspectives, respectively. The product carbon intensity of Scenario 2 was reported as 2.43 kg CO_2/kg H_2, which was the lowest among all scenarios. On the other hand, Scenario 3 exhibited the lowest economic carbon intensity among all scenarios, i.e., 10.91 kg CO_2/USD.

Product and economic carbon intensity values provide useful insights to the environmental impact of hydrogen production processes. The value of these insights is approximately equivalent to that of a traditional LCA study, but our analysis can be conducted during the conceptual design stage, when relevant data is scarce. Moreover, the idea presented by this work is highly adaptable once the process carbon emissions are determined. This methodology is readily applicable to a wide range of conceptual design problems. A lack of concrete data is a common problem for all projects during the conceptual design phase. The many possible reaction pathways based on various feed tock types also contribute to the large number of potential solutions that need to be vetted to identify high potential design options. The newly proposed indices, i.e., product and economic carbon intensity, only require the process emissions from a limited scope (feed stock to product, excluding emission during feed stock acquisition and after product formation) as input. Through simplification of carbon emission analysis, some leeway is made available with regard to the quality of data required to conduct the relevant study. Both of these are key factors that allow carbon footprint analysis to be conducted more efficiently. Future work should focus on using the methodology to identify where data acquisition efforts should focus to give better footprint estimates. Evaluating decarbonization efforts based on the hybrid use of renewable energy, solar electrolyser, and/or combined heat and power to reduce the carbon footprint of the power consumption is another promising direction. Exploration into improving the economic analysis and factoring in the wider scale carbon trading market and governmental subsidies are also valid avenues for improvement of this work.

Supplementary Materials: The following supporting information can be downloaded at: https://www.mdpi.com/article/10.3390/pr13041254/s1, Figure S1: Full PFD for base case, Figure S2: Simulation flowsheet of base case design, Figure S3: Full PFD for scenario 1 (with CC unit), Figure S4: Simulation flowsheet of scenario 1 (with CC unit), Figure S5: Full PFD for scenario 2 (with pre-reformation unit), Figure S6: Simulation flowsheet of scenario 2 (with pre-reformation unit), Figure S7: Full PFD for scenario 3 (with methanation units), Figure S8: Simulation flowsheet of scenario 3 (with methanation units). Table S1: Scope 1 Emission for each design alternative case (kg CO_2/h), Table S2: Scope 2 Emission for each design alternative (kg CO_2/h), Table S3: Detailed carbon footprint reduction of each case (kg CO_2/h), Table S4: Detailed capital cost breakdown of each design (Mil USD).

Author Contributions: Conceptualization, Z.A.P. and D.C.Y.F.; methodology, E.Y.H.C.; software, E.Y.H.C.; validation, Y.K.W. and D.C.Y.F.; formal analysis, E.Y.H.C.; writing—original draft preparation, E.Y.H.C.; writing—review and editing, Y.K.W., Z.A.P., R.R.T. and D.C.Y.F.; visualization, E.Y.H.C.; supervision, Y.K.W. and D.C.Y.F. All authors have read and agreed to the published version of the manuscript.

Funding: This research received no external funding.

Data Availability Statement: The original contributions presented in this study are included in the article/Supplementary Materials. Further inquiries can be directed to the corresponding author(s).

Acknowledgments: The support and guidance provided by Joshua YL Liew is gratefully acknowledged.

Conflicts of Interest: The authors declare no conflict of interest.

Nomenclature

CAGR	Compound annual growth rate
UNFCCC	United Nations Framework Convention on Climate Change
CCS	Carbon capture and storage
LCA	Life cycle analysis
IOA	Input–output analysis
CEPA	Carbon emission pinch analysis
EIA	Environmental impact assessment
ISO	International Organization for Standardization
SMR	Steam methane reformation
GHG	Greenhouse gas
E_1	Scope 1 emission
E_2	Scope 2 emission
C_G	CO_2 generated by process unit
P_c	Total power consumption
C_E	Carbon intensity of electricity purchased
C_N	Net carbon footprint
CI_P	Product carbon intensity
E_{CCUS}	CO_2 emissions captured, utilized, or reduced
H	Hydrogen produced
OP	Operating cost
C_w	Water cost
C_{Fuel}	Fuel cost
C_{Feed}	Feed cost
C_m	Materials cost
P_N	Net power consumption
P_G	Electrical power generated by the system
T	Electric tariff rate

MARR	Minimum acceptable rate of return
AW	Annual cost
C_A	Annualized cost
CI_E	Economic carbon intensity
P_T	Theoretical amount of hydrogen recoverable
P_A	Actual production rate
HDS	Hydrodesulfurization
WGS	Water gas shift
PSA	Pressure swing adsorption
CC	Carbon capture
AOT	Annual operating time
MEA	Monoethanolomine

References

1. Power Generation Market. Precedence Research. 2023. Available online: https://www.precedenceresearch.com/power-generation-market#:~:text=The%20global%20power%20generation%20market%20size%20was%20accounted%20at%20USD,8.04%25%20from%202023%20to%202032 (accessed on 28 January 2023).
2. World Energy Balances: Overview. IEA. Paris. 2023. Available online: https://www.oecd.org/about/#:~:text=The%20Organisation%20for%20Economic%20Co,%20and%20well-being%20for%20all (accessed on 5 November 2023).
3. Ritchie, H.; Roser, M. CO_2 Emissions. Our World in Data. 2020. Available online: https://ourworldindata.org/co2-emissions (accessed on 28 January 2023).
4. United Nations. For a Liveable Climate: Net-Zero Commitments Must Be Backed by Credible Action, United Nations. 2023. Available online: https://www.un.org/en/climatechange/net-zero-coalition#:~:text=to%2010%20levels.-,To%20keep%20global%20warming%20to%20no%20more%20than%201.5%C2%B0,reach%20net%20zero%20by%202050 (accessed on 26 January 2024).
5. United Nations Framework Convention on Climate Change. COP 28: What Was Achieved and What Happens Next? United Nations Framework Convention on Climate Change. 2024. Available online: https://unfccc.int/cop28/5-key-takeaways#end-of-fossil-fuels (accessed on 26 January 2024).
6. United Nations Framework Convention on Climate Change. Record Number of National Adaptation Plans Submitted in 2023, But More Are Needed. 2023. Available online: https://unfccc.int/news/record-number-of-national-adaptation-plans-submitted-in-2023-but-more-are-needed (accessed on 26 January 2024).
7. Energy Institute. Statistical Review of World Energy. Energy Institute: London, UK, 2023.
8. Ezhiljenekkha, G.B.; MarsalineBeno, M. Review of Power Quality Issues in Solar and Wind Energy. *Mater. Today Proc.* **2020**, *24*, 2137–2143. [CrossRef]
9. Wang, Z. Identifying green hydrogen produced by grid electricity. *Int. J. Hydrogen Energy* **2024**, *81*, 654–674. [CrossRef]
10. National Academies of Science Engineering and Medicine. *New Directions for Chemical Engineering*; National Academies Press: Washington, DC, USA, 2022. [CrossRef]
11. IRENA. Hydrogen, Internaitonal Renewable Energy Agency. 2022. Available online: https://www.irena.org/Energy-Transition/Technology/Hydrogen (accessed on 4 December 2023).
12. Park, C.; Koo, M.; Woo, J.; Hong, B.I.; Shin, J. Economic valuation of green hydrogen charging compared to gray hydrogen charging: The case of South Korea. *Int. J. Hydrogen Energy* **2022**, *47*, 14393–14403. [CrossRef]
13. Noussan, M.; Raimondi, P.P.; Scita, R.; Hafner, M. The role of green and blue hydrogen in the energy transition—A technological and geopolitical perspective. *Sustainability* **2021**, *13*, 298. [CrossRef]
14. Kakoulaki, G.; Kougias, I.; Taylor, N.; Dolci, F.; Moya, J.; Jäger-Waldau, A. Green hydrogen in Europe—A regional assessment: Substituting existing production with electrolysis powered by renewables. *Energy Convers. Manag.* **2021**, *228*, 113649. [CrossRef]
15. Yue, M.; Lambert, H.; Pahon, E.; Roche, R.; Jemei, S.; Hissel, D. Hydrogen energy systems: A critical review of technologies, applications, trends and challenges. *Renew. Sustain. Energy Rev.* **2021**, *146*, 111180. [CrossRef]
16. Lebrouhi, B.E.; Djoupo, J.J.; Lamrani, B.; Benabdelaziz, K.; Kousksou, T. Global hydrogen development—A technological and geopolitical overview. *Int. J. Hydrogen Energy* **2022**, *47*, 7016–7048. [CrossRef]
17. Marocco, P.; Gandiglio, M.; Audisio, D.; Santarelli, M. Assessment of the role of hydrogen to produce high-temperature heat in the steel industry. *J. Clean. Prod.* **2023**, *388*, 135969. [CrossRef]
18. Falfari, S.; Cazzoli, G.; Mariani, V.; Bianchi, G.M. Hydrogen Application as a Fuel in Internal Combustion Engines. *Energies* **2023**, *16*, 2545. [CrossRef]

19. Zhang, Y.; Wang, Y.; Cui, H.; Wang, F.; Li, Z.; Sun, C.; Jia, Y.; Xie, Y.; Song, F.; Wang, J.; et al. Co-pyrolysis of biomass with Red Mud: An efficient approach to improving bio-oil quality and resourceful utilization of the iron in Red Mud. *Fuel* **2023**, *355*, 129482. [CrossRef]
20. Thomas, G.; Pidgeon, N.; Henwood, K. Hydrogen, a less disruptive pathway for domestic heat? Exploratory findings from public perceptions research. *Clean. Prod. Lett.* **2023**, *5*, 100047. [CrossRef]
21. Martire, M.; Kaya, A.F.; Morselli, N.; Puglia, M.; Allesina, G.; Pedrazzi, S. Analysis and optimization of a hybrid system for the production and use of green hydrogen as fuel for a commercial boiler. *Int. J. Hydrogen Energy* **2024**, *56*, 769–779. [CrossRef]
22. Ali, A.; Houda, M.; Waqar, A.; Khan, M.B.; Deifalla, A.; Benjeddou, O. A review on application of hydrogen in gas turbines with intercooler adjustments. *Results Eng.* **2024**, *22*, 101979. [CrossRef]
23. IRENA. *Hydrogen: A Renewable Energy Perspective*; International Renewable Energy Agency: Abu Dhabi, United Arab Emirates, 2019.
24. Voldsund, M.; Reyes-Lúa, A.; Fu, C.; Ditaranto, M.; Nekså, P.; Mazzetti, M.J.; Brekke, O.; Bindingsbø, A.U.; Grainger, D.; Pettersen, J. Low carbon power generation for offshore oil and gas production. *Energy Convers. Manag. X* **2023**, *17*, 100347. [CrossRef]
25. Alghool, D.; Haouari, M.; Trucco, P. It is not the same blue: A comparative LCA study of blue hydrogen supply network pathways. *Int. J. Hydrogen Energy* **2024**, *81*, 214–224. [CrossRef]
26. Bauer, C.; Treyer, K.; Antonini, C.; Bergerson, J.; Gazzani, M.; Gencer, E.; Gibbins, J.; Mazzotti, M.; McCoy, S.T.; McKenna, R.; et al. On the climate impacts of blue hydrogen production. *Sustain. Energy Fuels* **2022**, *6*, 66–75. [CrossRef]
27. Hydrogen Council. *Hydrogen Decarbonization Pathways*; Hydrogen Council: Brussels, Belgium, 2021; p. 14. Available online: https://hydrogencouncil.com/wp-content/uploads/2021/01/Hydrogen-Council-Report_Decarbonization-Pathways_Executive-Summary.pdf (accessed on 4 December 2023).
28. Incer-Valverde, J.; Korayem, A.; Tsatsaronis, G.; Morosuk, T. "Colors" of hydrogen: Definitions and carbon intensity. *Energy Convers. Manag.* **2023**, *291*, 117294. [CrossRef]
29. Burchart, D.; Gazda-Grzywacz, M.; Grzywacz, P.; Burmistrz, P.; Zarębska, K. Life Cycle Assessment of Hydrogen Production from Coal Gasification as an Alternative Transport Fuel. *Energies* **2022**, *16*, 383. [CrossRef]
30. Zang, G.; Graham, E.J.; Mallapragada, D. H_2 production through natural gas reforming and carbon capture: A techno-economic and life cycle analysis comparison. *Int. J. Hydrogen Energy* **2023**, *49*, 1288–1303. [CrossRef]
31. Dong, Q.; Zhong, C.; Geng, Y.; Dong, F.; Chen, W.; Zhang, Y. A bibliometric review of carbon footprint research. *Carbon Footpr.* **2024**, *3*, 3. [CrossRef]
32. Tan, R.R.; Foo, D.C. Pinch analysis approach to carbon-constrained energy sector planning. *Energy* **2007**, *32*, 1422–1429. [CrossRef]
33. Tjan, W.; Tan, R.R.; Foo, D.C. A graphical representation of carbon footprint reduction for chemical processes. *J. Clean. Prod.* **2010**, *18*, 848–856. [CrossRef]
34. Yap, J.Y.; Tan, J.; Foo, D.C.; Tan, R.R.; Papadokonstantakis, S.; Badr, S. A graphical method for carbon dioxide emissions reduction in multi–product plants. *Process Saf. Environ. Prot.* **2020**, *133*, 51–63. [CrossRef]
35. Tan, R.R.; Aviso, K.B.; Foo, D.C. Carbon emissions pinch analysis of economic systems. *J. Clean. Prod.* **2018**, *182*, 863–871. [CrossRef]
36. Ramanath, T.; Foo, D.C.; Tan, R.R.; Tan, J. Integrated enterprise input-output and carbon emission pinch analysis for carbon intensity reduction in edible oil refinery. *Chem. Eng. Res. Des.* **2023**, *193*, 826–842. [CrossRef]
37. Nita, A.; Fineran, S.; Rozylowicz, L. Researchers' perspective on the main strengths and weaknesses of Environmental Impact Assessment (EIA) procedures. *Environ. Impact Assess. Rev.* **2022**, *92*, 106690. [CrossRef]
38. Mandelik, Y.; Dayan, T.; Feitelson, E. Issues and dilemmas in ecological scoping: Scientific, procedural and economic perspectives. *Impact Assess. Proj. Apprais.* **2005**, *23*, 55–63. [CrossRef]
39. Morgan, R.K. Environmental impact assessment: The state of the art. *Impact Assess. Proj. Apprais.* **2012**, *30*, 5–14. [CrossRef]
40. Robin, S. *Chemical Process Design and Integration*, 2nd ed.; Wiley: Hoboken, NJ, USA, 2016.
41. National Grid. What Are Scope 1, 2 and 3 Carbon Emissions? 2023. Available online: https://www.nationalgrid.com/stories/energy-explained/what-are-scope-1-2-3-carbon-emissions (accessed on 28 November 2023).
42. United States Environmental Protection Agency. Scopes 1, 2 and 3 Emissions Inventorying and Guidance. 2024. Available online: https://www.epa.gov/climateleadership/scopes-1-2-and-3-emissions-inventorying-and-guidance (accessed on 15 September 2024).
43. Gas Processors Suppliers Association. *Engineering Data Book*; Gas Processors Suppliers Association: Tulsa, OK, USA, 2004; pp. 1–32.
44. Yang, Y.; Wang, G.; Zhang, L.; Zhang, S.; Lin, L. Comparison of Hydrogen Specification in National Standards for China. In Proceedings of the 2019 4th International Conference on Advances in Energy and Environment Research (ICAEER 2019), Shanghai, China, 16–18 August 2019; Volume 118, p. 03042. [CrossRef]
45. Energy Institute. UK's Largest Blue Hydrogen Production Hub Given Initial Go Ahead. 2024. Available online: https://knowledge.energyinst.org/new-energy-world/article?id=138543 (accessed on 29 August 2024).
46. Energy Commision. Malaysia Energy Statistics Handbook 2021. 2023. Available online: www.st.gov.my (accessed on 3 February 2023).

47. Pricing & Tariffs. Tenaga Nasional Berhad. 2014. Available online: https://www.tnb.com.my/commercial-industrial/pricing-tariffs1/ (accessed on 19 April 2023).
48. Moonmaw, W.; Burgherr, P.; Heath, G.; Lenzen, M.; Nyboer, J.; Verbruggen, A. Annex II: Methodology. In *Special Report on Renewabe Energy Sources and Climate Change Mitigation*; Edenhofer, O., Pichs-Madruga, R., Sokona, Y., Seyboth, K., Matschoss, P., Kadner, S., Zwickel, T., Eickemeier, P., Hansen, G., Schlomer, S., et al., Eds.; Cambridge University Press: Cambridge, UK, 2011.
49. Yan, F.; Simon, N.; McCurdy, M. Why Should We Care About Hydrogen Carbon Intensity? ICF. 2022. Available online: https://www.icf.com/insights/energy/care-about-hydrogen-carbon-intensity (accessed on 28 January 2023).
50. Yusuf, M. Selection of Amine in Natural Gas Sweetening Process for Acid Gases Removal: A Review of Recent Studies. *Pet. Petrochem. Eng. J.* **2017**, *1*, 1–7. [CrossRef]
51. Collins, L. It Is Now Almost 14 Times More Expensive to Drive a Toyota Hydrogen Car in California Than a Comparable Tesla EV. Hydrogeninsight. 2023. Available online: https://www.hydrogeninsight.com/transport/analysis-it-is-now-almost-14-times-more-expensive-to-drive-a-toyota-hydrogen-car-in-california-than-a-comparable-tesla-ev/2-1-1519315 (accessed on 2 September 2024).
52. Suruhanjaya Tenaga Malaysia. Regulated Piped Gas Prices. 2021. Available online: https://www.st.gov.my/contents/2021/Fuel%20Prices/11%20Nov/1-%20Regulated%20piped%20gas%20prices%20as%20of%20November%202021.pdf (accessed on 3 February 2023).

processes

MDPI

Article

Development of MEA-Based and AEP-Based CO_2 Phase Change Absorbent

Yongyan Wang [1], Fanghui Cheng [2], Jingsong Li [3], Yingshu Liu [1], Haihong Wang [4], Ziyi Li [1,*] and Xiong Yang [1,5,*]

1 College of Energy and Environmental Engineering, University of Science and Technology Beijing, Beijing 100083, China; m202220258@xs.ustb.edu.cn (Y.W.); ysliu@ustb.edu.cn (Y.L.)
2 Shaanxi Blower (Group) Co., Ltd., Xi'an 710000, China; cfh442952210@163.com
3 Research Institute of Technology of Shougang Group, Beijing 100041, China; 18810634180@163.com
4 R&D Center Department, Beijing District Heating Group, Beijing 100026, China; ustbwhh@163.com
5 Key Laboratory of Plateau Oxygen and Living Environment of Xizang Autonomous Region, College of Science, Xizang University, Lhasa 850000, China
* Correspondence: ziyili@ustb.edu.cn (Z.L.); yangx@ustb.edu.cn (X.Y.)

Abstract: In energy conservation and low-carbon environmental protection, separating and capturing CO_2 from blast furnace gas is a crucial strategy for the steel industry to achieve its dual carbon goals. This study conducts an experimental study on the phase change absorption of carbon dioxide for the low-energy capture of carbon dioxide in blast furnace gas in iron and steel enterprises. The experiment used 30%wt monoethanolamine (MEA) and 30%wt 1-(2-aminoethyl)piperazine (AEP) as a reference to blend different absorbents, and the CO_2 absorption effect of the absorbents was tested. The results indicated that the MEA system phase change absorbents have the best absorption effect when the mass ratio of additives to water is 5:5, and the AEP system has the best absorption effect at 7:3. The absorption effect of different phase separators is as follows: n-propanol > sulfolane > isopropanol. AEP/n-propanol/H_2O (7:3) has a maximum absorption load of 2.03 molCO_2·mol^{-1} amine, a relatively low rich phase ratio of 0.46, and low regeneration energy consumption. The load capacity of different absorbents was calculated based on the load experiment results, and it was found that the loading capacity of the MEA system was greater than that of the AEP system, with the maximum load capacity of MEA/n-propanol/H_2O (5:5) being 4.02 mol/L. Different types of absorbents exhibited an increase in rich phase density with the increase in additive quality. The regeneration performance of the absorbent indicated that at a temperature of 393.15 K, the desorption load of n-propanol aqueous solution rich phase in the absorbent was high, and the desorption speed was the fastest.

Keywords: phase change absorbent; CO_2 capture; MEA; AEP

1. Introduction

Countries worldwide are pursuing the goal of low-carbon emissions [1,2]. Vigorously promoting CCUS (carbon capture, utilization, and storage) technology is an important means of achieving the dual carbon goals of the steel industry. The steel production process in China is mainly based on the blast furnace converter method [3]. About 89% of energy input comes from coal, which has high carbon emissions. Steel enterprises' energy-saving and consumption reduction starts with the "three gases" (blast furnace gas, converter gas, and coke oven gas). Among them, blast furnace gas emissions are the highest. Therefore, the effective utilization of blast furnace gas is the key to energy conservation and consumption reduction in the steel industry [4].

Considering factors such as one-time investment, operating costs and energy consumption, and waste treatment, the chemical absorption method can fully leverage the advantages of comprehensive waste heat utilization in steel enterprises, effectively utilizing low-temperature heat. It is an ideal method for capturing carbon dioxide in blast furnace gas. Chemical absorbents are the key to the chemical absorption method. The traditional chemical absorbents mainly include ammonia absorbent [5], amino acid salt absorbent [6], potassium carbonate absorbent [7,8], and organic amine absorbent [9]. Among them, organic amine absorbent is the most representative chemical absorbent. Amine absorbent has the advantages of a fast reaction rate, high gas purification, good absorption effect, and low cost [10]. Monoethanolamine (MEA) [11], diethanolamine (DEA), and N-methyldiethanolamine (MDEA) [12] are commonly used. However, most traditional absorbents have their limitations, such as the volatilization of ammonia, slow absorption rate, low absorption capacity, and high energy consumption [8,13,14]. Han et al. [15] found that the energy consumption of absorbent regeneration can be reduced to 2.5 GJ/t, and the volume fraction of nitrogen in the tail gas leaving the absorption tower is as high as 0.02%. Mao et al. [16] reported that the average energy consumption of using monoethanolamine to capture CO_2 was 5.4 GJ/t. 30% MEA is one of the commonly used reference absorbents, but its high regeneration energy consumption (about 3.7–4.0 GJ/tCO_2) hinders its large-scale application [17]. The energy consumption of absorbent regeneration comprises thermal desorption reaction heat, sensible heat of absorbent rising temperature, and latent heat of water evaporation. P.H.M. Feron et al. [18] studied the thermal desorption energy consumption of 30% MEA absorbent. The results showed that the total energy consumption of 30% MEA absorbent was about 135–205 KJ$\cdot$mol^{-1} CO_2, and the heat of thermal desorption reaction accounted for 40–60% of the total energy consumption, the latent heat of water evaporation accounts for 30–40% of the total energy consumption. Although water is the most suitable co-solvent from the cost and environmental impact perspective, the high water content (>50%) in the absorbent and the strong interaction between MEA and CO_2 are the reasons for the high desorption energy consumption. Therefore, researchers have developed new mixed amine absorbents [19,20], such as MEA+MDEA [21], MDEA+piperazine(PZ) [22], and Butyl ethanolamine (BAE)+2-Dimethylaminoethanol (DMAE) [23]. Although mixed amine absorbents can improve the drawbacks of single absorbers, they cannot simultaneously have high absorption and desorption performance [24].

Phase change absorption is considered as a CO_2 capture technology with great potential because it can greatly reduce the amount of recycled liquid and thus reduce energy consumption [25]. Phase change absorbent absorbs CO_2 and divides it into rich and lean phases [26]. Only the rich phase is regenerated, while the lean phase is directly returned to the absorption tower for reuse, greatly reducing the sensible heat and latent heat of vaporization required for adsorption. At first, researchers mixed different types of amines to develop phase separation absorbents. Wange et al. [27] adopted the mixed solution of diethylenetriamine(DETA) and 2-(diethylamine)-ethanol (DEEA) as a new phase change adsorbent to absorb CO_2, and the rich phase accounted for 70% when the rich phase load reached 2.23 mol/L. Liu et al. [28] mixed 50 wt% DEEA and 2-((2-aminoethyl) amino) ethanol (AEEA, 25 wt%). When the rich phase load reached 2.95 mol/kg, the rich phase accounted for 80%. The total amine concentration of mixed amine absorbents is relatively high, usually exceeding 60%. High concentrations can cause amine volatilization and degradation [29]. Excessive wealth ratio can also increase desorption energy consumption.

To address the drawbacks of mixed amine phase change absorbents. In recent years, the research direction is to use organic solvents to replace water in phase change absorbents. The most widely used organic solvents in phase separation are sulfoxide [30–33] and alcohol [34–37]. They promote the phase separation of the absorbent and the absorption of CO_2 [38]. Wang et al. [33] used DETA and sulfolane to make phase change absorbent to simulate the absorption of CO_2. During the CO_2 absorption process, sulfolane, as an inert solvent, was separated from the absorption product instead of water, reducing the total heat load by 19%. Zhang et al. [36] prepared a CO_2 phase change adsorbent (CPCA) composed of MEA/alcohol/H_2O, and studied the species distribution, physical properties, circulation capacity, absorption, and regeneration capacity of CPCAs. Compared with 30 wt% MEA/H_2O, the CO_2 cycling capacity of CPCA increased by 62%. In the MEA–propanol system, propanol can significantly promote CO_2 mass transfer [39]. Adding organic solvents to the alcohol amine phase change absorbent can reduce the enrichment to 50%, and the desorption energy consumption is about 2.2–2.8 GJ/tCO_2 [29], which is an ideal phase change absorbent. It is also reported that there are primary amine/secondary amine + tertiary amine + organic physical solvent. In the mixed solution, primary amine/secondary amine is used as the absorption enhancer, while tertiary amine serves as the principal absorbents, and physical solvent act as phase separation agent. For example, Wang et al. [30] proposed a two-phase absorbent that can adjust the phase separation behavior of Triethylenetetramine (TETA)-DEEA through sulfolane without sacrificing high CO_2 absorption capacity, while reducing the volume ratio of the rich phase and increasing the CO_2 load. Wang et al. [40] proposed a tertiary amine-based biphasic solvent: N,N-dimethylethanolamine (DMEA)-PZ-1-Butanol for efficient CO_2 capture. The addition of 1-Butanol not only produced phase separation and reduced the activation energy of the catalyst, but also synergistically with PZ to increase the inherent slow absorption rate of DMEA to a level comparable to that of 5M MEA. Although the regenerative energy consumption can be greatly reduced in the circulation system, the phase change absorbent still has problems such as high viscosity, large volatility, and unsatisfactory phase separation ratio. Therefore, developing solutions with higher CO_2 load, smaller abundance ratio, and lower viscosity is the future direction of phase change absorbent.

This study aims to develop a new phase change absorbent with good absorption performance, low enrichment ratio, and suitable viscosity for carbon dioxide capture by the chemical absorption method. MEA and AEP were used as the main absorption components of the phase change absorbent, and n-propanol, isopropanol, and sulfolane were used as phase separation agents. The absorbent's absorption, desorption, and phase separation effects were observed.

2. Materials and Methods

2.1. Materials

MEA and AEP were selected as the main absorption components of the phase separation absorbent; n-propanol, isopropanol, and sulfolane were selected as the phase separation additives. Based on 30%wt MEA and 30%wt AEP, different kinds of absorbents were prepared by changing the ratios of water and additives, and the mixed-phase system of absorber components/additives/H_2O could be obtained.

MEA (AR) was obtained from InnoKai Technology Co., Ltd., Shenzhen, China; AEP (AR) was acquired from Bohr Chemical Reagent Co., Ltd., Shanghai, China. N-propanol (AR), Isopropanol (AR), and Sulfolane were purchased from Mindray Biochemical Technology Co., Ltd., Shenzhen, China. HCl (AR) was purchased from China National Pharmaceutical Group Chemical Reagent Beijing Co., Ltd., Beijing, China. CO_2 ($\geq$99.99%, volume fraction) was purchased from Beijing Huanyu Jinghui Jingcheng Gas Technology Co., Ltd., Beijing, China. Throughout the experiments, deionized water was used and all the materials for this study were used without further purification.

2.2. Experimental Methods

2.2.1. Absorption Experiment

The schematic diagram of the absorption experimental device is shown in Figure 1, which is composed of a rotameter, a gas washing bottle, a constant temperature water bath device, and a soap film flowmeter. CO_2 enters the gas washing bottle through the flow rate controlled by the rotameter and reacts with the absorbent. After being absorbed and saturated, it flows out from the outlet of the washing bottle. The soap film flowmeter connected to the outlet records the CO_2 flow rate. The absorption device is in a constant temperature water bath, ensuring an absorption temperature of 40 °C. Adjust the rotameter to set the intake flow rate to 200 mL/min and introduce CO_2 into the prepared absorbent. After the absorbent is introduced into the gas, CO_2 is gradually measured from the outlet, and the gas flow rate is measured every 2 min using a soap film flowmeter. After the outlet flow rate is equal to the inlet flow rate, it is considered close to absorption saturation. Continue ventilation for 10 min and complete a set of experiments.

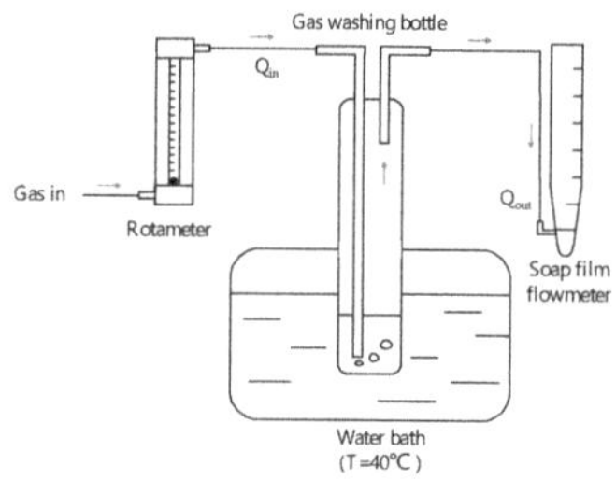

Figure 1. Schematic of absorption apparatus.

Evaluate the absorption performance of the absorbent through absorption rate and absorption load. According to Xiao et al. [41], the CO_2 absorption rate was defined as follows:

$$r_{co2,ab} = \frac{Q_{in} - Q_{out}}{n_{MEA} \times V_0}\left(\frac{P_{act}T_0}{P_0T_{act}}\right) \tag{1}$$

where $r_{co2,ab}$ ($molCO_2 \cdot mol^{-1} amine \cdot min^{-1}$) is the absorption rate, Q_{in} ($L \cdot min^{-1}$) is the intake flow rate, n_{MEA} (mol) is the molar amount in the washing gas cylinder, V_0 (22.4 $L \cdot mol^{-1}$) is the molar volume of gas at 1×10^5 Pa and 273.15 K, P_{act} (Pa) is the experimental gas pressure, and T_{act} (K) is the experimental gas temperature, P_0 (1×10^5 Pa), T_0 (273.15 K).

The absorption load can be calculated based on the absorption rate, and its formula is as follows:

$$\alpha = \int_0^t r_{co2,ab} dt \tag{2}$$

where α ($molCO_2 \cdot mol^{-1} amine$) is the absorption load, and t (min) is the reaction time.

After the absorption experiment, if the phase is separated, a measuring cylinder will be used to measure the volume of the lean phase and the rich phase to obtain the proportion of the rich phase volume. The proportion of rich phase volume was defined as follows:

$$\beta_1 = V_{rich}/(V_{rich} + V_{lean}) \tag{3}$$

where β_1 is the proportion of rich phase volume, V_{rich} (mL) is the volume of rich phase, and V_{lean} (mL) is the volume of the lean phase.

2.2.2. Desorption Experiment

The schematic diagram of the desorption experimental device is shown in Figure 2 and is composed of a condenser, a flask, a magnetic stirrer, and a soap film flowmeter. The phase change absorbent after CO_2 absorption is divided into two phases, and the rich phase is taken for desorption research. The high-temperature CO_2 produced after heating the rich phase is cooled by the condenser tube and then enters the soap film flowmeter connected to the air outlet to measure the flow rate at this time. Set up a heating magnetic stirrer to heat to a desorption temperature of 120 °C, take 10 mL of saturated absorbent, and put it into a constant temperature kerosene while ensuring the rotor is evenly stirred in the solution. After cooling and drying in the condenser pipe, the released gas is passed into the soap film flowmeter, and the outlet gas flow rate is measured every 2 min until it cannot be measured from the outlet to the gas overflow.

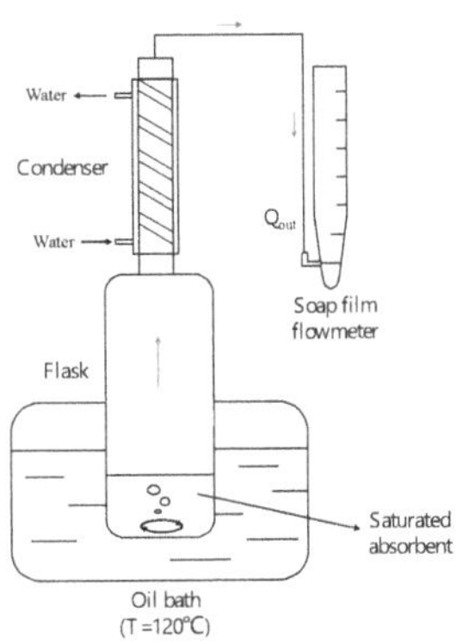

Figure 2. Schematic of desorption apparatus.

The desorption rate can be calculated by the gas outlet rate of the device, and is defined as follows:

$$r_{co2,des} = \frac{Q_{out}}{n_{MEA} \times V_0} \left(\frac{P_{act} T_0}{P_0 T_{act}} \right) \tag{4}$$

where $r_{co2,des}$ ($molCO_2 \cdot mol^{-1} amine \cdot min^{-1}$) is the desorption rate.

The desorption load can be obtained by integrating the desorption rate with time, and is defined as follows:

$$\alpha' = \int_0^t r_{co2,des} dt \tag{5}$$

where α' ($molCO_2 \cdot mol^{-1}$ amine) is the desorption load.

2.2.3. CO_2 Load Measurement Experiment

The CO_2 load measurement device is shown in Figure 3 and consists of a U-shaped tube, burette, agitator, and conical bottle. After the absorption of CO_2 by the absorbent is completed, let it stand for a while and observe stable phase separation. Separate the upper and lower liquid phases. Excessive hydrochloric acid is dripped into the liquid to be measured by a burette, and the CO_2 produced by the reaction enters the connected U-shaped tube for quantitative measurement. Record the initial reading V_1 when the U-tube is connected to the atmosphere, and record the amount of hydrochloric acid as V_2. Hydrochloric acid reacts with absorbent to release CO_2, creating a pressure difference in the U-shaped tube and a liquid level difference. After the reaction, adjust the position of the U-tube, and record the reading V_3 in the U-tube at this time.

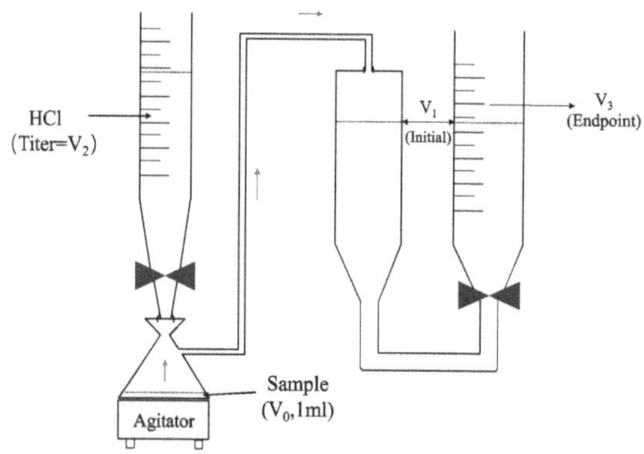

Figure 3. Schematic of CO_2 loading titration apparatus.

The load of CO_2 is defined as follows:

$$loading(\text{mol/L}) = \frac{V_3 - V_2 - V_1}{24.4 \times V_0} \tag{6}$$

where V_0 (mL) is the volume of the solution to be tested, and 24.4 refers to the volume of 1 mol gas at room temperature (298.15 K). V_1 (mL) is usually taken as the initial reading before the reaction in the U-shaped tube; V_2 is the titration amount of dilute hydrochloric acid, mL; V_3 (mL) is the reading of the end point after the completion of the reaction in the U-shaped tube.

3. Results

3.1. Phase Separation of Absorbent

A total weight of 100 g of absorbent was prepared by selecting n-propanol, isopropanol, and sulfolane as phase separators when the concentration of MEA was 30 wt% and the adsorption temperature was 313.15 K. That is, each absorbent includes 30 g of MEA. All ratios mentioned below are the mass ratio of phase separation additives to water under the premise of a total weight of 70 g. Table 1 shows the phase separation results at different mass ratios. It can be seen that the MEA/propanol/H_2O absorbents absorb CO_2 in the various phases until the mass ratio of 4:6, which is due to the large proportion of water in the system, which can completely dissolve the additive and absorb the reaction products.

Table 1. Phase separation results of MEA/phase separation agent/H_2O absorbent with different mass ratios.

Additive–Water	n-Propanol	Isopropanol	Sulfolane
1:9	x	x	x
2:8	x	x	x
3:7	x	x	√
4:6	√	√	√
5:5	√	√	√
6:4	√	√	√
7:3	√	√	√
8:2	√	√	√
9:1	√	√	√

According to the results of the phase separation of the phase-change absorber in the MEA system, when the weight ratio of phase separator–water reaches the extreme value, various absorption properties of phase-separated absorber are weak and the phase separation effect is not obvious. Therefore, three ratios of 3:7, 5:5, and 7:3 were selected for the absorption experiments with AEP as the main component of the phase change absorber, and the absorbent was uniformly separated after the absorption was completed.

Observe the phase change absorbent with phase separation, as shown in Tables 2 and 3. As the additive content increases, the volume of the rich phase gradually decreases, and the rich phase ratio decreases. At the same mass ratio, sulfolane brings the greatest rich phase ratio, and n-propanol and isopropanol are relatively small. The lower the proportion of the rich phase of the adsorbent after phase separation, the less the solution flows into the desorption tower for regeneration in the process, which is more conducive to reducing the energy consumption of regeneration. The 7:3 group of the AEP system absorbent added with n-propanol has an excellent absorption capacity and a relatively low phase-rich volume ratio of 0.46, which is beneficial for reducing regeneration energy consumption.

Table 2. Rich phase volume ratio of MEA/additive/H_2O absorbent.

Reagent Components	n-Propanol	Isopropanol	Sulfolane
3:7	Unseparated phase	Unseparated phase	0.88
5:5	0.67	0.66	0.68
7:3	0.47	0.65	0.67

Table 3. Rich phase volume ratio of AEP/additive/H_2O absorbent.

Reagent Components	n-Propanol	Isopropanol	Sulfolane
3:7	0.81	0.85	0.94
5:5	0.63	0.60	0.74
7:3	0.46	0.44	0.69

3.2. Phase Separation Mechanism

MEA and AMP can undergo phase separation because the solubility of MEA-CO_2 and AMP-CO_2 in water is higher than in alcohol after the reaction. CO_2 is introduced into the MEA/additive/water system, and the reaction is shown in Equation (7) [42]. The phase separation situation of MEA/n-propanol/H_2O absorbent is shown in Figure 4. This reaction is a salinization reaction, generating carbamate and protonated MEA, which were formed and could be regarded as MEA salt. Water will first solvate these components. With the addition of salt, phase separation would happen for a water-miscible organic liquid from its aqueous solution. MEA salts cause water clusters to evolve around them, forming alcohol clusters gradually. When the size of water and alcohol groups reaches the macroscopic scale, two phases are formed. After phase separation, the upper liquid phase mainly contains additives and related components and is a lean liquid phase. The lower liquid phase mainly contains CO_2 absorption products and MEA, making it a rich liquid phase.

$$CO_2 + 2MEA \leftrightarrow MEAH^+ + MEACOO^- \tag{7}$$

When the concentration of propanol in the absorbent increases, the CO_2 load in the lower phase increases, and the concentration of propanol slightly decreases. The CO_2 load in the upper phase remains unchanged. While the concentration of propanol increases, water content decreases [43]. The concentration of salt can affect the solubility of organic solvents, thereby affecting the phase separation of absorbents [44,45]. The increase in the concentration of propanol also represents a decrease in the concentration of water. After the absorption reaction is completed, the concentration of salt increases, but the amount of water that can interact with propanol decreases. As a result, more propanol rises to the lean phase, leading to a change in concentration.

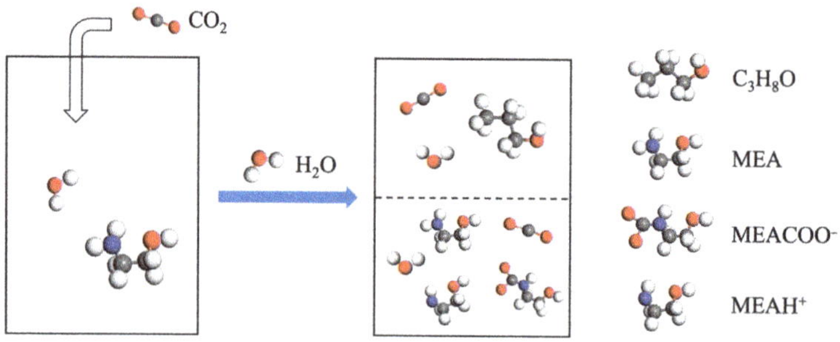

Figure 4. Phase separation diagram of MEA/n-propanol/H_2O absorbent.

The phase change behavior of absorbents is also related to the hydrophobicity of additives, and the higher the hydrophobicity, the more obvious the phase separation phenomenon. Sulfolane is a highly hydrophobic solvent, while the alcohol amines and reaction-related products used in the experiment are hydrophilic solutes. Therefore, as CO_2 is introduced, the concentration of the products gradually increases. Sulfolane is gradually discharged from the aqueous phase, and the solution is divided into two phases from a homogeneous phase. After the absorbent of the AEP/additive/water system absorbs CO_2, a reaction occurs, as shown in Equation (8). The product ($AEPH^+$ and $AEPCOO^-$ as shown in Figure 5) and additive are immiscible and migrate to the upper liquid phase to form a rich phase with amines, while the remaining additives and related components remain in the lower liquid phase as a lean solution.

$$CO_2 + 2AEP \leftrightarrow AEPH^+ + AEPCOO^- \tag{8}$$

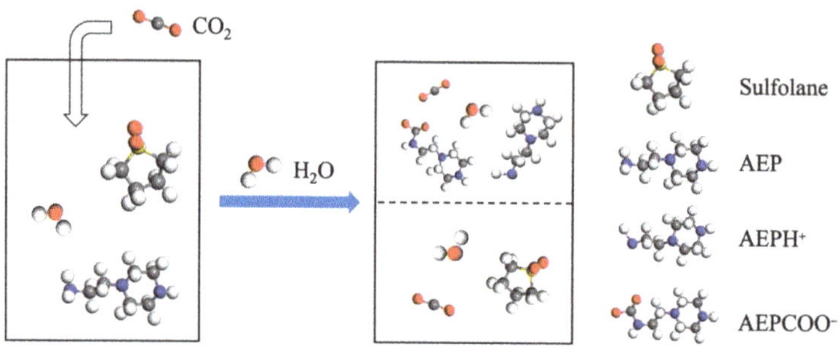

Figure 5. Diagram of the relationship between the phase separation phenomenon and the dipole moment of AEP/sulfolane/H_2O absorbent (Debye is the dipole moment).

The phase separation behavior of phase change absorbents may also be related to the polarity of the components. Typically, the larger the dipole moment of a molecule, the stronger its polarity [46]. The dipole moments of AEP and sulfolane are small, while the reaction products with CO_2 have high polarity. It can be inferred from this that after absorbing CO_2, the rich liquid phase contains highly polar products, which cannot be miscible with sulfolane, resulting in a phase transition, as shown in Figure 5.

In summary, the essence of the phase change absorbent adding organic solvent as a phase separation agent is to replace the water in the phase change absorbent with organic solvent to achieve the phase separation effect while reducing the volume proportion of the rich phase. The reaction of MEA and AEP in contact with CO_2 is similar, and the difference in phase separation ratio is mainly due to the characteristics of the added organic solvent.

3.3. Research on CO_2 Absorption by Phase Change Absorbent

3.3.1. Phase Change Absorbent with MEA as the Main Component

The mixed solvent absorption loads of MEA and different additives are shown in Figure 6. The absorption load curves all showed that the rate of absorption load increased rapidly in the early stage of adsorption, and the increased rate of absorption load slowed down with the extension of time. Overall, the phase change absorbents with MEA as the main absorption component have the highest absorption load when the mass ratio of the phase separation agent to water is 5:5. Among them, MEA/n-propanol (0.65 molCO_2·mol^{-1}amine) > MEA/sulfolane (0.61 molCO_2·mol^{-1}amine) > MEA/isopropanol (0.45 molCO_2·mol^{-1}amine). Liu et al. [28] mixed DEEA and MEA aqueous solution as phase change absorbents, and the CO_2 absorption load at 40 °C was 0.49 molCO_2·mol^{-1}amine. Zhang et al. [47] studied the optimal ratio and performance of MEA/n-propanol/H_2O and adjusted the mass ratio of n-propanol to water to maintain the MEA with a mass fraction of 30%. The results showed that the optimal ratio of m (MEA):m (n-propanol):m (H_2O) = 3:3:4 system has a circulating CO_2 load of 1.70 mol/kg (absorbent), which is 1.3 times that of 30% MEA. In this paper, the maximum absorption load at m (MEA):m (n-propanol):m (H_2O)= 3:3.5:3.5 was obtained, which is close to the known results.

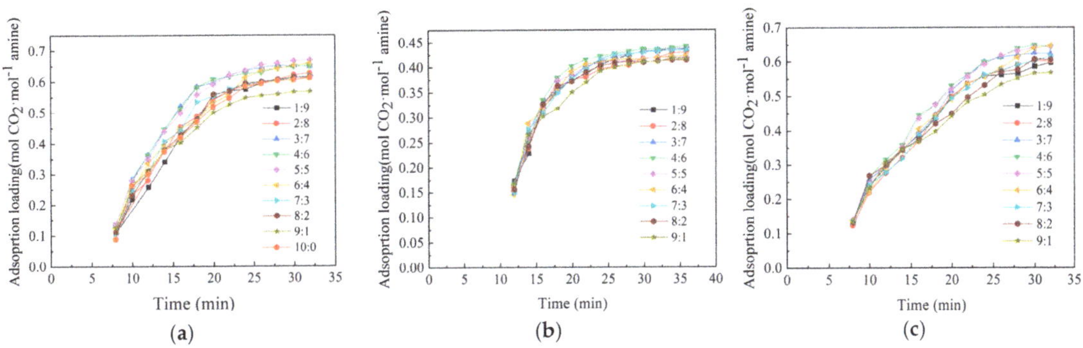

Figure 6. Absorption loads of MEA-based absorbents with different additives at different concentrations: (**a**) MEA/n-propanol; (**b**) MEA/isopropanol; (**c**) MEA/sulfolane.

The absorption rates of mixed solvents of MEA and different additives are shown in Figure 7. As shown in the figure, the absorption rate of each group shows a similar trend. In the initial stage of gas introduction, the absorption rate rapidly decreases. As the ventilation time increases, the downward trend of the absorption rate slows down until it approaches 0. Overall, there is not much difference in the absorption rate of each group of absorbents under different weight ratios. At the same 5:5 ratio, initial absorption rate was sulfolane (0.020 molCO_2·mol^{-1}amine·min^{-1}) > N-propanol (0.016 molCO_2·mol^{-1}amine·min^{-1}) > Isopropanol (0.014 molCO_2·mol^{-1}amine·min^{-1}). The 5:5 group, which performed excellently in absorbing load, was found to have a moderate absorption rate among different mass ratio groups. Zhang et al. [47] found that in the MEA/n-propanol/H_2O system, the initial absorption rate of the system with the optimal ratio of m (MEA):m (n-propanol):m (H_2O) = 3:3:4 was 5 times that of the mass fraction of 30% MEA. The presence of propanol will increase the diffusion coefficient of CO_2 and thus increase the absorption rate of the absorbent, but the high concentration of propanol will inhibit the MEA and CO_2 reaction.

The addition of sulfolane has a more significant effect on increasing the CO_2 absorption rate, because sulfolane has a strong affinity with acidic gases.

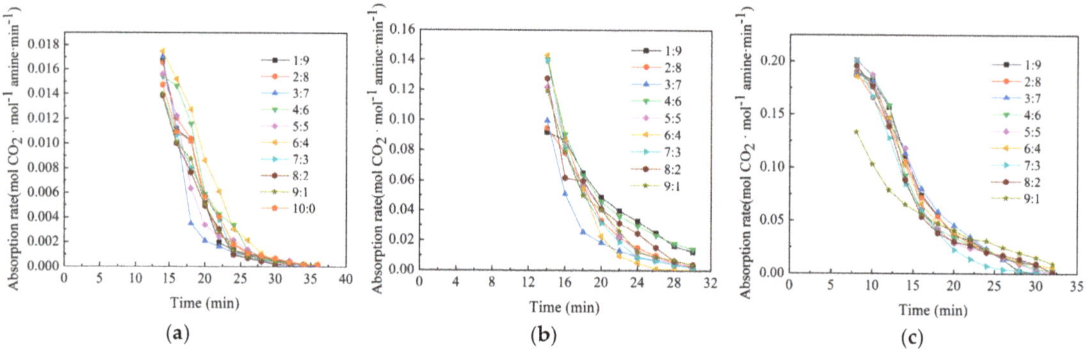

Figure 7. Absorption rate of MEA-based absorbents with different additives at different concentrations: (**a**) MEA/n-propanol; (**b**) MEA/isopropanol; (**c**) MEA/sulfolane.

3.3.2. Phase Change Absorbent with AEP as the Main Component

The mixed solvent absorption loads of AEP and different additives are shown in Figure 8. The absorption load of the 7:3 group in each absorbent was significantly higher than the other two groups, with AEP/n-propanol (2.03 $molCO_2 \cdot mol^{-1}amine$) > AEP/sulfolane (1.60 $molCO_2 \cdot mol^{-1}amine$) > AEP/isopropanol (1.56 $molCO_2 \cdot mol^{-1}amine$). The absorption rates of mixed solvents of AEP and different additives are shown in Figure 9. The rate trend is the same as that of the MEA system absorbent. The 7:3 group performed best in absorbing load but the slowest absorption rate. Li et al. [37] optimized the ratio of AEP and 1-propanol, and the results showed that under the optimal ratio of m (AEP):m (1-propanol):m (H_2O) = 2:4:4, the CO_2 loading capacity was 2.3 times that of 30% MEA.

The absorption rates of AEP and different additives in mixed solvents are shown in Figure 9. At the same 7:3 ratio, the initial absorption rates of different phase separation agents were as follows: isopropanol (0.016 $molCO_2 \cdot mol^{-1}amine \cdot min^{-1}$) > sulfolane (0.015 $molCO_2 \cdot mol^{-1}amine \cdot min^{-1}$) > n-propanol (0.014 $molCO_2 \cdot mol^{-1}amine \cdot min^{-1}$). The absorption saturation time of AEP-based absorbent is longer than that of MEA-based absorbent, and it is generally completed 45 min after ventilation, but the absorption load is greater.

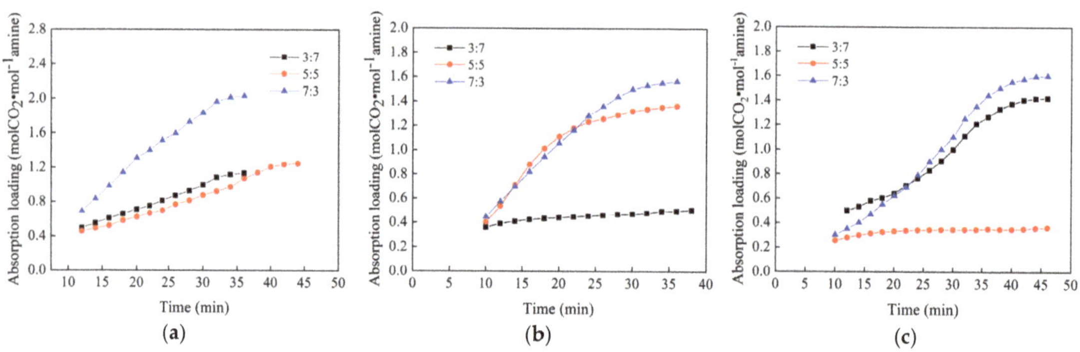

Figure 8. Absorption load of AEP-based absorbents with different additives at different concentrations: (**a**) AEP/n-propanol; (**b**) AEP/isopropanol; (**c**) AEP/sulfolane.

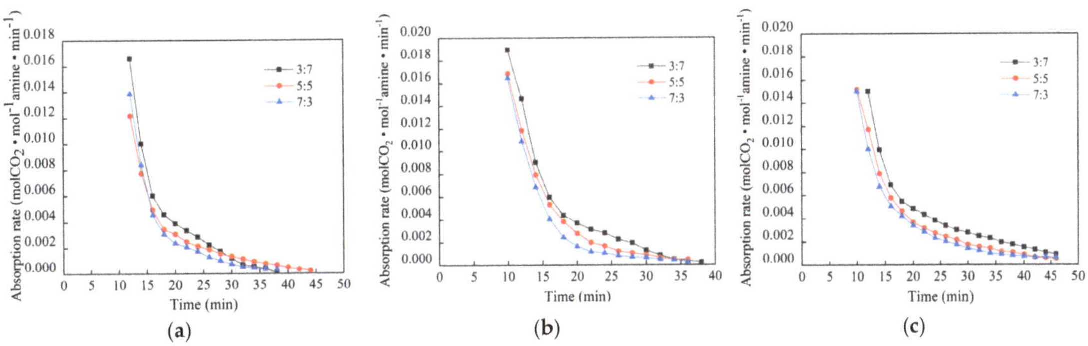

Figure 9. Absorption rate of AEP-based absorbents with different additives at different concentrations: (**a**) AEP/n-propanol; (**b**) AEP/isopropanol; (**c**) AEP/sulfolane.

Table 4 shows the key performance parameters for MEA and AEP to obtain the optimal absorption load. The phase change absorbent of the MEA system has the best absorption effect when the mass ratio of the phase separation agent to water is 5:5. The phase change absorbents of the AEP system have the best absorption effect when the mass ratio of the phase separation agent to water is 7:3. The absorption load is higher than that of the MEA system. Compared to MEA, the primary (secondary) amino group of AEP can add a strong absorb effect. Its steric hindrance effect is also beneficial for the regeneration cycle, and propanol plays a major role in phase separation in the system. The absorption rate of the absorbents of the two systems decreased significantly within 15 min, and the absorption rate curves were similar, which may be because the sample mass was small and the difference was not obvious.

Table 4. Comparison of optimal absorption effect points for MEA-based and AEP-based absorbents.

	MEA	AEP
Phase separator	n-propanol	n-propanol
Mass ratio	5:5	7:3
Highest absorption load	0.65 $molCO_2 \cdot mol^{-1}amine$	2.03 $molCO_2 \cdot mol^{-1}amine$

In terms of CO_2 absorption performance, the experiments in this paper show that the absorber with n-propanol added possesses the maximum absorption load. On the one hand, the volatilization rate of n-propanol is smaller than that of isopropanol, and the high volatility of the absorbent components is not favorable to the overall reaction balance of the absorption process, and also leads to the loss of absorbent, which in turn affects the absorption effect. On the other hand, MEA reacts with CO_2 to produce salts with different solubility in different solvents, and the polarity of organic solvents has a great influence on the solubility of substances. Overall, the solubility and absorption rate of CO_2 when propanol is used as a solvent is better than that of water [48]. Compared with alcohols, sulfolane has a stronger affinity for CO_2, so it can effectively improve the CO_2 absorption performance, but the melting point of sulfolane is extremely low, and the density of the phase change absorber of the sulfolane system is much larger, and the amine is required to have a certain degree of hydrophilicity [49]; otherwise, it is possible to have solid–liquid partitioning, which to a certain extent increases the uncontrollable factors and influences the actual absorption effect.

3.4. Research on CO_2 Desorption of Phase Change Absorbents

3.4.1. Phase Change Absorbent with MEA as the Main Component

The phase change absorber splits into two phases after absorbing carbon dioxide, which has the advantage of reducing the consumption of renewable energy by simply heating the rich phase for desorption. The rich phase of MEA/additive aqueous solution with a mass ratio of 5:5 was investigated under the most suitable desorption conditions (desorption temperature of 120 °C) and the desorption curves were tested, as shown in Figure 10. It can be seen that with the increase in desorption time, the desorption rates of the three adsorbents showed the change rule of increasing and then decreasing. At the beginning of desorption, the temperature of the solution increased rapidly due to the effect of 120 °C oil bath, and the desorption rate also increased, and reached the maximum value at the measurement point of 2 min; thereafter, due to the gradual decrease in the concentration of carbon dioxide in the product, the desorption rate began to decrease rapidly, but the three kinds of solutions slowed down after 10 min, 14 min, and 12 min, and the gas could not be detected from the outlet at 28 min, until it was almost 0. The desorption curve of the three adsorbents showed a change rule of increasing and then decreasing. The desorption load generally reached a high level at 10 min. At this point, the desorption of MEA/n-propanol was 77% of the final desorption (0.44 mol molCO_2·mol^{-1}amine), MEA/isopropanol was 84% of the final desorption (0.30 molCO_2·mol^{-1}amine), and MEA/sulfolane was 93% of the final desorption (0.40 molCO_2·mol^{-1}amine).

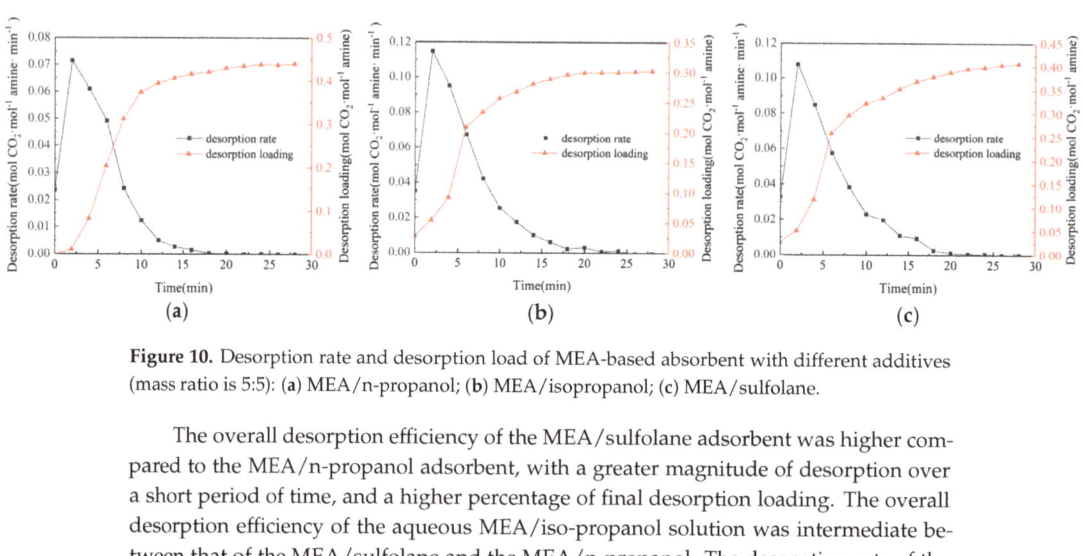

Figure 10. Desorption rate and desorption load of MEA-based absorbent with different additives (mass ratio is 5:5): (**a**) MEA/n-propanol; (**b**) MEA/isopropanol; (**c**) MEA/sulfolane.

The overall desorption efficiency of the MEA/sulfolane adsorbent was higher compared to the MEA/n-propanol adsorbent, with a greater magnitude of desorption over a short period of time, and a higher percentage of final desorption loading. The overall desorption efficiency of the aqueous MEA/iso-propanol solution was intermediate between that of the MEA/sulfolane and the MEA/n-propanol. The desorption rate of the aqueous MEA/iso-propanol solution was faster than that of the MEA/sulfolane and the MEA/isopropanol adsorbent over the period of 2–10 min. MEA/n-propanol adsorbents, but the final desorption was not as high as these two adsorbents because the absorption load was smaller than them.

3.4.2. Phase Change Absorbent with AEP as the Main Component

AEP/additive aqueous solutions with a mass ratio of 7:3 were investigated under optimum desorption conditions (desorption temperature of 120 °C), and its desorption profile was tested as shown in Figure 11. The trends in desorption rate and desorption load profiles were similar to those of the MEA/additive aqueous solution adsorbent, but the desorption time was shorter. As the desorption time increased, the desorption load increased and the rate of increase slowed down with time until saturation. The desorption load generally reached basic saturation at 6 min and the final desorption load was reached at 12 min when the rate of resolution was almost zero. AEP/n-propanol (1.00 $molCO_2{\cdot}mol^{-1}amine$) > AEP/iso-propanol (0.73 $molCO_2{\cdot}mol^{-1}amine$) > AEP/sulfolane (0.66 $molCO_2{\cdot}mol^{-1}amine$). The 30 wt% AEP/n-propanol aqueous solutions of the three adsorbents with AEP as the main component had higher rich phase desorption loadings and faster desorption rates at a desorption temperature of 393.15 K. The results showed that the three adsorbents had higher rich phase desorption loadings than those of the three adsorbents.

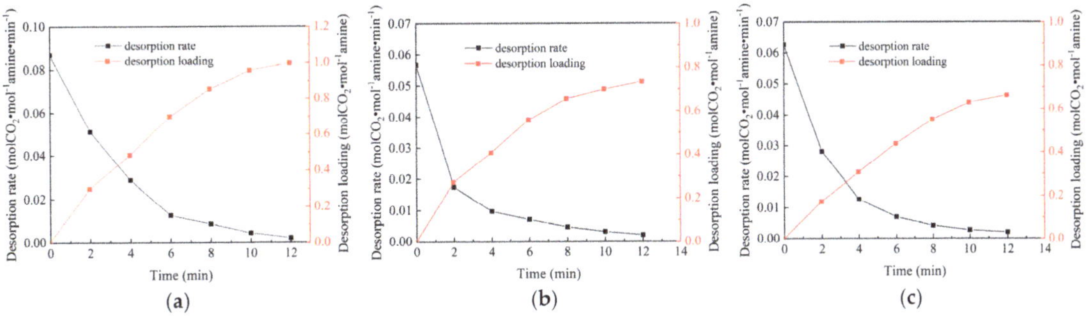

Figure 11. Desorption rate and desorption load of AEP-based absorbent with different additives (mass ratio is 7:3): (**a**) AEP/n-propanol; (**b**) AEP/isopropanol; (**c**) AEP/sulfolane.

Table 5 shows a comparison of the desorption loadings obtained for MEA-based and AEP-based absorbent at mass ratios of 5:5 and 7:3, respectively. The aqueous solution of n-propanol as the additive in both MEA-based and AEP-based absorbents had the highest desorption load and the fastest desorption rate among the three additive solutions. Because of the higher absorber load in the AEP-based absorbent, the desorption load is also higher than in the MEA-based absorbent. However, the absorber desorption rate is slower in the AEP-based absorbent. After absorption, the absorbent-rich phase is separated by static separation. The rich absorber phase of the AEP-based absorbent has a higher viscosity than the MEA-based absorbent. After placing the solution in the desorption unit, the rotor takes some time to stir smoothly, which affects the initial desorption rate.

Table 5. Comparison of desorption loads of MEA and AEP systems with different additives.

		n-Propanol	Isopropanol	Sulfolane
MEA	Mass ratio	5:5	5:5	5:5
	desorption loads ($molCO_2{\cdot}mol^{-1}amine$)	1.00	0.73	0.66
AEP	Mass ratio	7:3	7:3	7:3
	desorption loads ($molCO_2{\cdot}mol^{-1}amine$)	1.00	0.73	0.66

3.5. CO_2 Loading Experiment of Phase Change Absorbent

Absorption load is the total amount absorbed as the absorber reaches saturation with increasing absorption time. All measurements in this section are rich phase loads, indicating the amount of CO_2 absorption per liter of the rich phase of the absorbent, which is an indicator to test the absorption performance. The homogeneous absorbent is left to stand after absorbing CO_2 and is divided into liquid–liquid phases, i.e., the lean phase and the rich phase.

3.5.1. Phase Change Absorbent with MEA as the Main Component

The results of the loading size of rich phase CO_2 for MEA/additives and the relationship between rich phase density and mass ratio are shown in Figure 12. The rich phase density of different absorbents increased with additive quality.

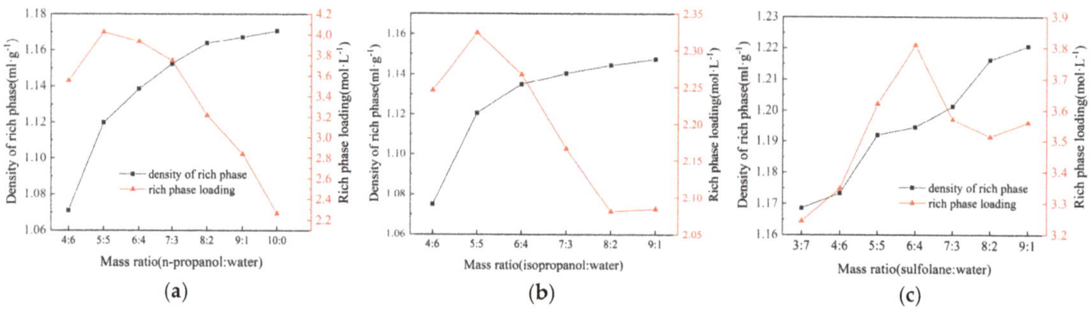

Figure 12. Density and load of MEA-based absorbent rich phase (mass ratio is 5:5): (**a**) MEA/n-propanol; (**b**) MEA/isopropanol; (**c**) MEA/sulfolane.

From the experimental results shown in Figure 12, it can be seen that different solvent systems have a great influence on the CO_2 loading capacity. After the passage of sufficient CO_2, the rich phase loading of MEA/iso-propanol ranges from 2.0 to 2.4 mol/L, while the rich phase loading of MEA/sulfolane and MEA/n-propanol is above 2 mol/L. The rich phase loading of MEA/sulfolane and MEA/n-propanol is above 2 mol/L. The rich phase loading did not increase as the proportion of n-propanol increased, with the highest value occurring around 5:5 at 4.02 mol/L. Although the adsorption loadings of the solutions did not change much, the rich phase loading decreased due to the increase of CO_2 in the lean phase. The MEA/iso-propanol system was similar to the MEA/n-propanol system; the rich phase load decreases after the peak value (2.32 mol/L) when the mass ratio is 5:5, but the reduction amplitude is smaller than that of MEA/n-propanol system. The rich phase CO_2 loading reached a maximum value of 3.81 mol/L at a mass ratio of 6:4 and then decreased. The mass ratio of sulfolane to water did not affect the loading capacity of the rich phase, which may be because some of the CO_2 products went into the lean phase.

In the MEA/n-propanol system, although the fastest absorption rate was observed at a mass ratio of 7:3, the highest rich phase loading was observed at 5:5, which had the highest rich phase loading of the three solutions. The MEA/n-propanol solution with a mass ratio of 7:3 had a relatively large phase-rich ratio, although the phase-rich loading was also larger. In addition, the solidification temperature range of sulfolane between 21.8 °C and 27.3 °C may affect the adsorption effect. Therefore, combining the above factors, the MEA/n-propanol system with 5:5 had the best adsorption effect with a suitable rich phase ratio and ideal loading capacity.

3.5.2. Phase Change Absorbent with AEP as the Main Component

The results of the magnitude of the phase-rich CO_2 loading of AEP/additives and the phase-rich density versus mass ratio are shown in Figure 13. The rich phase densities of the different absorbents all increased with the increase in additive mass. It was found that the maximum rich phase CO_2 loading was achieved at a mass ratio of 5:5, and the minimum loading of AEP/n-propanol was 0.71 mol/L. In contrast, the minimum loadings of AEP/iso-propanol and AEP/sulfolane were 0.49 mol/L and 1.12 mol/L, respectively. The maximum loading of AEP/sulfolane system at 7:3 was 1.56 mol/L, which was higher than that of AEP/n-propanol at 0.99 mol/L. However, the rich phase ratio was too large, the viscosity was high, and the sulfolane, which was a solid at room temperature, had to be melted before use, which consumed more energy and was unsuitable for recycling in the actual process.

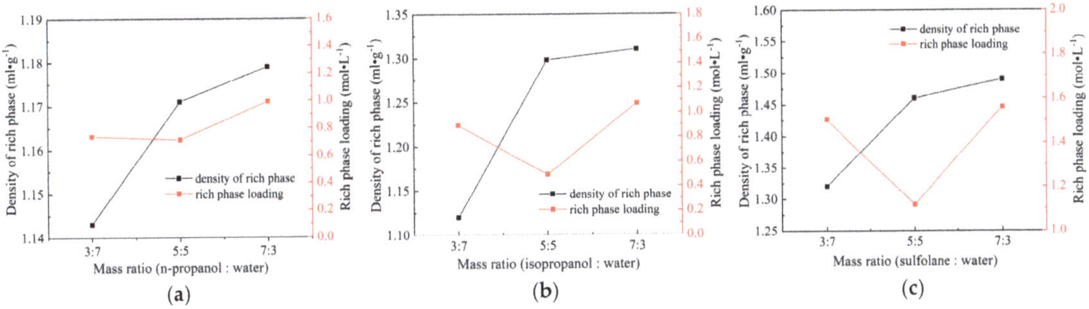

Figure 13. Density and load of AEP-based absorbent rich phase (mass ratio is 7:3): (**a**) AEP/n-propanol; (**b**) AEP/isopropanol; (**c**) AEP/sulfolane.

Table 6 shows a comparison of the maximum points of the rich phase loading capacity of the MEA and AEP systems with different additives. In general, the phase rich load of the MEA system is larger than that of the AEP system. After phase separation, the rich phase viscosity of the AEP system absorbent is obviously greater than that of the MEA system, and the high viscosity is not conducive to material transfer in solution and also affects the chemical reaction in the loading experiment. In practical application, when the rich phase viscosity of the absorbent increases, the heat transfer area of the rich and lean liquid needs to be changed, and the heat load of the reboiler increases, thus increasing the generation energy consumption.

Table 6. Comparison of rich phase loads of MEA and AEP systems with different additives.

		n-Propanol	Isopropanol	Sulfolane
MEA	Mass ratio	5:5	5:5	6:4
	Rich phase load (mol/L)	4.02	2.32	3.81
AEP	Mass ratio	7:3	7:3	7:3
	Rich phase load (mol/L)	0.99	1.07	1.56

4. Discussion

The CO_2 adsorption performance of the MEA-based system had a maximum value of 0.65 molCO_2·mol^{-1}amine with n-propanol as an additive and a phase separator to water ratio of 5:5, while the CO_2 absorption performance of the AEP-based system had a maximum value of 2.03 molCO_2·mol^{-1}amine with n-propanol as an additive and a phase separator to water ratio of 7:3. Among the similar results reported so far, Zhang et al. [47] reported that the cyclic capacity of the CPCA with 30% MEA + 30% 1-propanol + 40% H_2O was 1.70 mol/kg, and Li et al. [37] proposed a novel biphasic solvent of AEP/1-propanol/H_2O solution (under the optimal ratio of m (AEP):m (1-propanol):m (H_2O) = 2:4:4), and the absorption capacity was 1.26 molCO_2·mol^{-1}amine. The experiments in this paper contain less water components in the dosage of the absorbent, which theoretically has a lower energy consumption while maintaining the existing excellent CO_2 absorption performance because the latent heat of vaporization of water is reduced.

The lower the rich phase of the absorber after phase separation, the lower the flow rate of the solution into the desorber for regeneration in the process, and the more favorable to reduce the regeneration energy consumption. In addition, the AEP/n-propanol solution has a relatively low rich phase ratio of 0.46 at a ratio of 7:3, which is a relatively excellent performance in reducing energy consumption.

Phase change absorbers have higher viscosities compared to typical chemical absorbers. Xu et al. [50] measured the viscosity of a formulated 4M DEEA/2M BDA absorber at an absorption temperature of 40 °C, which was increased by a factor of 6 compared to the commonly used 30% WMEA single-phase absorber. The viscosity of the phase-rich phase of the AEP system absorber was significantly higher than that of the MEA system absorber after phase separation. After phase separation, the viscosity of the rich phase of the AEP system absorber is significantly higher than that of the MEA system absorber, which may be inconvenient for the transfer of substances in solution and affect the chemical reactions in the loading experiments. In practice, when the viscosity of the absorbent rich phase increases, the rich-rich heat exchange area needs to be changed because the heat exchange of the heat exchanger will be affected [51], and the above series of reactions increase the heat load of the reboiler, which in turn increases the regeneration energy consumption. However, high viscosity is beneficial for better phase stabilization of the absorber after CO_2 absorption, so that a separation state can be maintained between the CO_2 absorbed rich liquid and the unreacted absorber-lean liquid, which is helpful for the subsequent separation operation. Moreover, a larger viscosity allows the absorbent to flow more slowly during the reaction process, thereby prolonging the contact time between the CO_2 and the absorbent and potentially improving the efficiency of CO_2 absorption.

Considering the one-time investment, operation cost, energy consumption, waste disposal, and other factors, the chemical absorption method can effectively bring out the advantages of comprehensive utilization of waste heat in iron and steel enterprises. It can make good use of low-temperature thermal mass, minimize investment, and improve profitability. Compared with the traditional absorption method, the phase change absorption method can reduce the regeneration energy consumption of the system, which is considered a CO_2 capture technology with great development potential. However, problems such as high enrichment ratio, high solution viscosity, and high volatility still exist. Currently, research in this field focuses on screening better absorbers by changing the composition, but systematic studies on the kinetics of phase separation time after absorption and how to reduce volatility and viscosity are still lacking and need to be carried out in depth to ensure the maximization of benefits and further industrialization and promotion.

5. Conclusions

In this work, based on 30%wt MEA and 30%wt AEP, different absorbents were prepared to test the absorption effect of CO_2, and the regeneration performance of absorbents was tested using the thermal desorption method. The load test results calculated the load capacity of different absorbents. In this paper, the phase separation mechanism of different absorbent systems is also analyzed, and the changes in rich phase density and rich ratio of each phase transition absorbent after phase separation are studied. Finally, the phase change absorbent suitable for CO_2 absorption was selected based on the results. The main conclusions are as follows:

1. Different absorbents were mixed based on 30%wt MEA and 30%wt AEP. The phase separation mechanisms of the different absorbent systems were analyzed, and the changes in physical properties, such as rich phase density and rich phase ratio of each phase change absorbent after phase separation were investigated. It was found that, with the increase in additive content, different types of absorbents showed the law that the volume of the rich phase gradually decreased, the ratio of the rich phase gradually decreased, and the density of the rich phase gradually increased.

2. The CO_2 phase change absorption effect was experimentally tested. The results showed that the MEA phase change absorbent had the best absorption effect when the mass ratio of additive to water was 5:5. In contrast, the AEP system had the best absorption effect at 7:3. The absorption effects of different phase separators were as follows: n-propanol > sulfolane > iso-propanol, and AEP/n-propanol/H_2O (7:3) had a maximum absorption load of 2.03 molCO_2·mol^{-1}amine. The AEP/n-propanol absorbent at 7:3 had excellent absorption load and a relatively low rich phase volume ratio of 0.46.
3. The loading capacities of different absorbents were calculated based on the results of the loading tests. It was concluded that the rich phase loading capacity of the MEA system is greater than that of the AEP system, and the maximum loading capacity of the MEA/n-propanol/H_2O (5:5) system was 4.02 mol/L.
4. The regeneration performance of the absorbent was tested by the thermal desorption method, and the results showed that the desorption load of the n-propanol aqueous-rich phase in the absorbent was higher, and the desorption rate was fastest at the temperature of 393.15 K. Compared with the MEA system absorbent, the AEP system absorbent had a higher desorption load but a slower desorption rate.

Author Contributions: Conceptualization, Y.W. and F.C.; methodology, Y.L., X.Y. and Z.L.; software, F.C.; validation, Y.W.; formal analysis, J.L. and H.W.; investigation, J.L.; resources, Z.L. and X.Y.; data curation, J.L. and H.W.; writing—original draft preparation, F.C.; writing—review and editing, Y.W.; visualization, Y.W. and F.C.; supervision, Z.L. and X.Y.; project administration, Y.L., Z.L. and X.Y.; funding acquisition, X.Y. All authors have read and agreed to the published version of the manuscript.

Funding: This research was funded by the National Key R&D Program of China, grant number 2022YFC3005803, and Xizang Autonomous Region Science and Technology Plan, grant number XZ202401YD0004.

Data Availability Statement: The original contributions presented in this study are included in the article; further inquiries can be directed to the corresponding authors.

Conflicts of Interest: Author Fanghui Cheng was employed by the Shaanxi Blower (Group) Co., Ltd. Author Jingsong Li was employed by the Research Institute of Technology of Shougang Group. Author Haihong Wang was employed by the company R&D Center Department, Beijing District Heating Group. The remaining authors declare that the research was conducted in the absence of any commercial or financial relationships that could be construed as a potential conflict of interest.

References

1. Mon, M.T.; Tansuchat, R.; Yamaka, W. CCUS Technology and Carbon Emissions: Evidence from the United States. *Energies* **2024**, *17*, 1748. [CrossRef]
2. Gupta, N.; Tanwar, R.; Dipesh; Kaushik, A.; Singh, R.; Patra, A.; Sar, P.; Khakharia, P. Perspectives on CCUS deployment on large scale in India: Insights for low carbon pathways. *Carbon Capture Sci. Technol.* **2024**, *12*, 100195. [CrossRef]
3. Zhang, Y.J.; Gao, F.; Wang, Z.H. Updated resource depletion characterization factors for life cycle assessment-case studies on iron and steel production in China. *Mater. Sci. Forum* **2016**, *847*, 358–365. [CrossRef]
4. Jin, P.; Jiang, Z.; Bao, C.; Lu, Y.; Zhang, J.; Zhang, X. Mathematical Modeling of the Energy Consumption and Carbon Emission for the Oxygen Blast Furnace with Top Gas Recycling. *Steel Res. Int.* **2016**, *87*, 320–329. [CrossRef]
5. Yang, N.; Yu, H.; Li, L.; Xu, D.; Han, W.; Feron, P. Aqueous ammonia (NH_3) based post combustion CO_2 capture: A review. *Oil Gas Sci. Technol.–Rev. IFP Energ. Nouv.* **2014**, *69*, 931–945. [CrossRef]
6. Aronu, U.E.; Hessen, E.T.; Haug-Warberg, T.; Hoff, K.A.; Svendsen, H.F. Vapor–liquid equilibrium in amino acid salt system: Experiments and modeling. *Chem. Eng. Sci.* **2011**, *66*, 2191–2198. [CrossRef]
7. Ayittey, F.K.; Saptoro, A.; Kumar, P.; Wong, M.K. Parametric study and optimisation of hot K_2CO_3-based post-combustion CO_2 capture from a coal-fired power plant. *Greenh. Gases Sci. Technol.* **2022**, *10*, 631–642. [CrossRef]
8. Luo, H.; Chioyama, H.; Thürmer, S.; Ohba, T.; Kanoh, H. Kinetics and structural changes in CO_2 capture of K_2CO_3 under a moist condition. *Energy Fuels* **2015**, *29*, 4472–4478. [CrossRef]

9. Luo, Q.; Cao, Y.; Liu, Z.; Feng, B.; Zhou, Q.; Li, N. A feasible process for removal and utilization of CO_2 in thermal power plants by MDEA+ DMSO scrubbing and Cu/TiO_2 photocatalytic reduction. *Appl. Therm. Eng.* **2019**, *153*, 369–378. [CrossRef]
10. Sun, Q.; Liu, Y.; Liu, A.; Guo, X.; Yang, L. Separation of CO_2 from Flue Gas via Absorption of Alcohol Amines. *Univers. J. Chem.* **2016**, *4*, 25–30. [CrossRef]
11. Lv, B.; Guo, B.; Zhou, Z.; Jing, G. Mechanisms of CO_2 capture into monoethanolamine solution with different CO_2 loading during the absorption/desorption processes. *Environ. Sci. Technol.* **2015**, *49*, 10728–10735. [CrossRef] [PubMed]
12. Donaldson, T.L.; Nguyen, Y.N. Carbon dioxide reaction kinetics and transport in aqueous amine membranes. *Ind. Eng. Chem. Fundam.* **1980**, *19*, 260–266. [CrossRef]
13. Yu, H.; Qi, G.; Xiang, Q.; Wang, S.; Fang, M.; Yang, Q.; Wardhaugh, L.; Feron, P. Aqueous ammonia based post combustion capture: Results from pilot plant operation, challenges and further opportunities. *Energy Procedia* **2013**, *37*, 6256–6264. [CrossRef]
14. Zhang, Z.; Li, Y.; Zhang, W.; Wang, J.; Soltanian, M.R.; Olabi, A.G. Effectiveness of amino acid salt solutions in capturing CO_2: A review. *Renew. Sustain. Energy Rev.* **2018**, *98*, 179–188. [CrossRef]
15. Han, K.; Ahn, C.K.; Lee, M.S. Performance of an ammonia-based CO_2 capture pilot facility in iron and steel industry. *Int. J. Greenh. Gas Control* **2014**, *27*, 239–246. [CrossRef]
16. Mao, Y.L.; Qu, Y.L.; Li, B.; Jing, X.; Zhu, J.B. Development and application potential analysis of carbon dioxide capture technology from flue gas in steel works. *IronSteel* **2016**, *51*, 6–10. [CrossRef]
17. Gao, W.; Liang, S.; Wang, R.; Jiang, Q.; Zhang, Y.; Zheng, Q.; Xie, B.; Toe, C.Y.; Zhu, X.; Wang, J.; et al. Industrial carbon dioxide capture and utilization: State of the art and future challenges. *Chem. Soc. Rev.* **2020**, *49*, 8584–8686. [CrossRef]
18. Feron, P.H.M.; Jansen, A.E. Capture of carbon dioxide using membrane gas absorption and reuse in the horticultural industry. *Energy Convers. Manag.* **1995**, *36*, 411–414. [CrossRef]
19. Gautam, A.; Monday, M.K. Review of recent trends and various techniques for CO_2 capture: Special emphasis on biphasic amine solutions. *Fuel* **2023**, *334*, 126616. [CrossRef]
20. Liu, F.; Qi, Z.; Fang, M.; Ding, H. Pilot test of water lean solvent of 2- (ethylamine) ethanol, 1-methyl-2-pyrrolidinone, and water for post combustion CO_2 capture. *Chem. Eng. J.* **2023**, *459*, 141634. [CrossRef]
21. Idem, R.; Wilson, M.; Tontiwachwuthikul, P.; Chakma, A.; Veawab, A.; Aroonwilas, A.; Gelowitz, D. Pilot plant studies of the CO_2 capture performance of aqueous MEA and mixed MEA/MDEA solvents at the University of Regina CO_2 capture technology development plant and the boundary dam CO_2 capture demonstration plant. *Ind. Eng. Chem. Res.* **2006**, *45*, 2414–2420. [CrossRef]
22. Khan, A.A.; Halder, G.N.; Saha, A. Experimental investigation on efficient carbon dioxide capture using piperazine (PZ) activated aqueous methyldiethanolamine (MDEA) solution in a packed column. *Int. J. Greenh. Gas Control* **2017**, *64*, 163–173. [CrossRef]
23. Gautam, A.; Monoj, K.M. Novel aqueous amine blend of 2-(Butylamino) ethanol and 2-Dimethylaminoethanol for CO_2 capture: Equilibrium CO_2 loading, RSM optimization, desorption study, characterization and toxicity assessment. *Sep. Purif. Technol.* **2023**, *322*, 124279. [CrossRef]
24. Archane, A.; Gicquel, L.; Provost, E.; Fürst, W. Effect of methanol addition on water–CO_2–diethanolamine system: Influence on CO_2 solubility and on liquid phase speciation. *Chem. Eng. Res. Des.* **2008**, *86*, 592–599. [CrossRef]
25. Ye, Q.; Wang, X.; Lu, Y. Screening and evaluation of novel biphasic solvents for energy-efficient post-combustion CO_2 capture. *Int. J. Greenh. Gas Control* **2015**, *39*, 205–214. [CrossRef]
26. Hu, Y.; Wang, Q.; Hu, D.; Zhang, Y.; Furqan, M.; Lu, S. Experimental study on CO_2 capture by MEA/n-butanol/H_2O phase change adsorbent. *RSC Adv.* **2024**, *14*, 3146–3157. [CrossRef] [PubMed]
27. Wang, L.; An, S.; Yu, S.; Zhang, S.; Zhang, Y.; Li, M.; Li, Q. Mass transfer characteristics of CO_2 absorption into a phase-change solvent in a wetted-wall column. *Int. J. Greenh. Gas Control* **2017**, *64*, 276–283. [CrossRef]
28. Liu, F.; Fang, M.; Dong, W.; Wang, T.; Xia, Z.; Wang, Q.; Luo, Z. Carbon dioxide absorption in aqueous alkanolamine blends for biphasic solvents screening and evaluation. *Appl. Energy* **2019**, *233*, 468–477. [CrossRef]
29. Li, Q.; Huang, X.; Li, N.; Qi, T.; Wang, R.; Wang, L.; An, S. Energy-efficient biphasic solvents for industrial CO_2 capture: Absorption mechanism and stability characteristics. *Energy* **2024**, *293*, 130710. [CrossRef]
30. Wang, L.; Liu, S.; Wang, R.; Li, Q.; Zhang, S. Regulating phase separation behavior of a DEEA–TETA biphasic solvent using sulfolane for energy-saving CO_2 capture. *Environ. Sci. Technol.* **2019**, *53*, 12873–12881. [CrossRef]
31. Luo, W.; Guo, D.; Zheng, J.; Gao, S.; Chen, J. CO_2 absorption using biphasic solvent: Blends of diethylenetriamine, sulfolane, and water. *Int. J. Greenh. Gas Control* **2016**, *53*, 141–148. [CrossRef]
32. Wang, R.; Jiang, L.; Li, Q.; Gao, G.; Zhang, S.; Wang, L. Energy-saving CO_2 capture using sulfolane-regulated biphasic solvent. *Energy* **2020**, *211*, 118667. [CrossRef]
33. Wang, L.; Yu, S.; Li, Q.; Zhang, Y.; An, S.; Zhang, S. Performance of sulfolane/DETA hybrids for CO_2 absorption: Phase splitting behavior, kinetics and thermodynamics. *Appl. Energy* **2018**, *228*, 568–576. [CrossRef]
34. Wang, R.; Yang, Y.; Wang, M.; Lin, J.; Zhang, S.; An, S.; Wang, L. Energy efficient diethylenetriamine–1-propanol biphasic solvent for CO_2 capture: Experimental and theoretical study. *Appl. Energy* **2021**, *290*, 116768. [CrossRef]

35. Wang, N.; Peng, Z.; Gao, H.; Sema, T.; Shi, J.; Liang, Z. New insight and evaluation of secondary Amine/N-butanol biphasic solutions for CO_2 Capture: Equilibrium Solubility, phase separation Behavior, absorption Rate, desorption Rate, energy consumption and ion species. *Chem. Eng. J.* **2022**, *431*, 133912. [CrossRef]
36. Zhang, W.; Jin, X.; Tu, W.; Ma, Q.; Mao, M.; Cui, C. Development of MEA-based CO_2 phase change absorbent. *Appl. Energy* **2017**, *195*, 316–323. [CrossRef]
37. Shen, L.; Liu, F.; Shen, Y.; Sun, C.; Zhang, Y.; Wang, Q.; Li, S.; Li, W. Novel biphasic solvent of AEP/1-propanol/H_2O for CO_2 capture with efficient regeneration performance and low energy consumption. *Sep. Purif. Technol.* **2021**, *270*, 118700. [CrossRef]
38. Wang, L.; Zhang, Y.; Wang, R.; Li, Q.; Zhang, S.; Li, M.; Liu, J.; Chen, B. Advanced monoethanolamine absorption using sulfolane as a phase splitter for CO_2 capture. *Environ. Sci. Technol.* **2018**, *52*, 14556–14563. [CrossRef]
39. Wang, R.; Liu, S.; Wang, L.; Li, Q.; Zhang, S.; Chen, B.; Jiang, L.; Zhang, Y. Superior energy-saving splitter in monoethanolamine-based biphasic solvents for CO_2 capture from coal-fired flue gas. *Appl. Energy* **2019**, *242*, 302–310. [CrossRef]
40. Wang, R.; Zhao, H.; Qi, C.; Yang, X.; Zhang, S.; Li, M.; Wang, L. Novel tertiary amine-based biphasic solvent for energy-efficient CO_2 capture with low corrosivity. *Energy* **2022**, *260*, 125045. [CrossRef]
41. Xiao, L.; Qiu, Z.; Feng, S.; Duan, X.; Zhao, Z.; Liu, Y.; Ma, L. Carbon dioxide absorption and desorption experiments based on MDEA. *Chem. Eng. Process.-Process Intensif.* **2024**, *204*, 109931. [CrossRef]
42. Danckwerts, P.V. The reaction of CO_2 with ethanolamines. *Chem. Eng. Sci.* **1979**, *34*, 443–446. [CrossRef]
43. Energy. New Findings from Beijing University of Chemical Technology Update Understanding of Energy (Development of MEA based CO_2 phase change adsorbent). *Energy Wkly. News* **2017**, 366.
44. Malinowski, J.J.; Daugulis, A.J. Salt effects in extraction of ethanol, 1-butanol and acetone from aqueous solutions. *AIChE J.* **1994**, *40*, 1459–1465. [CrossRef]
45. Marcilla, A.; Ruiz, F.; García, A.N. Liquid-liquid-solid equilibria of the quaternary system water-ethanol-acetone-sodium chloride at 25 °C. *Fluid Phase Equilibria* **1995**, *112*, 273–289. [CrossRef]
46. Liu, F.; Fang, M.; Yi, N.; Wang, T. Research on Alkanolamine-Based Physical–Chemical Solutions As Biphasic Solvents for CO_2 Capture. *EnergyFuels* **2019**, *33*, 11389–11398. [CrossRef]
47. Zhang, W.; Jin, X.; Tu, W.; Ma, Q.; Mao, M.; Cui, C. A novel CO_2 phase change absorbent: MEA/1-propanol/H_2O. *Energy Fuels* **2017**, *31*, 4273–4279. [CrossRef]
48. Wanderley, R.R.; Pinto, D.D.D.; Knuutila, H.K. Investigating opportunities for water-lean solvents in CO_2 capture: VLE and heat of absorption in water-lean solvents containing MEA. *Sep. Purif. Technol.* **2020**, *231*, 115883. [CrossRef]
49. Liu, S.; Ling, H.; Lv, J.; Gao, H.; Na, Y.; Liang, Z. New insights and assessment of primary alkanolamine/sulfolane biphasic solutions for post-combustion CO_2 capture: Absorption, desorption, phase separation, and technological process. *Ind. Eng. Chem. Res.* **2019**, *58*, 20461–20471. [CrossRef]
50. Xu, Z.; Wang, S.; Chen, C. Solubility of N_2O in and Density and Viscosity of Aqueous Solutions of 1, 4-Butanediamine, 2-(Diethylamino)-ethanol, and Their Mixtures from (298.15 to 333.15) K. *J. Chem. Eng. Data* **2013**, *58*, 1633–1640. [CrossRef]
51. Li, L.; Voice, A.K.; Li, H.; Namjoshi, O.; Nguyen, T.; Du, Y.; Rochelle, G.T. Amine blends using concentrated piperazine. *Energy Procedia* **2013**, *37*, 353–369. [CrossRef]

processes

MDPI

Article

The Palaeocene Lista Shale: A Planned Carbon Capture and Storage Top Seal for the East Mey CO_2 Storage Site

Nourah AlNajdi *, Richard H. Worden * and James E. P. Utley

Department of Earth, Ocean, and Ecological Sciences, University of Liverpool, Jane Herdman Building, 4 Brownlow Street, Liverpool L69 3GP, UK; james.utley@liverpool.ac.uk
* Correspondence: n.alnajdi@liverpool.ac.uk (N.A.); r.worden@liverpool.ac.uk (R.H.W.)

Abstract: Top seals and overburden above reservoirs at geological carbon capture and storage (CCS) sites can be major concerns when they are at risk of being mineralogically and texturally unstable in the presence of high-pressure CO_2. Here we report on the pore systems, mineralogy, and surface area attributes of the Palaeocene Lista Shale, the caprock to the Mey Sandstone at the UK's planned East Mey CCS site. The core was logged, and then mineral quantification was undertaken with X-ray powder diffraction mineralogy, light optics and electron microscopy analyses. Laser particle size analysis was used for grain size determination. Porosity, pore throat diameter, surface area and pore body size were measured via mercury intrusion porosimetry and nitrogen adsorption analyses. The mudstone facies from the Lista Shale are dominated by smectite-rich matrix and silt-grade quartz, with small quantities of chlorite and sodic-plagioclase. Chlorite, sodic-plagioclase, and even smectite are known to be capable of reacting with, and potentially leading to mineral sequestration of CO_2. The mean pore throat and pore body diameters are 17 and nearly 18 nm, respectively, showing that the Lista is mesoporous; the similarity of pore body and pore throat dimensions reveals a predominance of plate and slit pores. Gas adsorption analyses revealed that the overall pore structure is complex, with a high tortuosity of fluid movement through a complex clay-rich matrix (this equates to a mean fractal dimension D2 value of 2.67). Gas adsorption analyses have also shown that grain surfaces are moderately complex (rough) due to the dominance of clay aggregates (this equates to a mean fractal dimension D1 value of 2.56). D2 being higher than D1 suggests that there is a relatively low potential to physically store CO_2 gas on grain surfaces. Conversely, the ability of the CO_2 to react with minor quantities of chlorite and sodic plagioclase, or even with smectite, could lead to increasing surface area of the remaining shale minerals with newly exposed reactive silicates leading to further enhanced mineral trapping of the injected CO_2. The restricted pore throat size linked to small grain size and poor sorting, and reflected by the high fractal D2 value, plus limited grain surface complexity, reflected by the low fractal D1 value, collectively suggest that mineral trapping of the injected CO_2 would be relatively slow (on the order of 1000s of years) if CO_2 penetrated the top seal.

Keywords: caprock; top seal; carbon capture; smectite; surface area; pore throat diameter; pore body size; porosity; permeability

1. Introduction

Mudstones represent a major control of fluid flow in sedimentary basins and in near surface environments. Mudstones routinely function as aquicludes or aquitards in sedimentary basins, restricting water movement patterns [1]. In petroleum systems, mudstones act as top seals to many hydrocarbon reservoirs [1,2]. Mudstones are now recognised as important controls on the vertical flow and trapping of CO_2 in carbon capture and storage (CCS) sites [3].

The success of long-term carbon dioxide (CO_2) storage in geological formations is typically expected to require a strong containment system for the injected CO_2 [4]. An important element of the containment system is the top seal stratigraphically above the storage formation [4]. Unlike during the appraisal of oil or gas discoveries, where a successful top seal can be assumed, it is essential to characterise and understand the top seal at any potential CO_2 storage site to minimise the risk of CO_2 leakage. Top seal characterisation involves the use of a range of methods to assess the petrophysical, geochemical and geomechanical properties.

Petrophysical properties, including porosity, pore structure and permeability, are potentially affected when, or if, mudstone interacts with CO_2 [5,6]. It has been suggested that the storage capacity of CO_2 by mudstone can be reduced due to reaction with CO_2 [7]. Significant mineralogical alteration in mudstone has been reported in several studies of CO_2 injection [6–9]. The reported alterations in mudstone structure were partly caused by the dissolution of CO_2 in formation waters, that subsequently generates carbonic acid, with the latter leading to induced mineral dissolution and precipitation processes [6,10].

One of the essential parameters that control interaction processes at an aqueous fluid-solid interface is a specific surface area [11]. Clay minerals tend to have the highest specific surface areas of any mineral group in sediments, sedimentary rocks and soils due to their small grain size and complex crystal morphology [11]. Here we seek to characterise the specific surface area, mineral composition, and both the primary (depositional) and secondary (diagenetic) factors that influence essential properties such as pore structure, pore throat diameter, and grain size distribution, in the top seal of a proposed CCS project at the Palaeocene East Mey formation in the UK North Sea, as part of the larger Acorn project.

The Acorn project, one of several CCS initiatives currently progressing in the UK [12,13] aims to utilise the Palaeocene Mey Sandstone Member as a storage site within an open saline aquifer located in the Moray Firth Basin in the North Sea, offshore UK (Figure 1A) [14]. The Palaeocene Lista Shale serves as the top seal for the Mey Sandstone Member. Risk assessment of the East Mey storage site was conducted by the Acorn research and technical team to define leakage potential and threats of leakage from, for example, an undetected open fault or a leaky abandoned well [13]. Notably, in the study area, leakage via geological factors was perceived to be less likely to happen than leakage via old abandoned wells [13]. Nonetheless, to understand and minimise the risk of possible leakage potential of CO_2 from the underlying reservoir via the top seal, we have undertaken a full characterisation of the Lista Shale top seal.

A general question of relevance to all future CCS projects is what rock characteristics and specific aspects of a seal's geological history result in a good seal at a CCS site. This question will be addressed here by reference to the Lista Shale. A synthesis is presented at the end of the paper that draws on the Lista Shale results and interpretation as well as the outcomes of several related studies. Characterising the top seal involves a combination of core analysis, petrographic assessment, and mineral quantification with pore system analysis. In this study, fractal dimension analysis has been applied to examine the pore surface and pore structure of the Lista Shale. The specific research questions that we have answered for the Lista Shale as a potential CCS top seal are as follows:

What are the characteristic pore sizes?
What are the genetic origins of the pore types present?
What is the best way to determine a specific surface area?
What controls reactive surface area?
Are there compositional controls on pore structure and surface area?
Is the Lista Shale likely to be a good seal for the injected CO_2?

2. Geological Background

The turbidite-derived stratigraphic intervals of the Cenozoic Palaeogene cycles of the Central North Sea became economically important in northwest Europe after the first discovery of the hydrocarbon reservoirs at the Arbroath Field in 1969 and the Forties Field in 1970 [15,16]. The collection of production and exploration datasets, and variable access to the core, from the mature North Sea hydrocarbon province has afforded an opportunity to investigate regional scale deep water sedimentation and how it is related to tectonic framework and sea level changes during the Palaeogene in the central North Sea [15]. In this study, we will investigate the relationship between deep water sedimentation of the turbidite system with the mineralogical composition in the Lista Shale CCS caprock.

The Chalk, underlying the Cenozoic clastic succession of interest here, was deposited in the Upper Cretaceous and lowermost Palaeocene (Figure 1). The Ekofisk Formation of the Chalk Group immediately underlies the mudstones of the Palaeocene Montrose Group, which is subdivided between the Maureen and Lista Formations [15,17].

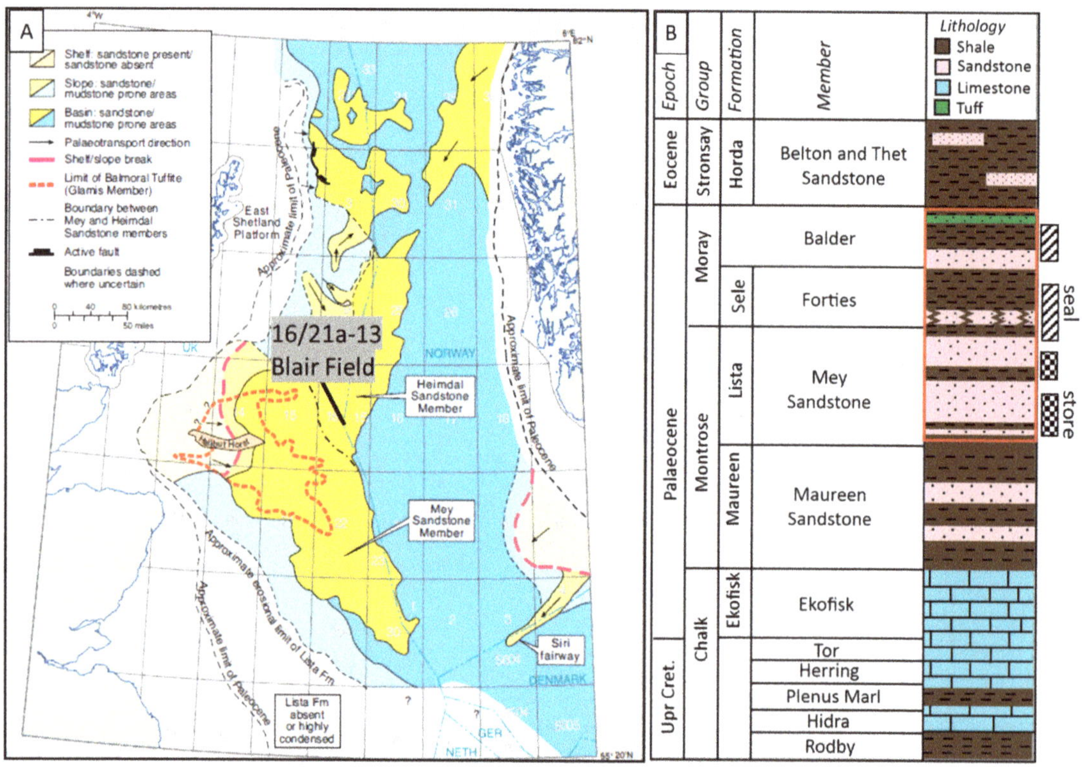

Figure 1. (**A**) Location map of the East Mey storage site in the Outer Moray Firth, Central North Sea, UK, adopted from the Millenium Atlas [18]. TheStudy area is within the Blair field, specifically at well 16/21a-13. (**B**) Lithostratigraphy of the East Mey CO_2 storage site, with the storage interval highlighted in red. The reservoir for CO_2 storage is the Mey Sandstone of the Palaeocene Lista Formation within the Montrose Group. This study focuses on the Lista Shale as the top seal for the Mey Sandstone Member.

The Montrose Group is composed of pelagic shales and turbidite sands [15,17] (Figure 1). The Maureen Formation contains discrete turbidite sandstone horizons (members) separated by pelagic shales. The Lista Formation sits above the Maureen Formation and contains pelagic shales and turbiditic sands. The Mey Sandstone Member, a planned CCS reservoir, sits within the Lista Formation [15,17]. The pelagic Lista Shale represents the top seal of the Mey Sandstone Member reservoir [14,19]. To be clear, the Mey Sandstone Member has also been locally named the Andrew Sandstone Member and the Balmoral Sandstone Member [14,15].

The Moray Group lies above the Montrose Group (Figure 1), consisting of the Sele Formation, associated with the Forties Member and the Balder Tuffaceous Member [15,17].

The thick Quaternary Nordland Group, at the top of the succession, is undifferentiated and dominated by mudstone, claystone and marls [14,19].

The burial and thermal histories of the East Mey site are relatively straightforward, with no significant uplift or erosion events identified since deposition commenced in the Palaeocene (66 to 56 Ma, Figure 2) [14]. The burial depth of the East Mey CCS site, which is close to the Blair Field, is approximately 6000 to 6600 ft (1800–2000 m), with a present-day temperature of about 60 to <70 °C [14]. Note that this relatively low maximum temperature, for a relatively short period of time, implies that mineral diagenesis and fabric alteration will not be highly advanced as it is only at the earliest stages of mesodiagenesis (burial diagenesis) [20] (Figure 2).

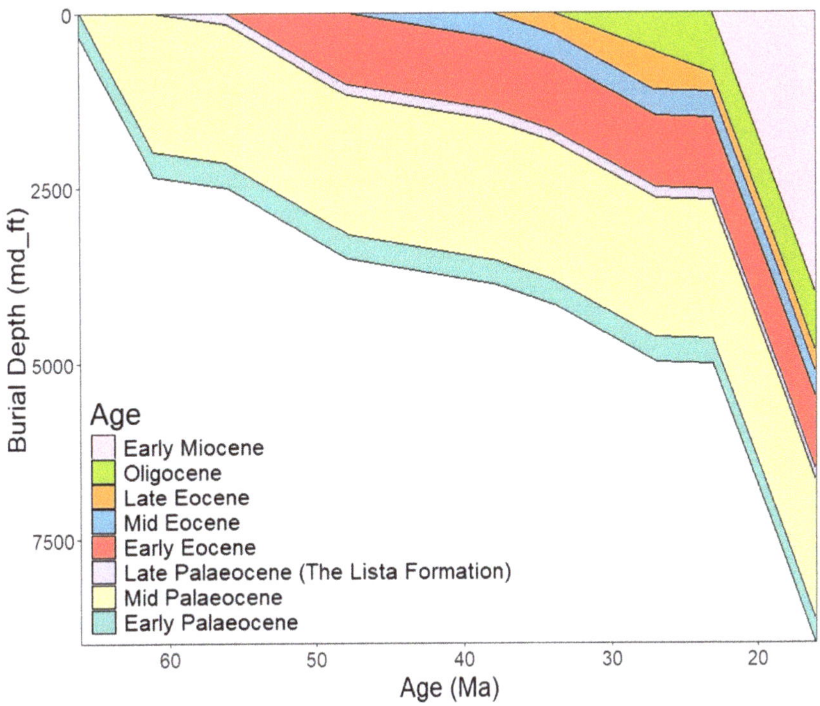

Figure 2. Burial history from the Early Palaeocene to Early Miocene Epochs of the Cenozoic Era in the Outer Moray Firth Basin, North Sea, UK. The Hordaland Group extends from the Early Miocene to Early Eocene age. The Rogaland Group, including the Balder and Sele Formation, are from the Late Palaeocene age. The Montrose Group including the Lista, Andrew, and Maureen Formation extends from the Late Palaeocene to Early Palaeocene age. The Ekofisk Formation from the Chalk Group belongs to the Early Palaeocene age. Based on available evidence, the burial curve for the Palaeocene is apparently straightforward, showing no significant uplift or erosion events [14]. Due to limited stratigraphic data for the post-Palaeocene and Neogene sections, a uniform deposition rate has been assumed for this interval.

3. Methodology

3.1. Characterisation Workflow

The characterisation workflow used in this study has been adopted from the Armitage workflow [21]. To start, the core was logged to investigate sedimentary features. Second, representative core samples were collected from the Lista Formation. Polished thin-section samples were used to carry out quantitative petrographic assessment via SEM/EDS and optical microscopy. Mineralogy was determined using XRD analysis. To relate mineralogy with porosity, mercury intrusion porosimetry analyses (MICP) and nitrogen gas adsorption experiments were undertaken to investigate the different pore types and mean pore size diameters, along with surface area determination (Figure 3). These data were then used for a risk assessment of the Lista Shale.

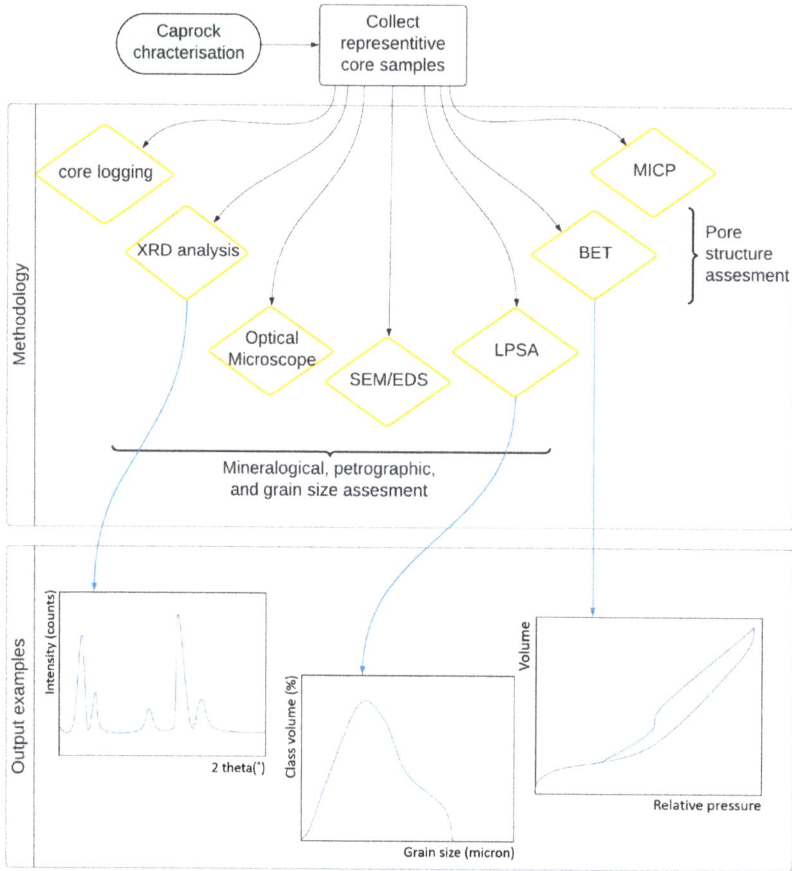

Figure 3. Characterisation workflow chart of the Lista Shale top seal.

3.2. Samples and Data

The Blair Field, within the planned CCS site for the East Mey project, is currently operated by Harbour Energy PLC (London, UK) [13,14,19]. Well 16/21a-13 has been selected for this study because, perhaps unusually, the core was taken through sealing lithologies during the assessment of local petroleum resources. Eleven representative samples were collected, made into 30 µm polished thin sections and were prepared for backscattered electron microscopy using standard methods (Table 1).

Table 1. Sampling of the Lista Shale from well 16/21a-13 including all methods undertaken in this study. The symbol (X) refers to samples taken from each technique and md refers to measured depth.

Depth (md-ft)	BET	MICP	XRD	Thin-Section	Optical Microscope	SEM	LPSA
6975	X	X	X	X	X	X	X
6981	X	X	X	X	X	X	X
6982	X		X				X

Table 1. *Cont.*

Depth (md-ft)	BET	MICP	XRD	Thin-Section	Optical Microscope	SEM	LPSA
6987	X		X	X	X	X	X
6989	X		X	X	X	X	X
6990	X	X	X				X
6992	X		X	X	X	X	X
6995.5	X		X	X	X	X	X
7001	X	X	X	X	X	X	X
7005	X	X	X	X	X	X	X
7008	X	X	X	X	X	X	X
7013	X		X				X
7015	X	X	X	X	X	X	X
7024.5	X	X	X	X	X	X	X

3.3. Lista Formation Core Logging

Approximately 62 feet of core was originally cut from well 16/21a-13. From this core, approximately 60% was in fine-grained top seal facies of the Lista Formation. For lithology, grain size, sedimentary structures, bioturbation index (BI), trace fossil and macrofossil identification, the core was logged at high resolution (mm-scale). The bioturbation index scale ranges from 0 to 6, where 0 is no trace of bioturbation and 6 represents total bioturbation [22]. Wireline logs were recorded through the reservoir and top seal at the time the well was drilled (by North Sea Sun Oil Company in 1984) but they proved to be of minimal help for detailed analysis of the lithology and rock properties during this study as neutron and shear compressional log data were not collected [14].

3.4. Microscopy

Optical microscopy was undertaken on polished thin sections using an Olympus BX51 microscope with an Olympus SC50 camera employing both plane polarised light (PPL) and crossed polarised light (XPL) imaging [21,23]. The Olympus BX51 microscope with the SC50 camera was manufactured by Olympus Corporation, Tokyo, Japan

Prior to electron microscopy, polished thin sections were carbon coated according to the method used by AlNajdi, et al. [21] to quantify pore shapes. Quantification was achieved using backscattered electron (BSE) and secondary electron (SE) microscopy on a Zeiss Gemini 450 SEM System at the Shared Research Facility (SRF) at the University of Liverpool.

Mineral proportions were determined and mineral maps were produced using secondary X-ray analysis in a dedicated scanning electron microscopy-energy dispersive spectroscopy SEM-EDS [14]; in this case, we used the University of Liverpool's FEI WellSite QEMSCAN used by [14,24].

3.5. Mineralogy via X-Ray Diffraction (XRD)

X-ray diffraction (XRD) analyses were undertaken to quantify the mineralogy of the core. Sample preparation was performed according to the standard methods applied at the University of Liverpool for mineral quantification and clay separation [14,21,25,26]. Mineral quantification was achieved using "HighScore Plus ®" software version 4.9A using the relative intensity ratio (RIR) method [27], with reference data from the International Centre for Diffraction Data's Powder Diffraction File-4+ Release 2022.

3.6. Laser Particle Size Analysis (LPSA)

The value of particle size determination during the analysis of top seals at carbon storage sites derives from its relation to pore volume and surface area [28]. Large particle sizes in top seals lead to high pore volumes, which potentially facilitate fluid flow and diffusion [28]. Linking laser particle size with N_2 adsorption (see Section 3.8) reveals the effect of particle size distribution on the gas adsorption process and provides unique insight into the relationship between surface area and pore volume measurements. Therefore, textural data from 14 core samples were acquired using a laser particle size analyser (LPSA) at the University of Liverpool; see AlNajdi, et al. [21] for details about sample preparation; the equipment used was a Beckman-Coulter LS13-320, manufactured by Danaher Corporation, California, USA. The data were processed using GRADISTAT version 9.1 to reveal particle size distributions [29,30].

3.7. Mercury Intrusion Capillary Pressure (MICP)

Mercury intrusion capillary pressure (MICP) analysis was used to characterise the pore system and pore size distribution for the Palaeocene Lista Shale [14]. Eight selected samples were analysed by MCA services (Cambridge, UK) [14]; see Peng, et al. [31] for details of the methodology and sample preparation. The relationship between pore throat diameter and applied pressure was defined by the Washburn equation [32].

$$\text{Washburn equation}: \ \mathrm{Dpt} = (4\gamma \cos(\theta))/\mathrm{Pc} \tag{1}$$

where Dpt is the pore throat diameter, Pc is capillary entry pressure, γ is the surface tension of the liquid, and θ is the contact angle in Radians between the specimen surface and mercury [14,32]. The contact angle varies as a function of fluid types and the specific mineral; however, the mercury-air contact angle (θ) is 141° and the surface tension of the mercury (γ) is 0.48 N/m [14,31,33,34].

3.8. N_2 Gas Adsorption via BET Analysis

Rock chips from the core were collected from the same depths as those used for LPSA and MICP analyses; see AlNajdi, et al. [21] and Luffel and Guidry [35] for details of the methods employed.

Gas adsorption measurements were performed using a Nova4200e surface area analyser, configured according to the static volumetric method to quantify gas adsorption. Prior to analysis, each sample underwent de-gassing in line with the ISO9277 [36].

The traditional pore diameter model was initially based on the Kelvin Equation (2), which was subsequently refined by Barrett, et al. [37] to account for multilayer adsorption, thus making it widely applicable for the calculation of pore diameter within the mesopore range [37,38]. The Barrett, Joyner, and Halenda (BJH) model describes the phenomena of capillary condensation within pores [37]:

$$\ln(\mathrm{P}/(\mathrm{P_o}) = -(2\gamma \mathrm{Vm})/\mathrm{RT}(\mathrm{r}_p - \mathrm{t_c}) \tag{2}$$

where $\ln(P/P_o)$ is the relative pressure of the nitrogen gas, γ is the surface tension of the bulk, V_m is the molar liquid nitrogen volume, R is the gas constant, T is the temperature at which the isotherm is calculated (77K), r_p is the pore radius, and t_c is the thickness of the adsorbed multilayer film [38,39].

3.9. Fractal Dimension Analysis for Complex Pore Structure and Surface

In recent years, fractal theory has been applied to characterise the heterogeneity of pore structures in rocks and various other materials [40–50]. AlNajdi, et al. [21] discussed how to interpret the results of fractal calculations in terms of top seal properties; the same approach will be employed in this study of the Lista Shale. The fractal dimension index (D), derived from the N_2 adsorption isotherm, ranges between 2 (regular/smooth surfaces) and 3 (irregular/rough surface) [6,51]. The Frenkel-Halsey-Hill (FHH) equation was used to estimate the D value from N_2 adsorption data [52]. Equation (3) presents the FHH model [51],

$$\ln \mathrm{V} = (\mathrm{D} - 3) \ln (\ln \mathrm{P_o}/\mathrm{P}) + \mathrm{C} \tag{3}$$

where V is the N_2 volume adsorbed at equilibrium pressure (cm^3), P_o is the saturated vapour pressure, P is the equilibrium pressure (MPa), C is a constant, and D is the fractal dimension. D is calculated from the slope of the linear regression correlation between lnV and $\ln(-\ln(P_o/P))$.

4. Results

In the following sections, the results from core logging, mineralogical assessment, petrographic assessment, gas adsorption, mercury intrusion, and particle size analyses are presented alongside the derived fractal dimensions. The results from each technique will be interpreted and compared in the discussion section.

4.1. Core Logging

Core samples from the Lista Shale Formation are predominantly mudstone, exhibiting fine lamination at scales of 1 to 5 mm, with localised turbidite sandstone beds that typically range from millimetres to centimetres scale (Figure 4). The cores are grey to dark grey and are locally bioturbated. The bioturbation index is approximately 3, dominated by trace fossils such as *Planolites* and *Zoophycos* (Figures 4 and 5). Traces of slickensides are present in the core and are associated with localised fractures (Figure 5). Centremetre-sized siderite nodules are locally present in parts of the core (Figures 4 and 5).

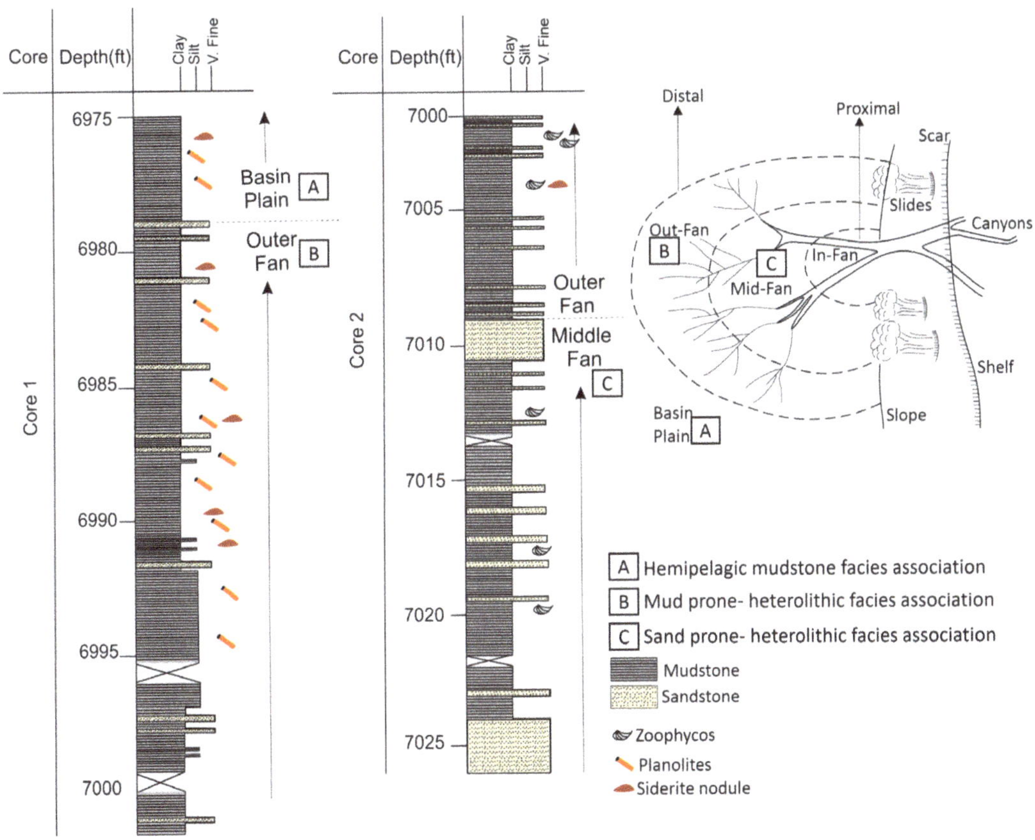

Figure 4. Comprehensive core logging of the Lista Shale for well 16/21a-13. The logging scale is based on 3 ft core sections. The core exhibits a colour range from dark grey to grey and is characterised by thin laminations at the 1 to 5 mm scale, with significant bioturbation and trace fossils such as *Planolites* and *Zoophycos*. At deeper depths, the grain size transitions to very fine sand. The core is

interbedded with sandstone laminae and contains siderite and carbonate nodules. Slickensides are visible within fractures. Three facies' associations were identified: facies association A, which is dominated by hemipelagic mudstone with thin (5 mm) sandstone beds, representing the basinal plain in a turbidite fan system; facies association B, characterised by mud-prone heterolithic facies that are distal and common in the outer fan system; and facies association C, consisting of sand-prone heterolithic facies, which is proximal and part of the middle fan in a turbidite fan system.

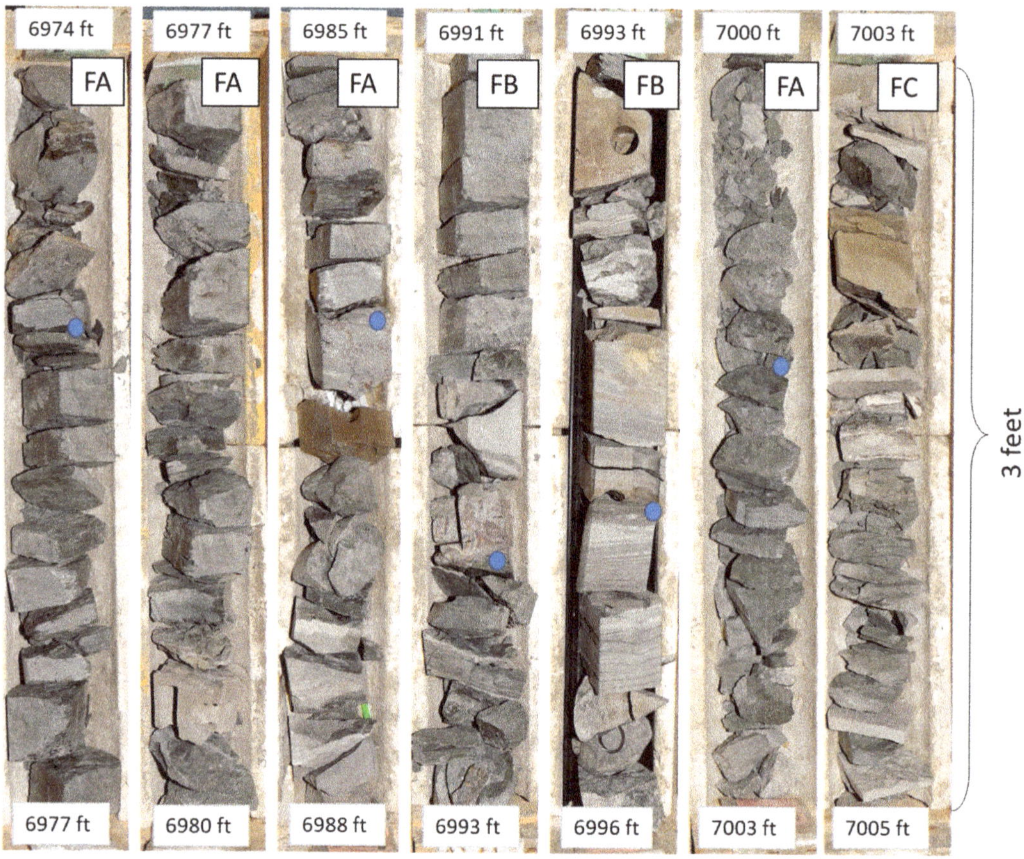

Figure 5. Core images with an indication of the Core sampling points are revealed by the blue circles. The three facies associations are FA, which is hemipelagic mudstone facies, FB which is mud-prone heterolithic facies, and FC which is sand-prone heterolithic facies. Each core box is 3 feet long.

4.2. Facies Associations

Six facies were identified in core: unbioturbated laminated mudstone; bioturbated mudstone; unbioturbated laminated silty mudstone; bioturbated silty mudstone; thin-bedded, very fine to fine, unbioturbated sandstones; and thicker bedded (10 cm), structureless unbioturbated sandstones (Figures 5–7).

Based on the facies classification for the Mey Sandstone Member of the Lista Formation published by Kilhams, et al. [15], three facies associations, linked to different environments in the turbidite fan system, are present in the Lista core (Figures 4 and 5).

Facies association A (hemipelagic mudstone) is from the basin plain beyond the turbidite fan system, dominated by bioturbated mudstones (Figure 7A). Facies association B (mud-prone heterolithic) is from the outer fan and is dominated by mudstone and silty mudstone with local thin-bedded sandstones. Facies association B is characterised by bioturbation, with traces of *Planolites* and *Zoophycos* (Figure 6A–C). Facies association C (sand-prone heterolithic) is from the mid-fan and is dominated by bioturbated mudstone with localised 10 cm-thick sandstone beds.

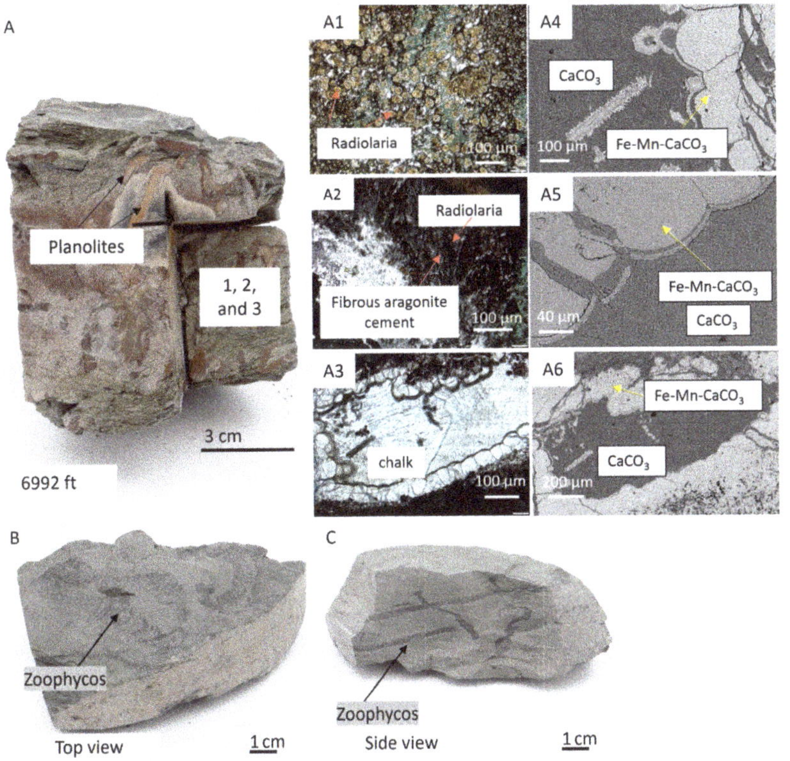

Figure 6. (**A**) is a core image from mud-prone heterolithic facies association at 6992 ft-md, with *Planolites* trace fossil. Images 1, 2 and 3 are optical thin sections of Image (**A**). In (**A1**), Radiolaria microfossils are observed. In (**A2**), radiolaria are localised around fibrous aragonite cement. In (**A3**), chalk is observed which is the milky white round clast shown in image (**A**). Images (**A4–A6**) are high-resolution BSE-EDX images of the chalk in image (**A3**) showing area of carbonate and iron-manganese-rich carbonate. (**B**,**C**) are core images of *Zoophycos* trace fossil at 7001 ft-md. (**B**) is a top view image of the trace fossil. (**C**) is a side view image of the *Zoophycos*.

4.3. Mineralogy and Petrography

Whole rock XRD analysis reveals that smectite is the dominant clay, comprising 20 to 47%, with minor amounts of mica and kaolinite, and trace amounts of chlorite (Figure 8 and Table 2). The Lista samples are silt-rich, containing between 18% and 53% quartz. Feldspar silt grains make up 8 to 10% of the samples, with plagioclase (specifically albite) slightly more prevalent than K-feldspar. Trace amounts of pyrite and calcite were also identified (Figure 8; Table 2).

Conventional light microscopy revealed the presence of detrital silt-grade quartz and trace fossils (Figure 9). Light microscopy also showed that radiolaria microfossils are localised around fibrous aragonite cement (Figure 6(A1,A2)).

Electron microscopy proved to be valuable for characterising the Lista Shale samples. SEM-EDS images confirmed that the shale is primarily composed of smectite, illite and quartz, aligning with the XRD results (Table 2). High-resolution backscattered electron microscopy (BSEM) images identified detrital quartz, K-feldspar and micas (Figure 10A–D).

Additionally, BSEM revealed the presence of pyrite framboids, which are common in the Lista Shale (Figure 10A,E), further validating the XRD findings (Table 2).

Figure 7. (**A**) is the core image of the hemipelagic mudstone facies at 6987 ft md, showing closed sutured fracture propagation. Image (**A1**) is an optical thin section of the fracture area. Images (**A2**,**A3**) are BSE-EDX images of the fracture indicating that the fractures are filled with Fe-Mg-rich carbonate solution between the original smectite matrix. (**B**) is the core image of the mud-prone heterolithic facies at 6995.5 ft-md. BSE images from (**B1**–**B3**) show the difference between the silt bed and coarse silt bed. BSE images (**B3**) show that there is localised carbonate cement around quartz grains in the coarse silt bed.

Illite is present (Figure 10F), confirming the XRD results (Table 2). The milky white sub-rounded clast in the BSEM image in Figure 6A is a fragment of chalk (Figure 6(A3)). BSEM images in Figure 6(A4–A6) show the composition of the chalk and the Fe-Mn-rich carbonate around the chalk. In Figure 7, core image A shows closed but sutured fractures filled with Fe-Mg-rich carbonate solution (Figure 7(A1–A3)). Core image B in Figure 7 is from mud-prone heterolithic facies association B and is laminated with coarse silt. However, the laminae are not only different in grain size but also in composition (Figure 7B). The rare coarse silt laminae are carbonate cemented, with cement localised around detrital quartz and K-feldspar (Figure 7(B1,B3)).

Table 2. Mineral analysis through X-ray diffraction (XRD) for well 16/21a-13.

Depth (ft)	Quartz	K-Feldspar	Plagioclase	Illite/Muscovite	Smectite	Chlorite	Kaolinite	Pyrite	Calcite
6975	18	8	6	30	31	3	4	0.5	0
6981	39	6	10	14	24	2	3	0	0
6982	28	9	5	24	27	3	4	0	0
6987	34	6	6	23	22	3	5	0	0.5
6989	33	7	7	21	22	4	5	1	0
6990	36	7	11	14	22	3	6	0	0
6992	22	7	6	23	36	3	3	0	0.5
6995.5	53	10	7	9	9	4	8	0.5	0
7001	27	6	9	16	35	2	5	0	0
7005.9	25	6	5	21	38	2	3	0.5	0
7008	52	5	9	9	20	2	2	0	0
7013	20	7	5	23	35	3	7	0	0
7015	30	7	6	11	36	3	6	0.5	0
7024.5	26	6	5	10	47	2	4	0.5	0

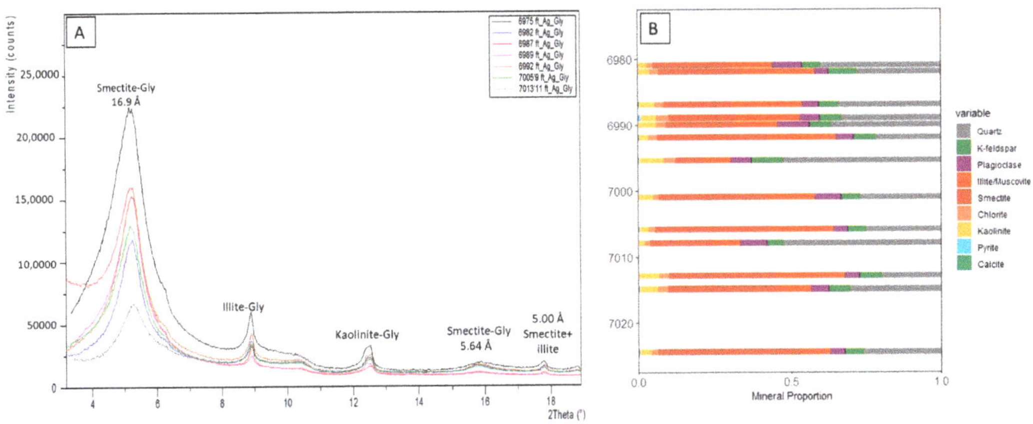

Figure 8. (**A**) XRD glycolated clay fraction pattern of the Lista Shale. (**B**) Mineral quantification of the Lista Shale from X-ray diffraction (XRD). Smectite and quartz are the dominant minerals in the Lista Shale.

4.4. Pore Types

Based on the classification for CCS top seals proposed by AlNajdi and Worden [23], two distinct pore types were observed (Figure 11). Interparticle pores are common between clay minerals (smectite) and silt grains composed of mica, quartz and feldspar (Figure 10A). Other types of interparticle pores are slit pores between the stack of clay aggregates (Figure 10B,D), or plate pores between clay platelets (Figure 11C,E). The intraparticle pores are primarily located within pyrite framboids (Figure 11F).

A schematic illustration is shown in Figure 11, linking the pore types to the adsorption process of liquid nitrogen and their ability to adsorb and store gases; this will be further addressed in the discussion section.

4.5. Particle Size Analysis (LPSA)

Particle size analysis reveals a range of unimodal, bimodal and polymodal distributions, with unimodal being dominant. Samples tend to be poorly sorted (Figure 12A; Table 3). The particle size varies from very fine medium silt to medium silt. The distribution of grain sizes ranges from sandy silt to silt. The mean grain size, also determined using the Folk and Ward method in GRADISTAT, is 7.08 μm (Figure 12B), representing very fine silt-grade material given that the boundary between clay and silt is 4 μm. The mean sorting, calculated using the Folk and Ward method, is 1.45 φ, indicating very poorly sorted primary sediment [24].

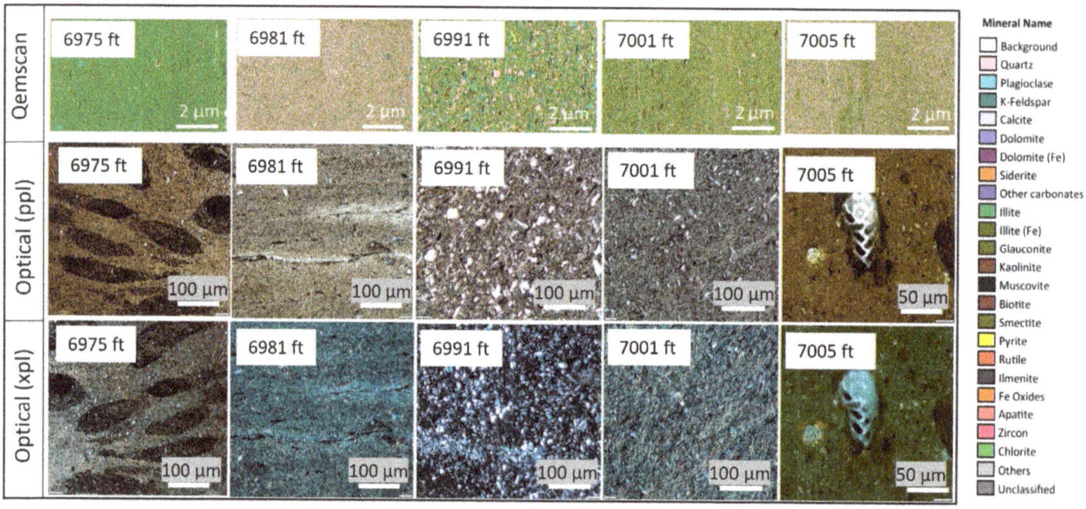

Figure 9. Optical and SEM-EDS images of the Palaeocene Lista Shale samples. The mineralogy quantification from SEM-EDS images at 2 μm resolution indicates that smectite and illite are the dominant minerals in the Lista Shale. However, at certain depths of the core such as 6981 ft, 6991 ft and 7005 ft (all measured depths), the quartz is notably dominant along with K-feldspar and plagioclase. Optical images of the Lista Shale in plain polarised light (ppl) and cross polarised light (xpl) show the distribution of detrital quartz grains within the fine grain matrix (smectite).

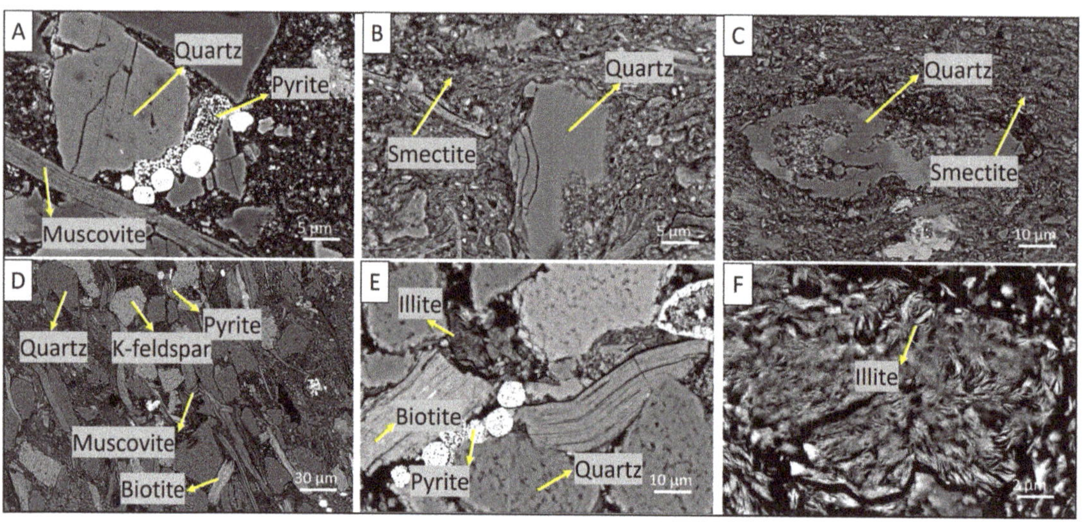

Figure 10. Backscattered electron images for well 16/21a-13. SEM-BSE Images: (**A**) polished section showing detrital quartz grains and pyrite framboids; (**B**,**C**) showing the presence of detrital quartz grain in the smectite matrix; (**D**,**E**) showing the compaction of the detrital quartz grain and micas within smectite matrix; (**F**) showing the presence of illite within the smectite matrix.

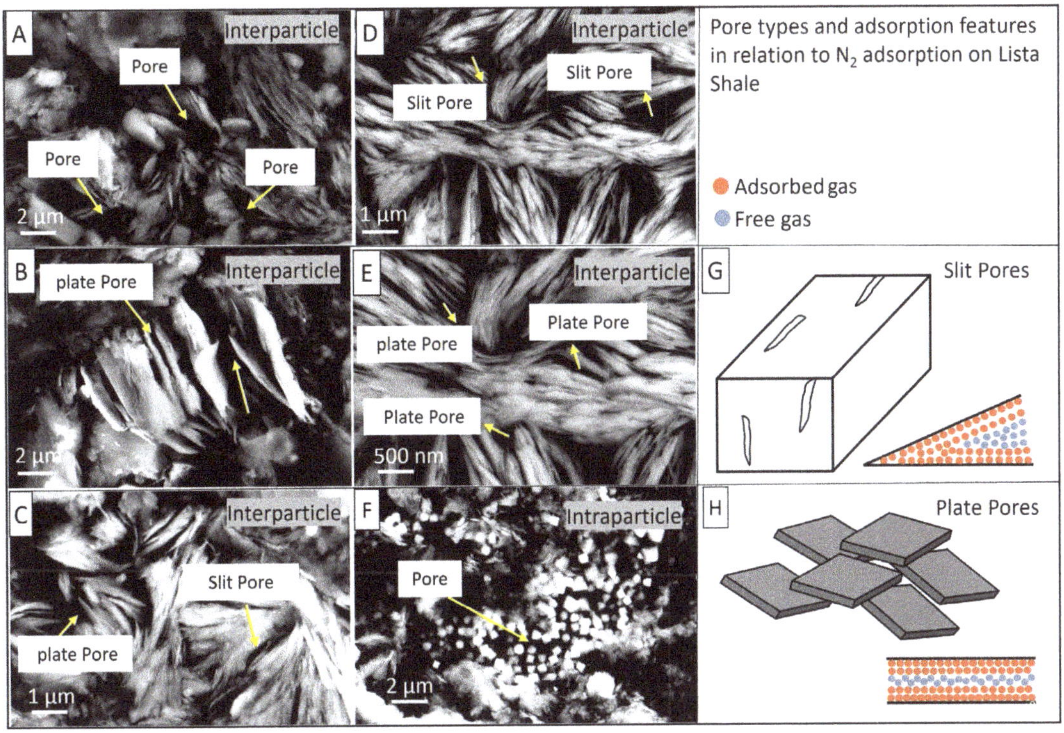

Figure 11. SEM images revealing pore type: (**A**) interparticle pores between grains; (**B**) interparticle slit pores in clay matrix; (**C**) interparticle plate pores between clay platelets; (**D**,**E**) SEM-BSE images of interparticle slit- and plate-pores; (**F**) intraparticle pores between pyrite framboids. The third column of the figure (**G**,**H**) represents a schematic illustration of the adsorption processes.

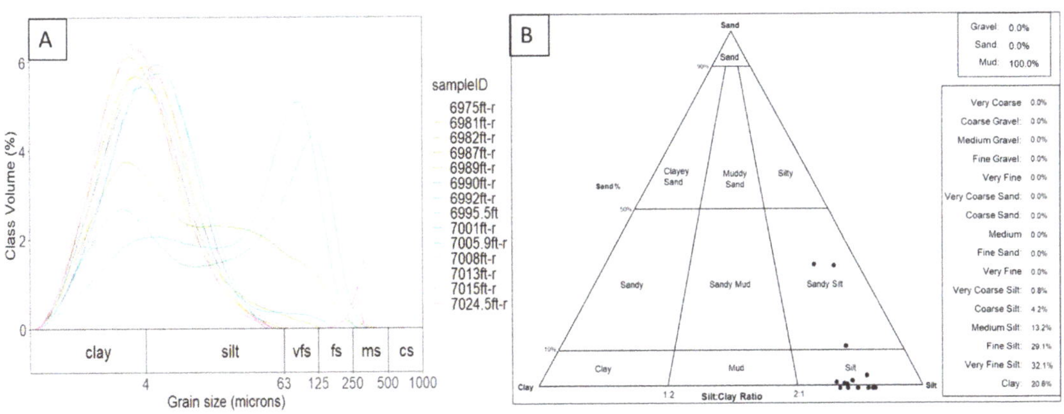

Figure 12. Grain size distribution from laser particle analysis (LPSA) for well 16/21A-13. (**A**) Data from LPSA show unimodal to bimodal type of the sediment, and the maximum grain size is medium sand (500 µm). (**B**) Display of proportions of sand, clay and silt from LPSA: interestingly for a unit formally named a shale, most samples fall in the silt category with some in the sandy silt category.

Table 3. Summary of results from nitrogen adsorption-desorption isotherms, MICP and LPSA, along with sorting and mean grain size (MGS) data. The surface area, pore volume and pore diameter measurements are obtained using the BET and BJH models for the Lista Shale from well 16/21a-13. Sorting and grain size data were analysed using GRADISTAT version 9.1, based on the Folk and Ward method [29]. ND: not determined for this sample.

	BET Measurements					MICP Measurements				LPSA Measurements	
Depth (ft)	Multi BET Surface Area (m^2/g)	BJH Surface Area (m^2/g)	BJH Pore Volume (cc/g)	BJH Pore Diameter (nm)	Average Pore Diameter (nm)	MICP Porosity	Median Pore Diameter (MICP)	Average Pore Throat Diameter (nm)	Total Pore Area (m^2/g)	Sorting Φ	MGN (μm)
6975	13.32	13.35	0.066	3.16	17.55	15.00	0.014	15.1	21.24	1.22	4.64
6981	11.41	17.21	0.071	3.16	18.68	15.71	0.013	14.07	23.99	1.19	4.22
6982	11.56	16.26	0.062	3.16	17.3	ND	ND	ND	ND	1.2	3.65
6987	7.06	10.93	0.048	3.14	18.49	ND	ND	ND	ND	1.33	4.12
6989	7.35	11.00	0.045	3.12	17.47	ND	ND	ND	ND	2.13	8.87
6990	4.11	7.34	0.031	3.40	17.84	14.15	0.018	23.53	12.56	2.51	19.46
6992	12.10	15.71	0.067	3.16	16.89	ND	ND	ND	ND	1.19	4.86
6995.5	4.56	7.66	0.029	3.16	17.68	ND	ND	ND	ND	2.17	23.63
7001	10.87	16.28	0.061	3.12	18.13	15.74	0.013	14.64	23.07	1.21	5.05
7005.9	11.05	16.32	0.062	3.16	18.23	16.33	0.014	14.74	24.12	1.37	4.47
7008	11.61	17.07	0.061	3.14	18.3	17.02	0.016	17.46	21.2	1.2	3.84
7013	11.22	16.17	0.058	3.12	18.19	ND	ND	ND	ND	1.31	5.06
7015	11.81	16.90	0.065	3.16	17.58	16.25	0.013	13.68	25.52	1.18	3.54
7024.5	9.87	15.00	0.064	3.14	17.53	15.45	0.012	13.78	23.96	1.16	3.76

4.6. Mercury Intrusion Porosimetry (MICP)

Mercury intrusion porosimetry (MICP) analysis was performed on eight samples (Figure 13B). The results, presented in terms of pore throat diameter and listed in Table 3, also include the derived mercury intrusion-derived shale porosity values.

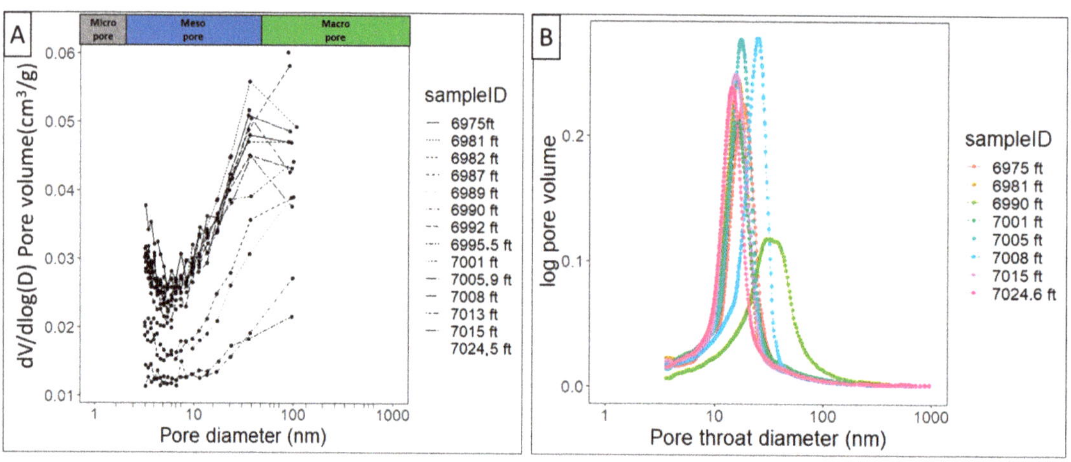

Figure 13. Part (**A**) shows the BJH pore size distribution for well 16/21a-13. The Lista Shale is primarily dominated by mesopores as indicated by the increase in partial pore volume, with minimal micropores and no evidence of macropores. Part (**B**) displays the pore throat distribution from mercury intrusion porosimetry (MICP) analysis for eight samples. The pore throat size distribution is unimodal, with an average size of 17 nm.

The average pore throat diameter of the Lista Shale is 17 nm. The porosity values determined by MICP range from 14% to 17% (Table 3), with the average porosity being approximately 16% (Table 3).

4.7. Pore Structure

Adsorption involves holding the gas molecules on a thin layer on the solid surfaces, while desorption is the process whereby adsorbed gas molecules are released from the surface [36]. According to the International Union of Pure and Applied Chemistry (IUPAC), the nitrogen gas adsorption isotherms observed in Figure 14 are classified as type IV, typical of mesopore adsorbents [39,53]. As the relative nitrogen pressure increases, multilayer adsorption occurs, resulting in a characteristic "knee-bend" shape on the adsorption isotherm. In Figure 14, this knee-bend shape indicates the transition from monolayer to multilayer adsorption. Gas in the mesopores and macropores begins to condense at higher relative pressures. As the relative pressure decreases, desorption occurs, and the amount of adsorbed gas reduces. Following desorption, the desorption curve coincides with the adsorption curve on the declining pressure path, forming a hysteresis loop. The shape of the hysteresis loop is type H3, which suggests the presence of open-type pores (e.g., between plate-like particles) or wedge-shaped slit pores [54,55] as shown in Figure 11B,C.

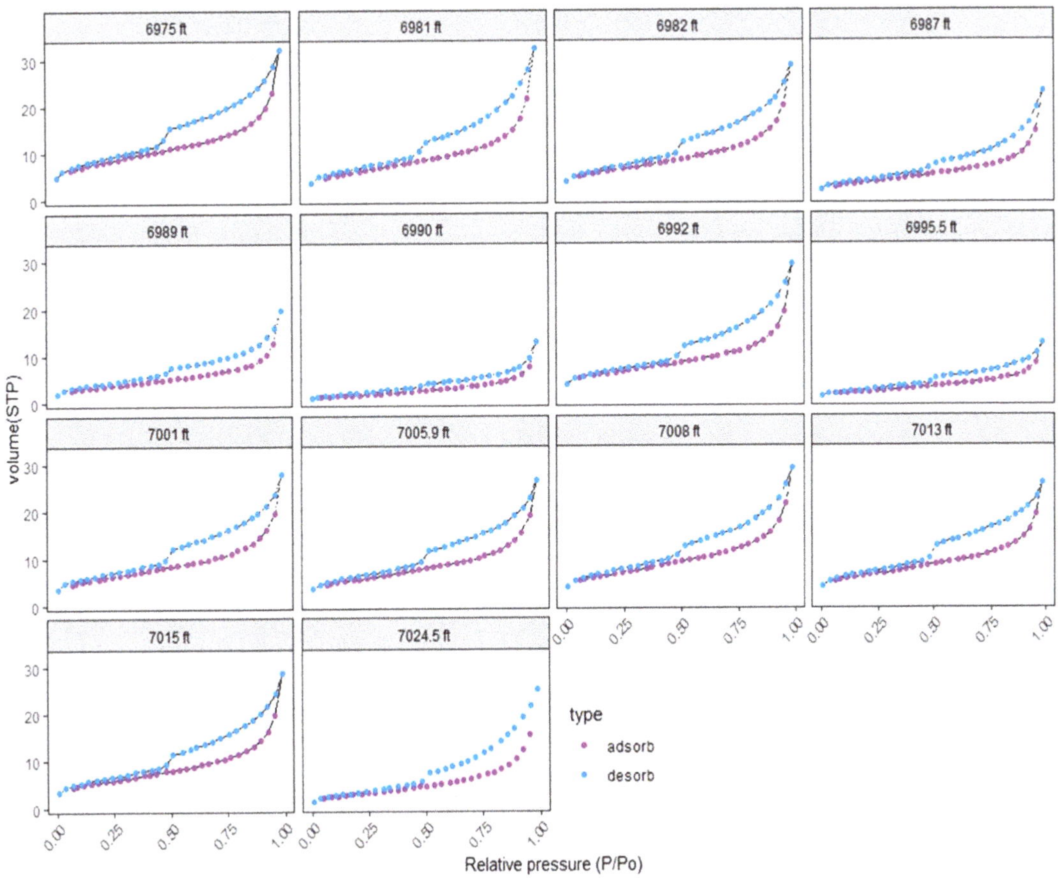

Figure 14. The N_2 adsorption-desorption isotherms at liquid nitrogen temperature (−197.3 °C) for gently crushed and sieved samples of the Lista Shale from well 16/21a-13 exhibits a type IV isotherm. This shape suggests that the pores span across micropores, mesopores and macropores.

Surface area and pore volume analyses, presented in Table 3, were calculated using the BET and BJH models. The bulk surface area ranges from 4.1 to 13.3 m^2/g, while the pore volume derived from nitrogen adsorption ranges from 0.029 to 0.071 cm^3/g (Table 3).

The pore body size diameters reported here are derived from the BJH model and based on the pore diameter classifications defined by IUPAC [53]. The partial volume of each pore diameter (dV/d(logD)) is plotted, and the area under the curve between two pore diameters can be used to assess the partial porosity for each pore size range [34]. The classification includes macropores, mesopores and micropores [35,56]. The results show that the dominant pore size falls within the mesopore range (Figure 13A), with an average overall pore body diameter of 17.84 nm (Figure 13A), which is similar to the pore throat diameter obtained from MICP analysis (Table 3).

4.8. Fractal Dimension Analysis Using Gas Adsorption Data

The adsorption isotherms were used to perform fractal dimension analysis using the Frenkel-Halsey-Hill (FHH) fractal model [6,21,51]. The standard adsorption and desorption isotherms typically merge in the relative pressure range of 0.40 to 0.45 (P/P_o) [6]. Consequently, the fractal dimension is derived from two distinct regions of the adsorption relative pressure (Figure 15) [6,21,51]. The first fractal region, referred to as D1 (Table 4), represents the pore surface and occurs at relative pressures below 0.45 [6,21,51]. The second fractal region, D2 (Table 4), represents the pore structure and is observed at relative pressures above 0.45 [6,21,51]. The fractal dimensions (D1 and D2) were calculated from the BET data, with R^2 (correlation coefficient) values for the fitted lines exceeding 0.9, confirming that the pores and surfaces possess fractal characteristics. The fractal dimension of the pore surface (D1) ranges from 2.4 to 2.6, indicating that some pore surfaces are smooth while others are irregular [21,51,56] (Table 4). The fractal dimension of the pore structure (D2) is around 2.7, suggesting rougher pore structures [21,51,56] (Table 4). Notably, the fractal dimension D2 is higher than D1 (Table 4).

Table 4. A summary of the FHH fractal model-fitted lines derived from the adsorption isotherms of representative samples, chosen based on their relative surface areas. The fractal dimension D1 characterises the pore surfaces for relative pressures (P/P_o) less than 0.45, while D2 characterises the pore structures for relative pressures greater than 0.45. All depths are measured as opposed to true vertical depths.

Sample	Fractal Dimension of Pore Surfaces ($P_o/P < 0.45$)			Fractal Dimension of Pore Structure ($P_o/P > 0.45$)		
	D1	Fitted Equation	R^2 (1)	D2	Fitted Equation	R^2 (2)
6975 ft	2.60	y = −0.2723x + 2.7876	0.999	2.72	y = −0.3924x + 2.7611	0.999
6981 ft	2.55	y = −0.3372x + 2.4631	0.998	2.66	y = −0.4469x + 2.4355	0.999
6982 ft	2.59	y = −0.3199x + 2.4856	0.999	2.68	y = −0.4052x + 2.4646	0.999
6987ft	2.48	y = −0.3777x + 1.8578	0.999	2.62	y = −0.5120x + 1.823	0.999
6989 ft	2.52	y = −0.3389x + 1.9949	0.998	2.66	y = −0.4703x + 1.9612	0.999
6990 ft	2.59	y = −0.3199x + 2.4856	0.999	2.68	y = −0.4052x + 2.4646	0.999
6992 ft	2.62	y = −0.3038x + 2.5429	0.996	2.69	y = −0.3701x + 2.5377	1.000
6995 ft	2.45	y = −0.3531x + 1.495	0.998	2.64	y = −0.5482x + 1.4054	0.999
7001 ft	2.57	y = −0.3095x + 2.4844	0.999	2.69	y = −0.4290x + 2.4449	0.999
7005 ft	2.56	y = −0.3221x + 2.4587	0.999	2.67	y = −0.4386x + 2.4218	0.999
7008 ft	2.57	y = −0.3120x + 2.5444	0.998	2.68	y = −0.4294x + 2.5148	0.999
7013 ft	2.60	y = −0.2955x + 2.5525	0.998	2.70	y = −0.3945x + 2.5204	0.999
7015 ft	2.57	y = −0.3285x + 2.4903	0.999	2.67	y = −0.4281x + 2.4619	0.999
7024 ft	2.43	y = −0.4500x + 1.9379	0.999	2.55	y = −0.5688x + 1.8917	0.999

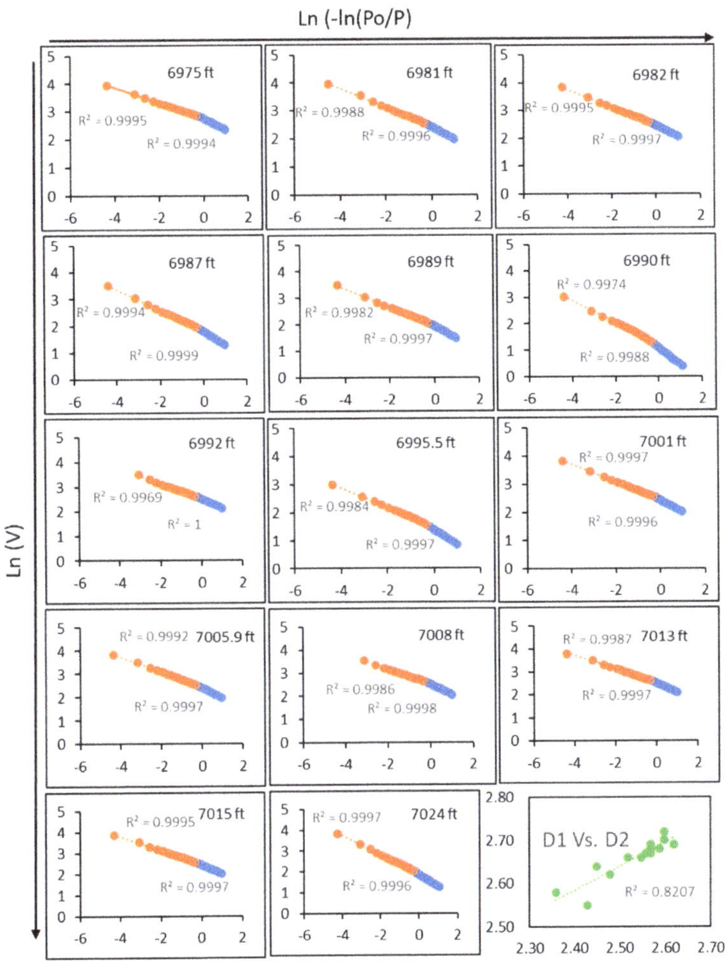

Figure 15. The FHH fractal [6,51] fitted curves for lnV vs. ln (−ln (P_0/P)) for the Lista Shale from well 16/21a-13. At relative pressures below 0.45 (highlighted in blue), a monolayer adsorption process occurs, dominated by Van der Waals forces. At relative pressures (highlighted in orange) above 0.45, the adsorption shifts to a capillary condensation process governed by surface tension, which reflects the pore structure. The highlighted green relative pressure points confirms that D1 and D2 have fractal dimensions with a correlation of 0.8. All depths are reported as measured depths.

5. Discussion

5.1. Mineralogy and Depositional Environment of the Lista Shale CCS Top Seal

The Lista Shale is finely laminated at the millimetre scale, suggesting fluctuating water flow velocities during its deposition (Figure 4). The presence of microfossils such as rare foraminifera (Figure 9), and radiolaria (Figure 6(A1)) and trace fossils such as *Planolites* (Figure 6A) and *Zoophycos* (Figure 6B,C) show that the Lista sediments were deposited in deep marine environmental settings [57,58]. *Zoophycos* is a trace fossil, feeding burrow with a worm-like shape commonly found in turbidite deep marine rock strata [59].

The core samples are dominated by smectite and quartz (Figure 8; Table 2). The quartz percentage varies throughout the core which relates to the shale interbedded with sandstone deposited as stacked lobes of middle and outer fan sediments (Figure 4). The variation in the thickness of sand input throughout is a characteristic of the Palaeocene turbidite system in the North Sea (Figure 4) [15,17,60]. Mudge and Bujak [60] indicated that the majority of sands associated with the Lista Formation are more widespread than the overlying Sele Formation, even though they are present in a similar area, suggesting a link to the original Mesozoic graben structure rather than the Cenozoic tectonic regime [15,61].

Three facies' associations were identified: (A) hemipelagic mudstone; (B) mud-prone heterolithic; and (C) sand-prone heterolithic (Figures 4 and 5). The term heterolithic particularly means alternating lithotype of layers, usually sand and mud [62]. The hemipelagic mudstone facies association is interpreted as basin floor deposits (Figure 4) without significant sand influx and the average grain size is clay (Figure 12) [15]. The term hemipelagic refers to a combination of gradual gravity sinking of very fine sediment, in this case with a mean grain size of 7 μm, and fine sediment delivered at the tail end of a flow event of material from a shelf towards an abyssal plain. Chalk debris flow occurs regularly as sub-rounded clasts within the mudstones (Figure 6A). The mud-prone heterolithic facies association is dominated by pelagic mudstone; the sandstone beds are usually thin (cm scale), and the average grain size is silt (Figure 12) [15]. The sand-prone heterolithic facies association is a mixture of sandstone and pelagic mud.

Mudstone and the grain size varies from clay to very fine sand (Figure 12) [15]. The sand-prone heterolithic facies are here interpreted as being marginal to the main sand channel in the turbidite fan system (Figure 4) [15].

In well 16/21a-13, core depths from 6974 ft to 6988 ft-md (measured depth) are part of a hemipelagic mudstone facies association and the core is dark grey, finely laminated, and contains trace fossils (Figures 4 and 5) along with traces of siderite nodules identified in thin sections and SEM-EDS (Figure 8). At core depth 6987 ft-md, a complex sutured fracture was observed (Figure 7A) suggesting that this part of the core is a damage zone related to an ancient and natural fault [63]. In general, a fracture damage zone is defined as a zone surrounded by fractures, showing grain breakage, concomitant reduction of grain size, change in clay mineral orientation, and increase in crack density [63]. The main fractures in the core are branching and localised in the clay-rich layers and are developed around more competent detrital quartz grains and pyrite (Figure 7(A1)) [63]. The fractures were initially open, possibly due to early tectonic activity linked to rifting and extension [15]; the fractures were then diagenetically filled with Fe-Mg-rich carbonate pieces of cement (Table 2) (Figure 7(A1–A3)). There are traces of slickensides associated with the damage zone suggesting some degree of shear as well as extension. Fractures cannot be ignored when it comes to assessing the integrity of the caprock for CCS projects, as fractures can play a major role in being conduits to flow movements and permeability pathways. Although the fractures evident in Figures 5 and 6A seem to be tightly cemented, they will have different geomechanical properties to the more clay-rich sections of the core and might be more susceptible to brittle failure under the new stress regime imposed when CO_2 is injected.

Core depths from 6991 to 6996 ft-md are part of the mud-prone heterolithic facies association, which is here attributed to outer fan deposits (Figure 5). At core depth 6995.5 ft-md, XRD data (Table 2) and BSE images show localised carbonate cement in the coarse silt beds, whereas the finer silt beds do not have calcite cement (Figure 7(B1–B3)). Calcite cementation of coarse silt is a common part of the diagenetic process in the Lista Formation; the origin of the calcite is unclear but it may be related to dissolved and reprecipitated calcite bioclasts or local redistribution favoured by the coarser grain of the cemented intervals [15]. The calcite in the coarse siltstone makes up 0.6 to 7.2% of the interval [15], which makes it potentially concerning when it comes to the integrity of the caprocks as part of CO_2-mineral interaction could lead to calcite dissolution [21].

5.2. The Genesis and Evolution of Pore Types in the Lista Shale CCS Caprock

A combination of primary sedimentary and secondary diagenetic processes leads to various pore types in mudstones [64]. Diagenetic processes during compaction and cementation influence the presence of both primary and secondary pores in these rocks. Clay-rich sediments like mudstones and shales are known to exhibit high porosity (e.g., up to 60 to 80%) at the time of deposition [2,23,65,66]. Currently, the porosity of the Lista Shale is approximately 16%, as indicated by MICP data (Table 3) and porosity data derived from density logs [14]. When compared to typical mudstone compaction curves, a porosity of 16% is common for rocks buried at around 6600 ft-true vertical depth.

Pores in the Lista Shale, after burial and diagenesis, include interparticle pores (Figure 11A–E), which are primary pores located between grains [23,64,67]. Interparticle pores between smectite platelets and between smectite and silt-grade materials are visible in SE images (Figure 11A,B,E). Intraparticle pores are observable in high-resolution images within pyrite framboids (Figure 11F). Primary interparticle porosity is the dominant pore type (Figure 11).

5.3. Pore Size Distribution in the Lista Shale CCS Caprock

Different aspects of pore size and structure are recorded by different analytical techniques. Different uses of the terms pore radius and pore diameter following mercury intrusion (MICP) and gas adsorption techniques in the literature tend to cause confusion [68,69]. For clarity, we will designate measurements from mercury intrusion as "pore throat diameter" and pore size measurements from N_2 adsorption as "pore body diameter" [21].

N_2 adsorption and MICP have different detection limits; MICP measurements can detect pore throats as small as 3.56 nm [68,70] (Figure 13B). At high mercury pressure, the rock fabric may be altered, potentially causing particle breakage and the opening of previously sealed pores [68,69]. The smallest pore body diameter detected by N_2 adsorption is 3.1 nm, while the largest detectable pore size using N_2 adsorption is 106.5 nm (Figure 13A). Despite the limitations of MICP and N_2 adsorption measurements, it has been suggested to combine both fluid invasion methods to characterise the full range of pore size distributions (Figure 16) [21,68,69,71].

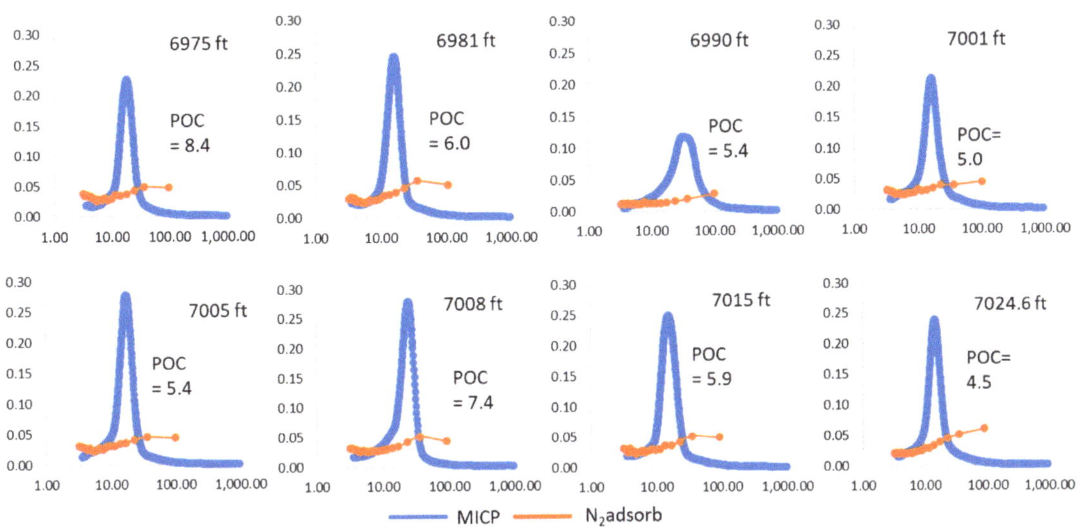

Figure 16. A combination of MICP and N_2 adsorption pore size distribution curves for well 16/21a-13 has been used to determine the point of connection (POC) between the two datasets. All depths listed are reported as measured depths.

From the overlapping porosity curves (Figure 16), there is a shared pore size range covered by the data from both MICP and BJH N_2 adsorption. This overlap enables the identification of the point of connection (POC) between the mercury intrusion and BJH N_2 adsorption curves, providing a complete representation of the pore size scale [21,38,69,70,72]. The overlap physically represents the filling mesopores (body and throats) leading to the formation of the N_2 monolayer.

The close correlation between the nitrogen adsorption-derived pore body diameter and mercury intrusion-derived pore throat diameter (Table 3) suggests that the pores predominantly resemble slits between grains, rather than narrow pores opening into larger pore bodies (Figure 11B). This is not surprising, given the H3 shape of the hysteresis loop which is characteristic of plate-like particles (Figure 14) [54] and considering that the Lista Shales are composed of plate-like sheet silicates such as smectite, kaolinite and mica (Figure 11D,E).

The average pore body diameter derived from gas adsorption is 17.84 nm (Figure 13A; Table 3). According to the IUPAC classification adopted by Rouquerol, et al. [71], the Lista Shale is predominantly composed of mesopores (2 to 50 nm) compared to micropores (<2 nm), with few pore bodies falling within the macropore classification (Figure 13A). Ding, et al. [55] concluded that micropores and mesopores provide the majority of surface area, facilitating gas adsorption on pore surfaces, while mesopores and macropores account for most of the pore volume, enabling gas storage as a discrete phase in shale [55]. Therefore, the dominance of mesopore bodies reflects the balance between free gas in macropore bodies and adsorbed gas in micropore bodies [21,55,73]. Schematic illustrations in Figure 11G,H show the occupation of free and adsorbed gas in slit pores and plate-like pores. Adsorption on pore surfaces is crucial for shale gas reservoir quality [74], although it may have a similarly significant impact on CO_2 adsorption on the pore surfaces of the Lista Shale.

The average pore throat diameter measured from MICP is about 17 nm (Figure 13B; Table 3). According to the classification scheme of pores proposed by [75] and adopted by Zhang, et al. [69] and AlNajdi and Worden [23], the Lista Shale pore throats (Figure 13B; Table 3) are classified as having a "transitional pore throat diameter" that ranges from 10 to 100 nm. Knowledge of the pore throat diameter, derived from MICP, is crucial for CCS projects, as it affects capillary entry pressure and the potential escape of CO_2 [23,76].

5.4. Compositional Influences on the Pore Structure, Surface Area and Fractal Dimensions of the Lista Shale

Correlation analysis with all the quantitative measurements has been undertaken, distinguishing negative and positive correlations (Figure 17). We have also derived *p*-value significance to identify meaningful correlations. Established statistical methods show that if the *p*-value is >0.05, the hypothesis (correlation) is insignificant, and if the *p*-value is $\leq$0.05, the hypothesis is significant [77]; increasingly low *p*-values are increasingly significant. In this research, adopting conventional statistical techniques, we have set the maximum *p*-value significance to equal 0.05.

There is a positive correlation between surface area (Table 3 and BJSA and MBSA in Figure 17) and pore volume (BJPV), which can be explained by the convoluted outlines of clusters of clay minerals leaving substantial aggregate-marginal, microporosity (that may be primary or secondary in origin) as well as elevated surface area (Figure 11). Xiong, et al. [78] reported a similar relationship.

The relationships between fractal dimensions and pore structure parameters are presented in Figure 17. The correlation matrix shows that fractal dimension D1 (pore surface) has a positive correlation with fractal dimension D2 (pore structure), which indicates that the Lista Shale has fractal attributes and that the fractal dimensions, D1 and D2, are not totally independent; a similar conclusion was reached by Chen, et al. [73]. As expected, there is a strong positive correlation between D1 and surface area (BJSA), reflecting the surface roughness (Figures 7B and 17) [51,73].

Clay mineral proportions have been reported to influence mudstone pore volume, pore structure and surface area [78–81]. There is a positive relationship between the dominant clay mineral, smectite and surface area (Figure 17), possibly due to a greater contribution of interparticle pores in smectite-rich, silt-poor, samples (Figure 10). According to Saidian, et al. [82], smectite and illite clays usually contribute the highest surface area in clay-rich rocks due to their sheet-like morphology (Table 2).

Complex clay morphology leads to challenging diffusion and gas flow in the pore network [83]. The correlation between the pore structure and fractal dimension indicates that the Lista Shale has a low potential for CO_2 flow and diffusion due to the complex and heterogeneous pore structure.

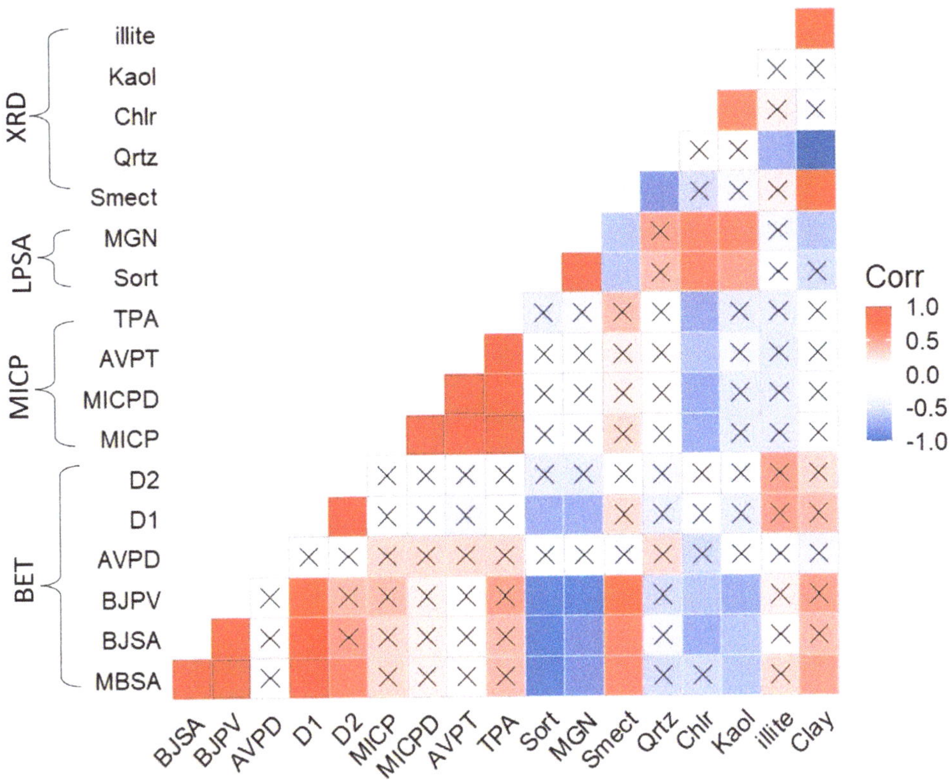

Figure 17. Correlogram plot for *p*-value significance tests from XRD, LPSA, fractals, MICP and BET data. *p*-values ≤ 0.05 are considered to be significant. *p*-values > 0.05 are considered to be insignificant and are marked with a cross ($\times$) in the figure. Positive correlations are displayed in blue and negative correlations in red. Colour intensity is proportional to the specific correlation coefficient. Multi Bet Surface area (MBSA), BJH surface area (BJSA), BJH pore volume (BJPV), average pore diameter (AVPD), MICP porosity (MICP), median pore diameter (MICPD), average pore throat (AVPT), total pore area (TPA), smectite content (Smect), chlorite content (Chlr), muscovite/illite content (illite), kaolinite content (Kaol), quartz content (Qrtz), sorting (Sort) and mean grain size (MGS).

5.5. The Impact of Reactive Surface Area on the Lista Shale Top Seal

The reactive surface area of minerals is a dynamic property that is expected to evolve over time as geochemical reactions develop; surface area varies from mineral to mineral [84,85]. The distribution of reactive surface area of minerals within a rock can significantly influence the progress of geochemical reactions during CO_2 injection [86], which in turn affects the integrity of the caprock.

Some minerals in shales, such as quartz or feldspars, can be compact (low surface area) silt- or even sand-grade grains. Some clay minerals, especially detrital smectite, can form irregular aggregates or pseudo-grain, which lead to high D1 fractal dimensions and commensurately high surface area (Figure 17). Other clay minerals, such as chlorite or kaolinite, can occur as detrital flakes or recrystallised or neomorphic version of detrital minerals; these tend to have relatively low surface area compared to smectite (Figure 17). Carbonate minerals [21], albeit not common in the Lista Shale, tend to occur as cement in pores and probably have fairly low surface area.

Aluminosilicate minerals, such as clay minerals, plus carbonate minerals and quartz (both detrital and biogenic), can act as pore-filling, or grain-coating agents, and in some cases, a cementing framework of grains [86–89].

Clay mineral- and feldspar-rich rocks have been reported to exhibit more complex relationship between CO_2/brine and porosity than carbonate rich rocks [86]. For example, Espinoza and Santamarina [90] described a decrease in total porosity after submerging kaolinite and montmorillonite in supercritical CO_2 (ScCO_2). In contrast, Smith, et al. [91] and Miller, et al. [92] reported an increase in caprock porosity upon exposure to CO_2/brine. Both Miller, et al. [92] and Smith, et al. [91] unexpectedly found that silicate minerals were more susceptible to reaction in the presence of CO_2 than carbonate minerals, possibly due to their relatively larger reactive surface areas. These reactions, in turn, resulted in the creation of new pores larger than 1 μm with notable increase in pore connectivity [86,91,92]. The inverse relationship between pore connectivity and porosity has been attributed to the roles of clay hydration or preferred mineral reprecipitation in pore throats following dissolution [86]. This is analogous to the phenomenon of redistributional secondary porosity in sandstones leading to reduced permeability [93]. XRD mineralogy data (Table 2) reveal chlorite and plagioclase, as wells as smectite, that may be reactive to CO_2 [94], exposing newly fresh pores which contribute to increased surface area changing the physical and chemical properties of the caprock.

6. Risk Assessment of the Lista Shale Top Seal

6.1. Seal Capacity and Its Relation to Depositional Environment

Worden, et al. [14] reported the CO_2 column height for the Rodby Shale top seal to the Captain Sandstone reservoir at the Acorn injection site, as well as the Lista Shale top seal based on MICP data. Gibson-Poole, et al. [95] also provided CO_2 column heights for localised intraformational seals within the Gippsland Basin in Australia using MICP data. According to Gibson-Poole, et al. [95], intraformational seals can span a range of depositional environments from deep marine to fluvial, with CO_2 column heights varying from 53 to 1191 m. Figure 18 combines the average CO_2 column heights published by Gibson-Poole, et al. [95] for different depositional environments, along with integrated Lista data from Worden, et al. [14]. Figure 18 demonstrates that the Lista Shale is classified as intermediate in terms of CO_2 column height that can be supported before capillary entry pressure is exceeded.

6.2. Lista Shale Seal Evaluation-Caprock Integrity and Risk of CO_2 Leaks

In this section, we examine the integrity of the Lista Shale and assess the leakage risks of CO_2 from the East Mey sandstone reservoir. The methods applied here have been adapted from those reported by Espinoza and Santamarina [96] and modified by AlNajdi, et al. [21] for evaluating the seals of target caprocks at various CCS sites, based on the petrophysical and mineralogical properties of the caprock, reservoir, and the anticipated CO_2 plume. The original approach by Espinoza and Santamarina [96] looked at bulk properties and so did not consider caprock depositional or diagenetic heterogeneity or structural discontinuities.

The escape of CO_2 from reservoir to the top seal plausibly might occur by either diffusion or advection of CO_2 [96,97]. Diffusion of CO_2 is a mechanism controlled by the difference in CO_2 concentration (or more accurately its fugacity) between the base and the top of a top seal [21]. A maximum permitted or acceptable leak rate was proposed by Espinoza and Santamarina [96] to be 3 kg/m^3/yr. The rate of diffusion ($q_{CO_2}^{diff}$) can be modelled by Fick's Law [98]:

$$q_{CO_2}^{diff} = -n \cdot D(\Delta x_{CO2})/t_h \quad (4)$$

where n represents fractional porosity (in the Lista, n = 0.16), D is the CO_2 diffusion coefficient (here assumed to be $D = 10^{-9}$ m^2/s [99]), x_{CO2} denotes difference in CO_2 concentration driving diffusion, with the maximum assumed value being ~0.044 kg/m^3, and t_h is the thickness of the caprock, which is measured as 15 m for single caprock unit and 100 m for the whole core section based on published wireline logs from the Lista Shale [14]. Given these assumptions and the data for the Lista Shale, the modelled CO_2 diffusion rate is about 1.5×10^{-5} kg/m^2/yr, which is considerably lower than the reference leak rate. Thus, diffusive loss of CO_2 seems to be unlikely to pose a concern for this top seal.

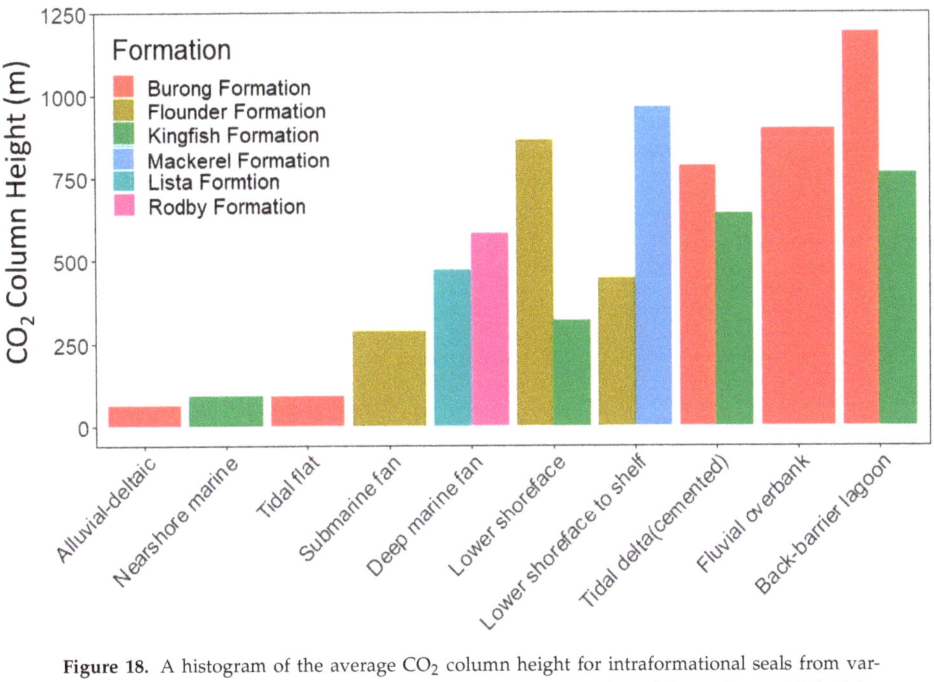

Figure 18. A histogram of the average CO_2 column height for intraformational seals from various formations across different depositional environments, adapted from data published by Gibson-Poole, et al. [95] and modified after [21]. According to Gibson-Poole, et al. [95], CO_2 column heights were categorised based on their ability to impede CO_2 flux. The most significant barriers to CO_2 flux are fine-grained mudstones from fluvial overbank and back-barrier lagoon deposits. Intermediate barriers are matrix-rich siltstones and fine-grained sandstones or mudstones from the lower shoreface to inner shelf depositional environments. The weakest barriers to CO_2 flux appear to be interbedded siltstones and fine-grained sandstones from tidal and nearshore marine settings. The Lista Shale, derived from deep marine fans, is similar to the calcareous Rodby Shale [21], and can be classified as an intermediate-strength barrier to CO_2 flux.

Advection of CO_2 through top seals takes place only if the capillary entry pressure of the caprock is surpassed [21,96]. The rate of CO_2 via advection ($q_{CO_2}^{adv}$) is affected by factors such as the pressure difference between the reservoir and the top of the top seal, the relative and absolute permeability, the density of CO_2, top seal thickness, and the viscosity of CO_2, as determined by Darcy's Law [100]:

$$q_{CO_2}^{adv} = (kk_{CO2}^{r} \cdot \rho_{CO2}) / \mu_{CO2} \cdot ((\Delta P - \rho_{CO2} \cdot g \cdot t_h)) / t_h \quad (5)$$

where k represents the absolute vertical permeability (in the Lista, this is 2.59×10^{-20} m^2, derived from MICP data and an assumption that $k_v/k_h = 0.01$). ρ_{CO2} denotes the mass density of CO_2 (assumed here to be 650 kg/m^3 [101]). μ_{CO2} is the viscosity of CO_2 (assumed here to be 0.05 mPa·s). t_h is the thickness of the caprock, which is assumed to be 15 m [14]. $(\Delta P - \rho_{CO2} \cdot g \cdot t_h)$ represent the total pressure head, assumed to be up to 10 MPa. The relative permeability of CO_2, k_{CO2}^r is assumed to be 0.64, 0.16 and 0.01, corresponding to near wellbore conditions (CO_2-saturation [S_{CO2}^r] of 0.8), mid plume (CO_2-saturation of 0.4), and far-field (CO_2-saturation of 0.1) assuming that (k_{CO2}^r) is equal to $(S_{CO2}^r)^2$ [96].

The modelled rate of advective flux of CO_2 is illustrated in Figure 19, in the event of capillary pressure of CO_2 exceeding the Lista Shale. The rate of advective CO_2 flux varies due to different CO_2 saturations and corresponding relative permeability scaling factors, which are influenced by distance from the injection well [21]. Near the injection well, with a 10 MPa CO_2 pressure difference, the post-breakthrough CO_2 flux could reach up to 4.2 kg/m^2/yr which exceeds the reference leak rate. At several hundred meters from the injection well, with a 5 MPa pressure difference, the flux may be as high as 0.55 kg/m^2/yr. Further away at many hundreds of meters from the injection well, with 1 MPa pressure difference, the flux rate could be up to 0.01 kg/m^2/yr. The highest risk is clearly in the wellbore region at 10 MPa pressure and the lowest which are below the reference leak rate are at 5 and 1 MPa. The greatest is clearly associated with high-pressure CO_2 injection in near-wellbore locations, but only if capillary entry pressure is exceeded.

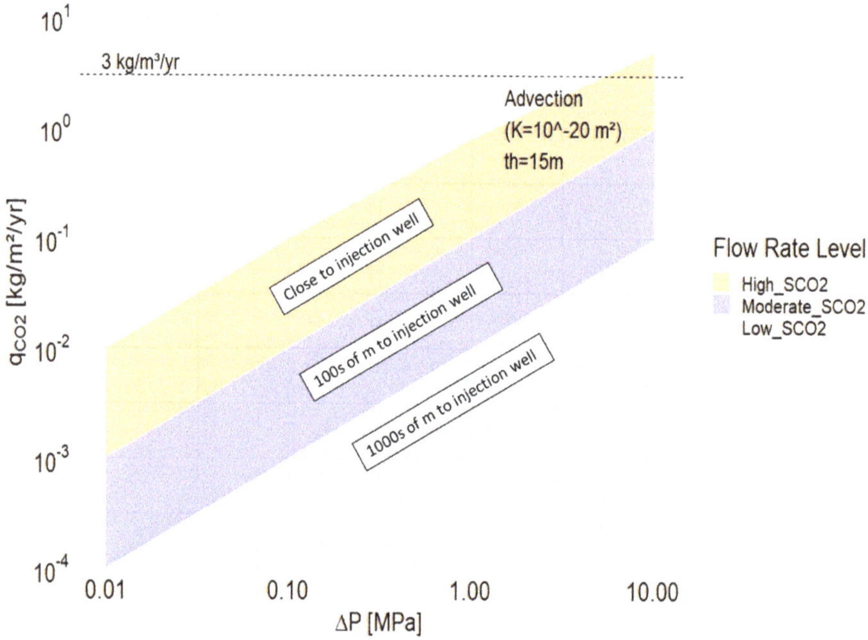

Figure 19. Analysis of the magnitude of advective flux rate if CO_2 capillary entry pressure has been exceeded, adapted from Espinoza and Santamarina [96] and modified after AlNajdi, et al. [21]. The calculation is based on equations 4 and 5 (refer to text). Here, n represents the fraction porosity (in the Lista, n is 0.16). D denotes the CO_2 diffusion coefficient (following convention, we assume D^* is

10^{-9} m^2/s [99], x_{CO2} is the difference in CO_2 concentration driving diffusion, with a maximum assumed value of approximately 0.044 kg/m^3 (representing the difference between maximum and minimum CO_2 concentration in water). k refers to the vertical permeability, which in the Lista is 2.59×10^{-20} m^2 (derived from MICP data, with an assumption that kv/kh = 0.01). ρ_{CO2} is the mass density of CO_2, here assumed to be 650 kg/m^3 [101]. μ_{CO2} is the viscosity of CO_2, which is taken to be 0.05 mPa·s. t_h is the thickness of the caprock, assumed here to be 15 m [14]. $(\Delta P - \rho_{CO2} \cdot g \cdot t_h)$ is the total pressure head, here set between 1 MPa and 10 MPa. k^r_{CO2} is CO_2 relative permeability; calculated for CO_2 saturations of 0.8, 0.4 and 0.1, representing near-wellbore (meters), mid-plume (100 s of meters) and far-field (1000 s of meters) conditions, and based on the assumption that (k^r_{CO2}) is equal to $(S^r_{CO2})^2$ [96]. The calculated CO_2 flux rate in the near-wellbore region, if capillary entry pressure is exceeded, approaches the maximum reference leak rate of 3 $kg/m^2/yr$, given a pressure difference (driving force) of 10 MPa. This rate is at the reference leak rate, indicating that advective loss of CO_2 through the top seal may be a concern in the near-wellbore region at the highest CO_2 saturations and pressures. In general, the risk of leakage diminishes with increasing distance from the near-wellbore region, as CO_2 saturation, and presumably pressure difference, decreases.

6.3. Seal Capacity

The effectiveness of geological formations for potential CO_2 storage can be assessed by using dimensionless sealing and stability numbers [96]. The sealing number represents the ratio of the capillary breakthrough pressure, Pc, to the buoyant pressure, ΔP, which is determined by the thickness of the continuous CO_2 plume, h [96],

$$\text{Sealing number} = \pi_1 = \left(P_c^*\right)/\Delta P = (\psi T_s \cos\theta\ S_s\ \rho_m)/(2e\ hg\ (\rho_w - \rho_{CO2})) \tag{6}$$

where ψ is the pore shape factor that ranges from 2 to 4; in the Lista case, it is assumed tobe 3 for slit pores. T_s denotes the interfacial tension between water and CO_2 (assumed to be 0.03 N/m after Bachu and Brant [102]). θ is the contact angle formed by the CO_2-water interface on the mineral surface (assumed to be 0.9185 Radians). Ss is the surface area, (assumed average surface area of the Lista is 9854.5 m^2/kg based on BET data, Table 3). The mineral bulk density ρ_m is assumed to be 2300 kg/m^3. e is the void ratio at breakthrough, here assumed to be 0.6. h is the thickness of the continuous CO_2 plume (assumed to be 15 m based on wireline log data). g is acceleration of gravity. ρ_w is the mass density of water assumed to be 1050 kg/m^3 based on regional formation water data in Warren and Smalley [103]. ρ_{CO2} is the density of CO_2 assumed to be 650 kg/m^3 [101].

A sensitivity test was conducted on the range of inputs to the calculated sealing number of Lista Shale, the dominant parameters as reported by AlNajdi, et al. [21] are shape factor, surface area and the thickness of CO_2 plume. The average sealing number of the Lista Shale is 39 (Figure 20). In comparison to other CCS case studies shown in Figure 20, the Lista Shale is similar to the Rodby Shale [21] and seems to offer a more effective seal than for example the Utsira Formation at Sleipner [104] and the regional seal at Rousse CCS pilot located in France [105]. The Lista Shale is slightly less effective than the In Salah (Krechba) seal [106].

6.4. Mechanical Stability of the Top Seal

Assessment of the relationship between vertical effective stress, σ'_{ZO}, at the reservoir depth, z, and the buoyant pressure, ΔP, initiated by CO_2 plume of thickness h can be measured via stability number (Equation (7)) [96]. Variation in pore pressure can lead to changes in vertical effective stress, which can later trigger fault reactivation and generate new fractures [96].

$$\text{stability number} = \pi^2 = \frac{\sigma'_{zo}}{\Delta P} = \frac{(Z_w\ g\rho_w + Zg\rho_{bulk}) - P_o}{hg(\rho_w - \rho_{CO2})} \tag{7}$$

where Z_w is the height of the water column above the seafloor (equal to 100 m, based on well reports). ρ_{bulk} is the bulk mass density (assumed to be 2300 kg/m^3). The sediment column height, Z, is assumed to be 1900 m. P_o is the initial fluid pressure at the reservoir-seal interface (assumed to be hydrostatic pressure at 20.6 MPa).

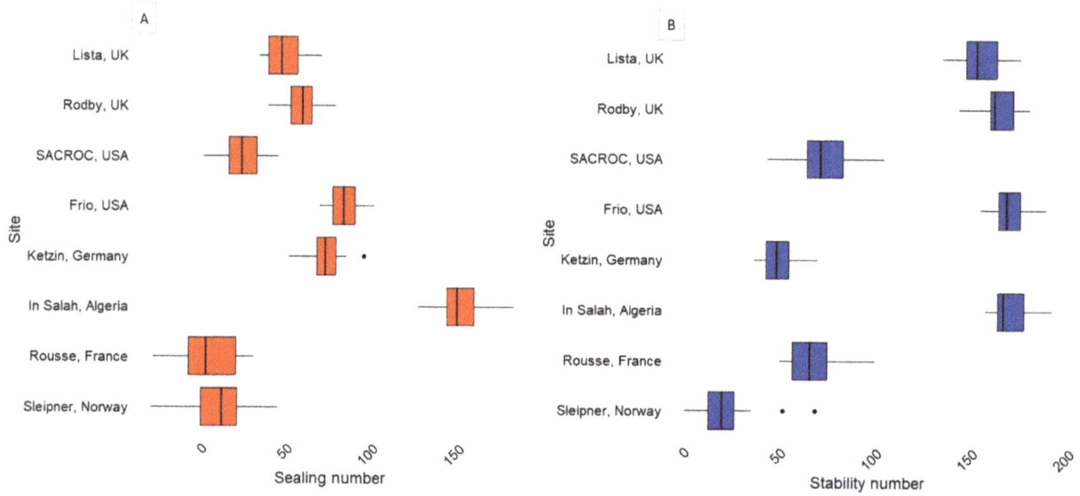

Figure 20. Analysis of the mechanical stability and sealing capacity of current and proposed CCS storage sites, including the Lista Shale [21,104–109] adapted from Espinoza and Santamarina [96] and modified after AlNajdi, et al. [21]. (**A**) The results from the dimensionless sealing number calculation, π^1, are plotted according to Equation (6). The sealing number for the Lista Shale has been calculated assuming a pore shape factor (ψ) of 3, representing slit-shaped pores. The interfacial tension (T_s) between water and CO_2 is taken to be 0.03 N/m [102]. The contact angle (θ) is 1.00 (Radians). Surface area (Ss) based on BET data is 9854.5 m^2/kg (Table 3), while mineral density (ρ_m) is assumed to be that of quartz, 2400 kg/m^3. The void ratio (e) at breakthrough is assumed to be 0.6. The thickness (h) of the continuous CO_2 plume is taken as 15 m based on wireline log data and the approach suggested by Espinoza and Santamarina [96]. g is acceleration of gravity. ρ_w is the mass density of water assumed to be 1050 kg/m^3 based on regional formation water data in [103]. ρ_{CO2} is the density of CO_2 assumed to be 650 kg/m^3 [101]. In terms of sealing number, the Lista Shale is comparable to the Rodby Shale [21], has a better sealing capacity than the Utsira Formation at Sleipner [104] and the Rousse CCS pilot site in France [105] but is slightly less effective than the seal at the In Salah (Krechba) former CCS site in Algeria [106]. (**B**) The dimensionless stability number calculation, π_2, is based on Equation (7). Z_w represents the water column height above the seafloor, assumed to be approximately 100 m for this part of the Moray Firth. Bulk density (ρ_{bulk}) is taken to be 2300 kg/m^3. The sediment column height (Z) is taken as 1900 m, based on wireline logs, and the initial fluid pressure (P_o) at the reservoir-seal interface is assumed to be hydrostatic, 20.6 MPa. For the stability number, the Lista Shale top seal performs comparably to the Rodby Shale [21], better than the Utsira Formation at Sleipner [104] and similarly to the seals at In Salah (Krechba) [106] and Frio [107]. These sealing and stability calculations seem to suggest that the Lista Shale has a high likelihood of serving as an effective seal. Note that the dots in the plot are outliners.

A sensitivity test was applied to the range of input parameters for the calculation of the stability number (Figure 20). The stability number is primarily controlled by three parameters: the bulk density of the sediment column, the height of the sediment column, and, most significantly, the maximum thickness of the CO_2 plume [21]. The average stability number of the Lista Shale is 361. When compared to other CCS field cases, the Lista Shale is similar to the Rodby Shale [21], In Salah (Krechba) [106] and Frio [107] in terms of stability and better than the Utsira Formation at Sleipner [104].

Drawing together the work published here (Figures 1–19) and other examples referred to in Figure 20A,B, we have produced a summary table that lists attributes of clastic top seal (Table 5). We have here specifically listed helpful and detrimental characteristics of each attribute. This table could serve as a checklist that could be used during the appraisal of any potential top seal to a CO_2 storage site.

7. Conclusions

1. The Palaeocene Lista Shale is smectite-rich and quartz silt-bearing. It was deposited on top of the Mey Sandstone as part of a linked turbidite fan-abyssal plain system in a deep marine environment.
2. The pores have been affected by both depositional and diagenetic processes; pore types include interparticle pores such as slit-shaped and plate-like pores, as well as intraparticle pores found within pyrite framboids.
3. The pore bodies are primarily mesopores, with small proportions of micropores and macropores also present. The prevalence of mesopores aligns with the observed pattern in which free gas predominantly occupies macropore bodies, while adsorbed gas is mainly found within micropore bodies.
4. The mean pore throat diameter measured by mercury intrusion is 17.00 nm, with a minimum of 3.56 nm and a maximum of 10,000 nm. For pore body diameters, BJH N_2 adsorption detected a minimum of 3.10 nm and a maximum of 106.5 nm. By combining two fluid invasion techniques, mercury intrusion and gas adsorption, the entire spectrum of pore body and pore throat sizes can be accurately represented. The connection at the inflection point, where mesopores begin to fill, marks the junction at which mercury intrusion pore throat measurements align with gas adsorption measurements.
5. The close similarity between the N_2 adsorption pore body diameter (average 17.84 nm) and the mercury intrusion pore throat diameter (average 17.0 nm) indicates that the pores in the Lista Shale are likely to be slit-shaped.
6. The complex heterogeneity of Lista Shale pore structure and mineral composition has strongly influenced the surface area. Greater quantities of smectite lead to higher surface area and pore volume, possibly due to a greater contribution of interparticle pores in the platy clay minerals that dominate in smectite-rich, silt-poor samples.
7. If smectite or other aluminosilicate minerals such as chlorite or plagioclase reacted with CO_2, newly exposed minerals would potentially lead to increased reactive surface area, possibly feeding back to increased reaction rates.
8. According to assessments of the diffusive flux, advection rate if capillary entry pressure is exceeded, and its overall stability and sealing characteristics, the Lista Shale shows strong potential to serve as an effective barrier against CO_2 leakage.

Table 5. Summary list of attributes for clastic top seal with listed beneficial and detrimental characteristics of each attribute for a potential CCS top seal.

Attribute	Helpful for a Good Top Seal	Detrimental to Being a Good Top Seal	Example	References
Pore body size	Preferably in the micro-pore range	When in meso- to macro-pore range	Lista Shale, Rodby Shale, Sichuan Basin	AlNajdi et al. [21], Chen et al. [51]
Pore throat size	Preferably in the micro-pore range	When in meso- to macro-pore range	Lista Shale, Rodby Shale, Mercia Mudstone	AlNajdi et al. [21], Armitage et al. [26]
Pore type	Slit pores preferred	Round pores	Lista Shale, Rodby Shale, Sichuan Basin	AlNajdi et al. [21], Chen et al. [51]
Pore volume	Low	High	Lista Shale, Rodby Shale, Sichuan Basin	AlNajdi et al. [21], Chen et al. [51]
Specific surface area	Low (to minimise reactivity)	High	Lista Shale, Rodby Shale, Sichuan Basin	AlNajdi et al. [21], Chen et al. [51]
Particle size	Low	High	Lista Shale, Rodby Shale	AlNajdi et al. [21], Pandey et al. [28]

Table 5. *Cont.*

Attribute	Helpful for a Good Top Seal	Detrimental to Being a Good Top Seal	Example	References
Sorting	Poorly sorted sediment preferred	Well sorted sediment preferred	Lista Shale, Rodby Shale	AlNajdi et al. [21]
Mineralogy—clay	Inert clay minerals (e.g., illite, kaolinite)	Reactive clay minerals (e.g., chlorite, some smectites)	Lista Shale, Rodby Shale, Mercia Mudstone	Worden et al. [14], AlNajdi et al. [21], Armitage et al. [26]
Mineralogy—carbonate	Zero carbonate preferred	Calcite-rich (most reactive carbonate)	Lista Shale, Rodby Shale	Worden et al. [14], AlNajdi et al. [21]
Mineralogy—silt minerals	Quartz preferred (least reactive)	Reactive feldspar (Ca-rich are worst of all)	Lista Shale, Rodby Shale, Mercia Mudstone	Worden et al. [14], AlNajdi et al. [21], AlNajdi and Worden [23], Armitage et al. [26]
Mineralogy—silt proportions	Low (minimum brittleness)	High (increasing brittleness)	Lista Shale, Rodby Shale, Mercia Mudstone	Worden et al. [14], AlNajdi et al. [21], AlNajdi and Worden [23], Armitage et al. [26]
Mineralogy—Ca-sulphates	Low (minimum risk of volume change)	High (elevated risk of fractures and veins)	Mercia Mudstone	Worden et al. [14], AlNajdi and Worden [23], Armitage et al. [26]
Brittleness	Low	High	Lista Shale, Rodby Shale	Worden et al. [14], Ingram and Urai [110]
Unconfined compressive strength	Low	High	Lista Shale, Rodby Shale	Worden et al. [14], Ingram and Urai [110]
Thickness of top seal	Thicker seals better to minimise transport out of the storage site and risk of fault leakage	Thin seals potentially poor as they would lead to risk of leakage	Lista Shale, Rodby Shale, Sleipner, Rousse, In Salah, Ketzin, Frio, SACROC	AlNajdi et al. [21], Espinoza and Santamarina [96]
Lateral extent of top seal	Wide lateral extent preferred to minimise spillage and upward CO_2 migration	Limited lateral extent would lead to risk of upwards leakage	Otway Basin	Kaldi et al. [4]
Fluid pressure relative to hydrostatic	Low fluid pressures preferred to minimise risk of capillary entry pressure being exceeded	High fluid pressures may lead to capillary entry pressure being exceeded	Lista Shale, Rodby Shale	Kaldi et al. [4], AlNajdi and Worden [23], Espinoza and Santamarina [96], Ingram and Urai [110]
Effective stress history (maximum palaeo-depth)	Sufficient to compact the mudstone, but low enough to prevent elevated brittleness	Negligible compaction is not helpful; excess compaction may not be helpful either	Lista Shale, Rodby Shale, Rousse, Mercia Mudstone, In Salah, Nordland Shale	AlNajdi and Worden [23], Ingram and Urai [110]
Thermal history (maximum palaeo-temperature)	Maximum palaeotemperature < 80 °C preferred to minimise mineral diagenesis and increasing brittleness	Maximum palaeotemperature > 120 °C as chemical compaction and mineral diagenesis lead to elevated brittleness	Lista Shale, Rodby Shale, Rousse, Mercia Mudstone, In Salah, Nordland Shale	AlNajdi and Worden [23]
Geological age of top seal	Younger rocks (e.g., Cenozoic) preferred to minimise diagenesis, but not too young to prevent sufficient compactional porosity-loss	Old rocks (e.g., Palaeozoic) have greatest chance of becoming brittle and strong (but will have lowest porosity and permeability)	Lista Shale, Rodby Shale, Rousse, Mercia Mudstone, In Salah, Nordland Shale	AlNajdi and Worden [23]

Author Contributions: Conceptualisation; N.A.; methodology; N.A. and J.E.P.U.; validation; R.H.W.; formal analysis; N.A. and R.H.W., investigation; N.A. and R.H.W., data curation; N.A., writing original draft preparation; N.A., reviewing and editing; R.H.W., supervision; R.H.W. All authors have read and agreed to the published version of the manuscript.

Funding: This research received no funding.

Data Availability Statement: Data will be available upon request.

Acknowledgments: We acknowledge Kuwait Institute for Scientific Research (KISR) for sponsoring the project.

Conflicts of Interest: The authors declare no conflicts of interest.

References

1. Aplin, A.C.; Fleet, J.A.; Macquaker, J.H.S. *Muds and Mudstones: Physical and Fluid Flow Properties*; The Geological Society of London: London, UK, 1999; p. 158.
2. Aplin, A.C.; Macquaker, J.H.S. Mudstone diversity: Origin and implications for source, seal, and reservoir properties in petroleum systems. *Am. Assoc. Pet. Geol. Bull.* **2011**, *95*, 2031–2059. [CrossRef]
3. Williams, G.A.; Chadwick, R.A. Influence of reservoir-scale heterogeneities on the growth, evolution and migration of a CO_2 plume at the Sleipner Field, Norwegian North Sea. *Int. J. Greenh. Gas Control* **2021**, *106*, 103260. [CrossRef]
4. Kaldi, J.; Daniel, R.; Tenthorey, E.; Michael, K.; Schacht, U.; Nicol, A.; Underschultz, J.; Backe, G. Containment of CO_2 in CCS: Role of caprocks and faults. *Energy Procedia* **2013**, *37*, 5403–5410. [CrossRef]
5. Fatah, A.; Bennour, Z.; Ben Mahmud, H.; Gholami, R.; Hossain, M. A Review on the Influence of CO_2/Shale interaction on shale properties: Implications of CCS in shales. *Energies* **2020**, *13*, 3200. [CrossRef]
6. Fatah, A.; Ben Mahmud, H.; Bennour, Z.; Gholami, R.; Hossain, M. The impact of supercritical CO_2 on the pore structure and storage capacity of shales. *J. Nat. Gas Sci. Eng.* **2022**, *98*, 104394. [CrossRef]
7. Pan, Y.; Hui, D.; Luo, P.; Zhang, Y.; Sun, L.; Wang, K. Experimental investigation of the geochemical interactions between supercritical CO_2 and shale: Implications for CO_2 storage in gas-bearing shale formations. *Energy Fuels* **2018**, *32*, 1963–1978. [CrossRef]
8. Cheng, Y.; Zeng, M.; Lu, Z.; Du, X.; Yin, H.; Yang, L. Effects of supercritical CO_2 treatment temperatures on mineral composition, pore structure and functional groups of shale: Implications for CO_2 sequestration. *Sustainability* **2020**, *12*, 3927. [CrossRef]
9. Qin, C.; Jiang, Y.; Luo, Y.; Xian, X.; Liu, H.; Li, Y. Effect of supercritical carbon dioxide treatment time, pressure, and temperature on shale water wettability. *Energy Fuels* **2016**, *31*, 493–503. [CrossRef]
10. Rochelle, C.A.; Czernichowski-Lauriol, I.; Milodowski, A.E. The impact of chemical reactions on CO_2 storage in geological formations: A brief review. *Geol. Soc. Lond.* **2004**, *233*, 87–106. [CrossRef]
11. Macht, F.; Eusterhues, K.; Pronk, G.J.; Totsche, K.U. Specific surface area of clay minerals: Comparison between atomic force microscopy measurements and bulk-gas (N_2) and -liquid (EGME) adsorption methods. *Appl. Clay Sci.* **2011**, *53*, 20–26. [CrossRef]
12. Jones, F. UK Government to Fund Acorn CCS in Second Phase of CCUS Development. Available online: https://www.power-technology.com/news/ccus-acorn-uk-government/#catfish (accessed on 14 September 2024).
13. Alcalde, J.; Heinemann, N.; Mabon, L.; Worden, R.H.; de Coninck, H.; Robertson, H.; Maver, M.; Ghanbari, S.; Swennenhuis, F.; Mann, I.; et al. Acorn: Developing full-chain industrial carbon capture and storage in a resource- and infrastructure-rich hydrocarbon province. *J. Clean. Prod.* **2019**, *233*, 963–971. [CrossRef]
14. Worden, R.H.; Allen, M.J.; Faulkner, D.R.; Utley, J.E.P.; Bond, C.E.; Alcalde, J.; Heinemann, N.; Haszeldine, R.S.; Mackay, E.; Ghanbari, S. Lower Cretaceous Rodby and Palaeocene Lista Shales: Characterisation and comparison of top-seal mudstones at two planned CCS sites, Offshore UK. *Minerals* **2020**, *10*, 691. [CrossRef]
15. Kilhams, B.; Hartley, A.; Huuse, M.; Davis, C. Characterizing the Paleocene turbidites of the North Sea: The Mey Sandstone Member, Lista Formation, UK Central Graben. *Pet. Geosci.* **2012**, *18*, 337–354. [CrossRef]
16. Evans, D.; Graham, C.; Armour, A.; Bathurst, P. *The Millennium Atlas: Petroleum Geology of the Central and Northern North Sea*; The Geological Society of London: London, UK, 2003.
17. Mudge, D.; Copestake, P. Lower Palaeogene stratigraphy of the northern North Sea. *Mar. Pet. Geol.* **1992**, *9*, 287–301. [CrossRef]
18. Allen, M.J.; Faulkner, D.R.; Worden, R.H.; Rice-Birchall, E.; Katirtsidis, N.; Utley, J.E.P. Geomechanical and petrographic assessment of a CO_2 storage site: Application to the Acorn CO_2 Storage Site, offshore United Kingdom. *Int. J. Greenh. Gas Control* **2020**, *94*, 102923. [CrossRef]
19. Ahmadi, Z.M.; Sawyers, M.; Kenyon-Roberts, S.; Stanworth, C.W.; Kugler, K.A.; Kristensen, J.; Bathurst, P. The Millenium Atlas: Petroleum Geology of the Central and Northern North Sea. *Geol. Soc. Lond.* **2003**, 548–597.
20. Gier, S.; Worden, R.H.; Krois, P.; Armitage, P.J.; Butcher, A.R.; Churchill, J.M.; Csoma, A.E.; Hollis, C.; Lander, R.H.; Omma, J.E.; et al. Comparing clay mineral diagenesis in interbedded sandstones and mudstones, Vienna Basin, Austria. In *Reservoir Quality of Clastic and Carbonate Rocks: Analysis, Modelling and Prediction*; Geological Society of London: London, UK, 2018; Volume 435.
21. AlNajdi, N.A.; Worden, R.H.; Utley, J.E.P. Compositional controls on the Lower Cretaceous Rodby Shale pore structure and surface area: A planned CCS top seal caprock for the Acorn storage site. *Int. J. Greenh. Gas Control* **2024**, *137*, 104219. [CrossRef]

22. Taylor, A.M.; Goldring, R. Description and analysis of bioturbation and ichnofabric. *J. Geol. Soc.* **1993**, *150*, 141–148. [CrossRef]
23. AlNajdi, N.; Worden, R.H. Porosity in mudstones and its effectiveness for sealing carbon capture and storage sites. In *Enabling Secure Subsurface Storage in Future Energy Systems*; Geological Society of London: London, UK, 2023; Volume 528, SP528-2022-2084. [CrossRef]
24. Worden, R.H.; Griffiths, J.; Wooldridge, L.J.; Utley, J.E.P.; Lawan, A.Y.; Muhammed, D.D.; Simon, N.; Armitage, P.J. Chlorite in sandstones. *Earth-Sci. Rev.* **2020**, *204*, 103105. [CrossRef]
25. Worden, R.H.; Charpentier, D.; Fischer, Q.; Aplin, A.C. Fabric development and the smectite to illite transition in Upper Cretaceous mudstones from the North Sea: An image analysis approach. *Geol. Soc. Lond. Spec. Publ.* **2005**, *249*, 103–114. [CrossRef]
26. Armitage, P.J.; Worden, R.H.; Faulkner, D.R.; Aplin, A.C.; Butcher, A.R.; Espie, A.A. Mercia Mudstone Formation caprock to carbon capture and storage sites: Petrology and petrophysical characteristics. *J. Geol. Soc.* **2013**, *170*, 119–132. [CrossRef]
27. Hillier, S. Accurate quantitative analysis of clay and other minerals in sandstones by XRD: Comparison of a Rietveld and a reference intensity ratio (RIR) method and the importance of sample preparation. *Clay Miner.* **2000**, *35*, 291–302. [CrossRef]
28. Pandey, J.S.; Daas, Y.J.; Karcz, A.P.; von Solms, N. Enhanced hydrate-Based geological CO_2 Capture and sequestration as a mitigation Strategy to address climate change. *Energies* **2020**, *13*, 5661. [CrossRef]
29. Blott, S.J.; Pye, K. GRADISTAT: A grain size distribution and statistics package for the analysis of unconsolidated sediments. *Earth Surf. Process. Landf.* **2001**, *26*, 1237–1248. [CrossRef]
30. Houghton, J.E.; Nichols, T.E.; Griffiths, J.; Simon, N.; Utley, J.E.P.; Duller, R.A.; Worden, R.H. Automated classification of estuarine sub-depositional environment using sediment texture. *J. Geophys. Res. Earth Surf.* **2023**, *128*. [CrossRef]
31. Peng, S.; Zhang, T.; Loucks, R.G.; Shultz, J. Application of mercury injection capillary pressure to mudrocks: Conformance and compression corrections. *Mar. Pet. Geol.* **2017**, *88*, 30–40. [CrossRef]
32. Washburn, E.W. The dynamics of capillary flow. *Phys. Rev.* **1921**, *17*, 273–283. [CrossRef]
33. Dewhurst, D.; Jones, R.; Raven, M. Microstructural and petrophysical characterization of Muderong Shale: Application to top seal risking. *Pet. Geosci.* **2002**, *8*, 371–383. [CrossRef]
34. Diamond, S. Pore size distributions in clays. *Clays Clay Miner.* **1970**, *18*, 7–23. [CrossRef]
35. Luffel, D.L.; Guidry, F.K. New core anlaysis methods for measuring reservoir rock properties of Devonian Shale. *J. Pet. Technol.* **1992**, *44*, 1184–1190. [CrossRef]
36. *ISO9277*; Determination of the Specific Surface Area of Solids by Gas Adsorption–BET Method. ISO: Geneva, Switzerland, 2010.
37. Barrett, E.P.; Joyner, L.G.; Halenda, P.P. The determination of pore volume and area distribution in porous substances. I. Computations from nitrogen isotherms. *J. Am. Chem. Soc.* **1951**, *73*, 373–380. [CrossRef]
38. Schmitt, M.; Fernandes, C.P.; da Cunha Neto, J.A.B.; Wolf, F.G.; dos Santos, V.S.S. Characterization of pore systems in seal rocks using nitrogen gas adsorption combined with mercury injection capillary pressure techniques. *Mar. Pet. Geol.* **2013**, *39*, 138–149. [CrossRef]
39. Thommes, M.; Kaneko, K.; Neimark, A.V.; Olivier, J.P.; Rodriguez-Reinoso, F.; Rouquerol, J.; Sing, K.S.W. Physisorption of gases, with special reference to the evaluation of surface area and pore size distribution (IUPAC Technical Report). *Pure Appl. Chem.* **2015**, *87*, 1051–1069. [CrossRef]
40. Ge, X.; Fan, Y.; Deng, S.; Han, Y.; Liu, J. An improvement of the fractal theory and its application in pore structure evaluation and permeability estimation. *J. Geophys. Res. Solid Earth* **2016**, *121*, 6333–6345. [CrossRef]
41. Wang, F.; Yang, K.; You, J.; Lei, X. Analysis of pore size distribution and fractal dimension in tight sandstone with mercury intrusion porosimetry. *Results Phys.* **2019**, *13*, 102283. [CrossRef]
42. Wang, X.; Jiang, Z.; Jiang, S.; Chang, J.; Zhu, L.; Li, X.; Li, J. Full-scale pore structure and fractal dimension of the Longmaxi Shale from the Southern Sichuan Basin: Investigations using FE-SEM, gas adsorption and mercury intrusion porosimetry. *Minerals* **2019**, *9*, 543. [CrossRef]
43. Huang, Y.; Zhang, P.; Zhang, J.; Tang, X.; Liu, C.; Yang, J.; Chen, Z. Fractal characteristics of pores in the Longtan Shales of Guizhou, Southwest China. *Geofluids* **2020**, *2020*, 8834758. [CrossRef]
44. Peng, J.; Han, H.; Xia, Q.; Li, B. Fractal characteristic of microscopic pore structure of tight sandstone reservoirs in Kalpintag Formation in Shuntuoguole area, Tarim Basin. *Pet. Res.* **2020**, *5*, 1–17. [CrossRef]
45. Tyler, S.W.; Wheatcraft, S.W. Fractal processes in soil retension. *Water Resour. Res.* **1990**, *26*, 1047–1054. [CrossRef]
46. Sahouli, B.; Blacher, S.; Pirard, R.; Brouers, F. Fractal analysis of mercury porosimetry data in the framework of the thermodynamic method. *J. Colloid Interface Sci.* **1999**, *214*, 450–454. [CrossRef]
47. Hunt, A.G.; Gee, G.W. Application of critical path analysis to fractal porous media: Comparison with examples from the Hanford site. *Adv. Water Resour.* **2002**, *25*, 129–146. [CrossRef]
48. Cai, J.; Yu, B. A discussion of the effect of tortuosity on the capillary imbibition in porous media. *Transp. Porous Media* **2011**, *89*, 251–263. [CrossRef]
49. Ghanbarian, B.; Hunt, A.G.; Ewing, R.P.; Skinner, T.E. Universal scaling of the formation factor in porous media derived by combining percolation and effective medium theories. *Geophys. Res. Lett.* **2014**, *41*, 3884–3890. [CrossRef]
50. Daigle, H.; Johnson, A.; Thomas, B. Determining fractal dimension from nuclear magnetic resonance data in rocks with internal magnetic field gradients. *Geophysics* **2014**, *79*, D425–D431. [CrossRef]

51. Chen, Y.; Tang, H.; Zheng, M.; Li, C.; Zhao, S.; Zhao, N.; Leng, Y. Fractal characteristics and significance of different pore types of the Wufeng-Longmaxi Formation, Southern Sichuan Basin, China, based on N(2) adsorption and image analysis. *ACS Omega* **2021**, *6*, 30889–30900. [CrossRef]
52. Anvir, D.; Jaroniec, M. An Isotherm equation for adsorption on fractal surfaces of heterogenous porous materials. *Langmuir* **1989**, *5*, 1431–1433.
53. Brunauer, S.; Emmett, P.H.; Teller, E. Adsorption of gases in multimolecular layers. *J. Am. Chem. Soc.* **1938**, *60*, 309–319. [CrossRef]
54. Sing, K.S.W. Reporting physisorption data for gas/solid systems with special reference to the determination of surface area and porosity (Recommendations 1984). *Pure Appl. Chem.* **1985**, *57*, 603–619. [CrossRef]
55. Ding, C.; He, J.; Wu, H.; Zhang, X. Nanometer pore structure Characterization of Taiyuan Formation Shale in the Lin-Xing Area based on nitrogen adsorption experiments. *Minerals* **2021**, *11*, 298. [CrossRef]
56. Wang, Y.; Liu, L.; Cheng, H. Gas adsorption characterization of pore structure of organic-rich shale: Insights into contribution of organic matter to shale pore network. *Nat. Resour. Res.* **2021**, *30*, 2377–2395. [CrossRef]
57. Gehrels, W.R. Sea Level Studies | Microfossil-based reconstructions of Holocene relative sea-level change. In *Encyclopedia of Quaternary Science*; Elsevier: Amsterdam, The Netherlands, 2013; pp. 419–428. [CrossRef]
58. Bradley, J. Zoophycos and Umbrella (Pennatulcacea): Their synthesis and identity. *Palaeogeography* **1973**, *13*, 103–128. [CrossRef]
59. Wetzel, A.; Werner, F. Morphology and ecological significance of Zoophycos in deep sea sediment off NW Africa. *Palaeogeography* **1980**, *32*, 185–212. [CrossRef]
60. Mudge, D.C.; Bujak, J.P. An integrated stratigraphy for the Paleocene and Eocene of the North Sea. *Geol. Soc. Lond. Spec. Publ.* **1996**, *101*, 91–113. [CrossRef]
61. O'Connor, S.J.; Walker, D. Paleocene reservoirs of the Everest trend. In Proceedings of the Petroleum Geology of Northwest Europe: Proceedings of the 4th Conference, London, UK, 2008; pp. 145–160.
62. Gupta, R.; Johnson, D.H. Characterization of heterolithic deposits using electrofacies analysis in the tide dominated lower. *Pet. Geosci.* **2001**, *7*, 321–330. [CrossRef]
63. Houben, M.E.; van Eeden, J.C.M.; Barnhoorn, A.; Hangx, S.J.T. Fracture-induced permeability in Whitby mudstone. *Environ. Sci. Technol.* **2020**, *54*, 9564–9572. [CrossRef]
64. Loucks, R.G.; Reed, R.M.; Ruppel, S.C.; Hammes, U. Spectrum of pore types and networks in mudrocks and a descriptive classification for matrix-related mudrock pores. *Am. Assoc. Pet. Geol. Bull.* **2012**, *96*, 1071–1098. [CrossRef]
65. Day-Stirrat, R.J.; McDonnell, A.; Wood, L.J. Diagenetic and seismic concernsassociated with interpretation of deeply buried 'mobile shales'. In *Shale Tectonics*; American Association of Petroleum Geologists Memoir: Tulsa, OK, USA, 2010; Volume 93, pp. 5–27.
66. Magara, K. Compaction and migration of fluids in Miocene Mudstone Nagaoka Plain, Japan. *Am. Assoc. Pet. Geol. Bull.* **1968**, *52*, 2466–2501.
67. Loucks, R.G.; Reed, R.M.; Ruppel, S.C.; Jarvie, D.M. Morphology, genesis, and distribution of nanometer-scale pores in siliceous mudstones of the Mississippian Barnett Shale. *J. Sediment. Res.* **2009**, *79*, 848–861. [CrossRef]
68. Kuila, U.; Prasad, M. Specific surface area and pore-size distribution in clays and shales. *Geophys. Prospect.* **2013**, *61*, 341–362. [CrossRef]
69. Zhang, J.; Liu, G.; Torsaeter, O.; Tao, S.; Jiang, M.; Li, G.; Zhang, S. Pore-throat structure characteristics and its effect on flow behavior in Gaotaizi tight siltstone reservoir, northern Songliao Basin. *Mar. Pet. Geol.* **2020**, *122*, 104651. [CrossRef]
70. Gao, Z.; Yang, X.; Hu, C.; Wei, L.; Jiang, Z.; Yang, S.; Fan, Y.; Xue, Z.; Yu, H. Characterizing the pore structure of low permeability Eocene Liushagang Formation reservoir rocks from Beibuwan Basin in northern South China Sea. *Mar. Pet. Geol.* **2019**, *99*, 107–121. [CrossRef]
71. Rouquerol, J.; Avnir, D.; Fairbridge, C.W.; Everett, D.H.; Haynes, J.H.; Pernicone, N.; Ramsay, J.D.F.; Sing, K.S.W.; Unger, K.K. Reccomendation for the characterisation of porous solids. *Pure Appl. Chem.* **1994**, *66*, 1739–1758. [CrossRef]
72. Cao, Z.; Liu, G.; Zhan, H.; Li, C.; You, Y.; Yang, C.; Jiang, H. Pore structure characterization of Chang-7 tight sandstone using MICP combined with N(2)GA techniques and its geological control factors. *Sci Rep* **2016**, *6*, 36919. [CrossRef] [PubMed]
73. Chen, L.; Jiang, Z.; Jiang, S.; Liu, K.; Yang, W.; Tan, J.; Gao, F. Nanopore structure and fractal characteristics of Lacustrine Shale: Implications for shale gas storage and production potential. *Nanomaterials* **2019**, *9*, 390. [CrossRef] [PubMed]
74. Zhang, N.; Zhao, F.; Guo, P.; Li, J.; Gong, W.; Guo, Z.; Sun, X. Nanoscale pore structure characterization and permeability of mudrocks and fine-grained sandstones in coal reservoirs by scanning electron microscopy, mercury intrusion porosimetry, and low-field nuclear magnetic resonance. *Geofluids* **2018**, *2018*, 2905141. [CrossRef]
75. Hodot, B. *Outburst of Coal and Coalbed Gas (Chinese Translation)*; China Coal Industry Press: Beijing, China, 1966; Volume 318.
76. Harding, F.C.; James, A.T.; Robertson, H.E. The engineering challenges of CO_2 storage. *Proc. Inst. Mech. Eng. Part A J. Power Energy* **2018**, *232*, 17–26. [CrossRef]
77. Di Leo, G.; Sardanelli, F. Statistical significance: P value, 0.05 threshold, and applications to radiomics-reasons for a conservative approach. *Eur. Radiol. Exp.* **2020**, *4*, 18. [CrossRef]
78. Xiong, J.; Liu, X.; Liang, L. Experimental study on the pore structure characteristics of the Upper Ordovician Wufeng Formation shale in the southwest portion of the Sichuan Basin, China. *J. Nat. Gas Sci. Eng.* **2015**, *22*, 530–539. [CrossRef]
79. Hu, J.; Tang, S.; Zhang, S. Investigation of pore structure and fractal characteristics of the Lower Silurian Longmaxi shales in western Hunan and Hubei Provinces in China. *J. Nat. Gas Sci. Eng.* **2016**, *28*, 522–535. [CrossRef]

80. Tian, H.; Pan, L.; Xiao, X.; Wilkins, R.W.T.; Meng, Z.; Huang, B. A preliminary study on the pore characterization of Lower Silurian black shales in the Chuandong Thrust Fold Belt, southwestern China using low pressure N2 adsorption and FE-SEM methods. *Mar. Pet. Geol.* **2013**, *48*, 8–19. [CrossRef]
81. Wang, Y.; Zhu, Y.; Chen, S.; Li, W. Characteristics of the nanoscale pore structure in northwestern Hunan Shale gas reservoirs using field emission scanning electron microscopy, high-pressure mercury intrusion, and gas adsorption. *Energy Fuels* **2014**, *28*, 945–955. [CrossRef]
82. Saidian, M.; Godinez, L.J.; Prasad, M. Effect of clay and organic matter on nitrogen adsorption specific surface area and cation exchange capacity in shales (mudrocks). *J. Nat. Gas Sci. Eng.* **2016**, *33*, 1095–1106. [CrossRef]
83. Ji, W.; Song, Y.; Jiang, Z.; Meng, M.; Liu, Q.; Chen, L.; Wang, P.; Gao, F.; Huang, H. Fractal characteristics of nano-pores in the Lower Silurian Longmaxi shales from the Upper Yangtze Platform, south China. *Mar. Pet. Geol.* **2016**, *78*, 88–98. [CrossRef]
84. Bolourinejad, P.; Shoeibi Omrani, P.; Herber, R. Effect of reactive surface area of minerals on mineralization and carbon dioxide trapping in a depleted gas reservoir. *Int. J. Greenh. Gas Control* **2014**, *21*, 11–22. [CrossRef]
85. Jia, W.; Xiao, T.; Wu, Z.; Dai, Z.; McPherson, B. Impact of mineral reactive surface area on forecasting geological carbon sequestration in a CO_2-EOR field. *Energies* **2021**, *14*, 1608. [CrossRef]
86. Akono, A.T.; Druhan, J.L.; Dávila, G.; Tsotsis, T.; Jessen, K.; Fuchs, S.; Crandall, D.; Shi, Z.; Dalton, L.; Tkach, M.K.; et al. A review of geochemical–mechanical impacts in geological carbon storage reservoirs. *Greenh. Gases Sci. Technol.* **2019**, *9*, 474–504. [CrossRef]
87. Peters, C.A. Accessibilities of reactive minerals in consolidated sedimentary rock: An imaging study of three sandstones. *Chem. Geol.* **2009**, *265*, 198–208. [CrossRef]
88. Bowen, B.B.; Ochoa, R.I.; Wilkens, N.D.; Brophy, J.; Lovell, T.R.; Fischietto, N.; Medina, C.R.; Rupp, J.A. Depositional and diagenetic variability within the Cambrian Mount Simon Sandstone: Implications for carbon dioxide sequestration. *Environ. Geosci.* **2011**, *18*, 69–89. [CrossRef]
89. Botto, J.; Fuchs, S.J.; Fouke, B.W.; Clarens, A.F.; Freiburg, J.T.; Berger, P.M.; Werth, C.J. Effects of mineral surface properties on supercritical CO_2 wettability in a siliciclastic reservoir. *Energy Fuels* **2017**, *31*, 5275–5285. [CrossRef]
90. Espinoza, D.N.; Santamarina, J.C. Clay interaction with liquid and supercritical CO_2: The relevance of electrical and capillary forces. *Int. J. Greenh. Gas Control* **2012**, *10*, 351–362. [CrossRef]
91. Smith, M.M.; Sholokhova, Y.; Hao, Y.; Carroll, S.A. Evaporite caprock integrity: An experimental study of reactive mineralogy and pore-scale heterogeneity during brine-CO_2 exposure. *Environ. Sci. Technol.* **2013**, *47*, 262–268. [CrossRef] [PubMed]
92. Miller, Q.R.S.; Wang, X.; Kaszuba, J.P.; Mouzakis, K.M.; Navarre-Sitchler, A.K.; Alvarado, V.; McCray, J.E.; Rother, G.; Bañuelos, J.L.; Heath, J.E. Experimental study of porosity changes in shale caprocks exposed to carbon dioxide-saturated brine II: Insights from aqueous geochemistry. *Environ. Eng. Sci.* **2016**, *33*, 736–744. [CrossRef]
93. Giles, M.R.; de Boer, R.B. Origin and significance of redistributional secondary porosity. *Mar. Pet. Geol.* **1990**, *7*, 378–397. [CrossRef]
94. Baines, S.J.; Worden, R.H. The long-term fate of CO_2 in the subsurface: Natural analogues for CO_2 storage. *Geol. Soc. Lond.* **2004**, *233*, 59–85. [CrossRef]
95. Gibson-Poole, C.M.; Svendsen, L.; Watson, M.N.; Daniel, R.F. Understanding stratigraphic heterogeneity: A methodology to maximize the efficiency of the geological storage of CO_2. In *Carbon Dioxide Sequestration in Geological Media-State of the Science*; Grobe, M., Pashin, J.C., Dodge, R.L., Eds.; AAPG Studies in Geology; 2009; Volume 59, pp. 347–364.
96. Espinoza, D.N.; Santamarina, J.C. CO_2 breakthrough—Caprock sealing efficiency and integrity for carbon geological storage. *Int. J. Greenh. Gas Control* **2017**, *66*, 218–229. [CrossRef]
97. Fleury, M.; Berne, P.; Bachaud, P. Diffusion of dissolved CO_2 in caprock. *Energy Procedia* **2009**, *1*, 3461–3468. [CrossRef]
98. Katopodes, N.D. Chapter 3–Diffusive mass transfer. In *Free-Surface Flow*; Katopodes, N.D., Ed.; Butterworth-Heinemann: Oxford, UK, 2019; pp. 184–270. [CrossRef]
99. Schloemer, S.; Krooss, B.M. Experimental characterisation of hydrocarbon sealing efficency of caprocks. *Mar. Pet. Geol.* **1997**, *14*, 565–580. [CrossRef]
100. Kirkham, M.B. 7—Water movement in saturated soil. In *Principles of Soil and Plant Water Relations*; Kirkham, M.B., Ed.; Academic Press: Burlington, MA, USA, 2005; pp. 85–100. [CrossRef]
101. Salem, A.M.; Shedid, S.A. Variation of petrophysical properties due to carbon dioxide (CO_2) storage in carbonate reservoirs. *J. Pet. Gas Eng.* **2013**, *4*, 91–102. [CrossRef]
102. Bachu, S.; Brant, B.D. Dependence of CO_2 -brine interfacial tension on aquifer pressure, temperature and water salinity. *Energy Procedia* **2009**, *1*, 3157–3164. [CrossRef]
103. Warren, E.A.; Smalley, P.C. *North Sea Formation Waters Atlas*; The Geological Society of London: London, UK, 1994.
104. Chadwick, R.A.; Zweigel, P.; Gregersen, U.; Kirby, G.A.; Holloway, S.; Johannessen, P.N. Geological reservoir characterization of a CO_2 storage site: The Utsira Sand, Sleipner, northern North Sea. *Energy* **2004**, *29*, 1371–1381. [CrossRef]
105. Tonnet, N.; Mouronval, G.; Chiquet, P.; Broseta, D. Petrophysical assessment of a carbonate-rich caprock for CO_2 geological storage purposes. *Energy Procedia* **2011**, *4*, 5422–5429. [CrossRef]
106. Armitage, P.J.; Worden, R.H.; Faulkner, D.R.; Aplin, A.C.; Butcher, A.R.; Iliffe, J. Diagenetic and sedimentary controls on porosity in Lower Carboniferous fine-grained lithologies, Krechba Field, Algeria: A petrological study of a caprock to a carbon capture site. *Mar. Pet. Geol.* **2010**, *27*, 1395–1410. [CrossRef]

107. Kharaka, Y.K.; Cole, D.R.; Hovorka, S.D.; Gunter, W.D.; Knauss, K.G.; Freifeld, B.M. Gas-water-rock interactions in Frio Formation following CO_2 injection: Implications for the storage of greenhouse gases in sedimentary basins. *Geology* **2006**, *34*, 577–580. [CrossRef]
108. Carey, J.W.; Wigand, M.; Chipera, S.J.; WoldeGabriel, G.; Pawar, R.; Lichtner, P.C.; Wehner, S.C.; Raines, M.A.; Guthrie, G.D. Analysis and performance of oil well cement with 30 years of CO_2 exposure from the SACROC Unit, West Texas, USA. *Int. J. Greenh. Gas Control* **2007**, *1*, 75–85. [CrossRef]
109. Boulin, P.F.; Bretonnier, P.; Vassil, V.; Samouillet, A.; Fleury, M.; Lombard, J.M. Sealing efficiency of caprocks: Experimental investigation of entry pressure measurement methods. *Mar. Pet. Geol.* **2013**, *48*, 20–30. [CrossRef]
110. Ingram, G.M.; Urai, J.L. Top-seal leakage through faults and fractures: The role of mudrock properties. *Geol. Soc. Lond. Spec. Publ.* **1999**, *158*, 125–135. [CrossRef]

processes

MDPI

Article

Real-Time Estimation of CO_2 Absorption Capacity Using Ionic Conductivity of Protonated Di-Methyl-Ethanolamine (DMEA) and Electrical Conductivity in Low-Concentration DMEA Aqueous Solutions

Sang-Jun Han, Joo Young Han and Jung-Ho Wee *

Department of Energy and Environmental Engineering, The Catholic University of Korea, 43 Jibong-ro, Wonmi-gu, Bucheon-si 14662, Gyeonggi-do, Republic of Korea; hsj8812@chungbuk.ac.kr (S.-J.H.); dbstjd5679@naver.com (J.Y.H.)

* Correspondence: jhwee@catholic.ac.kr or jhwee@korea.ac.kr; Tel.: +82-2-2164-4866

Abstract: The present study investigates the real-time estimation of CO_2 absorption capacity (CAC) based on the electrical conductivity (EC) of low-concentration di-methyl-ethanolamine (DMEA) solutions (0.1–0.5 M). CO_2 absorption experiments are conducted to measure the variation in CAC and EC during CO_2 absorption, revealing a strong correlation between the two properties. The ionic conductivity of $DMEAH^+$ formed during absorption is calculated to be 53.1 $S{\cdot}cm^2/(mol{\cdot}z)$, which is found to be larger than that of $TEAH^+$ and $MDEAH^+$. This can be attributed to the smaller molar mass and higher ionic mobility of $DMEAH^+$. A significant finding is that the measured EC (EC_M) of the DMEA solutions consistently demonstrates a lower value than the theoretically predicted value. This discrepancy is due to the larger ionic size of $DMEAH^+$, which results in a reduction in the real mean ionic activity coefficient. This effect becomes more pronounced with increasing DMEA concentration. Consequently, a higher CAC is required to produce the same change in EC at higher amine concentrations. Based on these findings, an empirical equation is devised to estimate CAC from EC_M in solutions of constant DMEA concentration. This equation will be employed as a practical approach for the in situ monitoring of CO_2 absorption using DMEA aqueous solution.

Keywords: carbon capture and storage; di-methyl-ethanolamine; CO_2 absorption; ionic conductivity; electrical conductivity; correlation

1. Introduction

Carbon dioxide (CO_2)—which is emitted from a variety of industrial processes, including fossil fuel combustion, cement production, and steel and chemical plants—constitutes a significant contributor to global warming. CO_2 constitutes approximately 76% of global greenhouse gas emissions, with annual emissions reaching 37.15 billion metric tons [1,2]. In July 2023, the UN Secretary-General underscored the urgency of the current climate change crisis, asserting that the era of global warming has concluded and that we have entered the period of "global boiling" [3]. To address climate change, ongoing research over the past 30 years has focused on the development of carbon dioxide capture and storage (CCS) technologies, which are characterized by high economic viability and performance. Various CCS methods, including bioconversion [4–7] and adsorption using novel materials [8–12], have been investigated; however, these approaches frequently fail to achieve the desired impact at a large scale. Among the various CCS technologies, chemical absorption using alkanolamine has been extensively studied over an extended period, and it is currently the only technology that has been successfully commercialized with the objective of addressing the urgent climate crisis [13–17].

Alkanolamines are a class of substances that contain both amino (-NH_2, -NHR, and -NR_2) and hydroxyl (-OH) functional groups. They are highly and selectively reactive with

CO_2, and they can be classified as primary, secondary, and tertiary amines based on the number of R groups they contain. An absorption process using mono-ethanolamine (MEA), a member of the primary amine group, has been commercialized due to its rapid reaction rate and low cost [18–22]. However, the reaction of MEA with CO_2 limits the theoretical maximum amine utilization to 0.5 mol CO_2/mol amine in high amine-concentrated solutions. Moreover, MEA is highly corrosive to equipment, requiring a significant amount of energy to strip the CO_2 from the absorbent (or to regenerate the solutions). An alternative approach is being investigated in which tertiary amines are utilized as absorbents. Although these amines, such as tri-ethanolamine (TEA), methyl-di-ethanolamine (MDEA), and di-methyl-ethanolamine (DMEA), have been reported to have relatively low reaction rates, they achieve a maximum amine utilization of 1.0 mol CO_2/mol amine, are relatively less corrosive, and require minimal energy for CO_2 stripping. Most studies using tertiary amines as the absorbent have concentrated on their blending effect with primary or secondary amines to enhance the performance of the absorbents [23–27]. However, there has been a relative lack of research into the specific absorption performance and properties of tertiary amine solutions, which would be expected to furnish crucial insights for the continued advancement of absorbents.

Recent advances in sensor technology have demonstrated the potential of low-cost, real-time monitoring systems to enhance data collection for environmental and industrial processes. For example, low-cost sensors have been employed to monitor spatial and temporal pollution in urban areas, providing a practical solution for large-scale exposure assessments [28]. One study highlighted that the use of a dispersion model enabled the testing of dust pollution distribution in a seaside town, thereby significantly enhancing pollution monitoring [29]. In water quality monitoring, electrochemical sensors have emerged as a crucial tool due to their capacity to transduce chemical interactions into electrical signals, thereby enabling real-time, multi-parameter monitoring, which is feasible even in resource-limited settings [30,31]. Similarly, electrolytic conductivity sensors, which are widely used in diverse industries, offer reliable and precise monitoring of electrolytes in various applications [32,33]. Given these advantages, we opted to employ electrical conductivity (EC) measurements as an effective and economical method for the real-time estimation of CO_2 absorption capacity in aqueous DMEA solutions. In this way, we utilize the advantages of EC sensors to provide a practical solution for the continuous monitoring of CO_2 absorption processes in industrial settings.

In this context, the present study investigated the CO_2 absorption performance and electrical characteristics of the CO_2 absorption system using DMEA aqueous solution with a relatively low concentration of DMEA (0.1–0.5 M). First, the ionic conductivity (IC) value of protonated DMEA ($DMEAH^+$) generated from the chemical absorption reaction between CO_2 and DMEA—which was achieved through a series of absorption experiments—was presented. The relationship between the CO_2 absorption capacity (CAC) (or the amount of CO_2 absorbed) and the EC of the solution at the arbitrary absorption time was also established for each solution, which finally resulted in the derivation of a general empirical equation for predicting the amount of CO_2 absorbed based on EC measurements of the solutions.

2. CO_2 Absorption Mechanism Using DMEA Aqueous Solutions

In the DMEA solutions prepared for the CO_2 absorption, a small amount of the DMEA molecule first receives H^+ from the water to become $DMEAH^+$ and OH^- in the solution, as expressed in Equation (1) [34,35].

$$\mathrm{DMEA} + \mathrm{H_2O} \leftrightarrow \mathrm{DMEAH^+} + \mathrm{OH^-} \tag{1}$$

The two ionic concentrations are determined from DMEA's base dissociation constant (pK_b) of 4.78 [36]. If CO_2 is then added to the DMEA solution to be absorbed, two reactions occur, as summarized in Equations (2) and (3) [34,35,37].

$$OH^- + CO_2 \rightarrow HCO_3{}^- \quad (2)$$

$$DMEA + H_2O + CO_2 \rightarrow DMEAH^+ + HCO_3{}^- \quad (3)$$

When CO_2 is injected into the DMEA solution, OH^-, which is generated via Equation (1), reacts with CO_2 according to Equation (2), thereby resulting in the formation of $HCO_3{}^-$. Subsequently, CO_2 is absorbed by reacting with OH^- in the H_2O activated by DMEA molecules to produce $DMEAH^+$ and $HCO_3{}^-$, as expressed in Equation (3). The CO_2 absorption mechanism in tertiary amine aqueous solution is designated as the "base-catalyzed reaction" for CO_2 hydration [38]. This is illustrated in Figure 1.

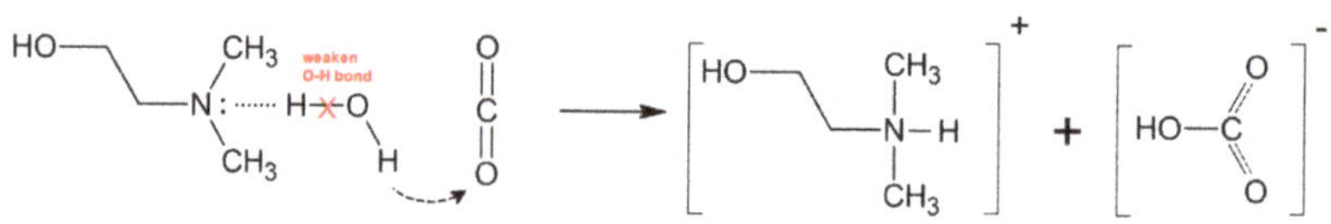

Figure 1. Reaction mechanism for CO_2 absorption in di-methyl-ethanolamine (DMEA) aqueous solutions.

In the absorption process, the DMEA molecule does not directly react with CO_2; instead, a lone pair of electrons on the nitrogen atom of DMEA forms hydrogen bonds with water molecules (illustrated by the dotted line in Figure 1), which activates the hydroxide ion (OH^-) on the opposite side of the water molecule. This enhances its reactivity with the injected CO_2 molecule. Consequently, the formation of $DMEAH^+$ and $HCO_3{}^-$, as well as their zwitterion ($DMEAH^+ \cdot HCO_3{}^-$), occurs in the solution during CO_2 absorption. Additionally, CO_2 can be physically absorbed into the water component, as described in Equation (4), which is referred to as the amount of physical absorption of CO_2; this is also included in the total CAC of the solutions [39,40].

$$H_2O(l) + CO_2(g) \rightarrow H_2CO_3(aq) \quad (4)$$

The amount of CO_2 physical absorption under experimental conditions (25 °C, 1 atm, and a CO_2 partial pressure of 0.33 atm) is 0.023 mol CO_2/L of pure distilled water [41]. The H_2CO_3 produced by physical absorption increases the concentration of either $HCO_3{}^-$ or $CO_3{}^{2-}$ according to the pH of the solution. However, both concentrations are very small compared to those of the ions produced by the chemical absorption reaction (Equations (2) and (3)), so they are not included in the calculation of the EC of the solutions.

3. Experimental Methodology and Calculations

3.1. Experimental Setup for CO_2 Absorption

Figure 2 shows a schematic diagram of the absorption experiment with DMEA solution.

The experimental setup, including the equipment and methods used in this study, was established identically to that used in our previous work on other amine-based solutions [41–43]. The details of the setup are described below.

A cylindrical semi-batch reactor (D: 110 mm; h: 80 mm; total volume: 0.76 L) maintained at 25 °C by a jacket connected to a water circulator was filled with 0.5 L of 0.1–0.5 M DMEA (99.5%, Sigma-Aldrich, St. Louis, MO, USA) aqueous solution. Next, the empty space in the reactor was filled with N_2 gas (99.9%) and completely sealed to confirm tightness. For CO_2 absorption, N_2 and CO_2 gasses (99.99%) were passed through a gas mixture (D: 90 mm; h: 400 mm; total volume: 2.55 L) at respective flow rates of 2.0 and 1.0 L/min using a mass flow controller (MFC; TSC-200, MKPrecision, Siheung-si, Republic of Korea). This resulted in a gas mixture with a CO_2 concentration of 33.3 vol%. Prior to absorption, this gas mixture was bypassed from the reactor through a three-way valve installed at the top of the reactor, after which its CO_2 concentration was checked using a CO_2 analyzer (maMos-200, Madur Electronics, madur Polska Sp. z o. o., Zgierz, Poland). The gas mixture was then used as the absorption gas.

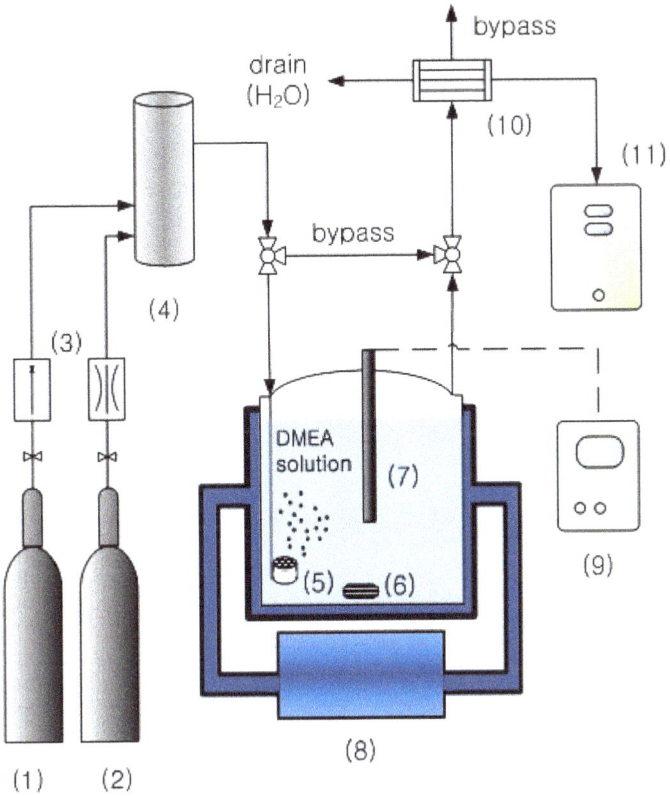

Figure 2. The experimental apparatus for CO_2 absorption using DMEA aqueous solutions. (1) N_2 cylinder, (2) CO_2 cylinder, (3) mass flow controller (MFC), (4) gas mixture, (5) bubbler, (6) magnetic stirrer, (7) pH/EC sensor, (8) water circulator, (9) pH/EC meter, (10) humidifier, (11) CO_2 gas analyzer.

When the gas mixture was injected into the reactor, a glass fiber bubbler with a pore size of 1 um was fixed at the end of the inlet gas line. This was carried out to ensure uniform contact between the solution and the gas. The solution was stirred at 380 rpm using a magnetic stirrer while absorbing. A pH/EC meter (Orion 4 Star, Thermo Scientific, Waltham, MA, USA) was also used to measure the variations in the pH and the EC of the solution caused by CO_2 absorption in real time at 5 s intervals. The CO_2 concentration in the outlet gas from the reactor was measured using a CO_2 analyzer. The absorption completion point was determined as the point at which the CO_2 concentration in the outlet gas was equal to the initial CO_2 concentration of the inlet gas.

3.2. Calculation of Electrical Conductivity (EC) of the CO_2-Absorbed DMEA Solutions

The EC of a CO_2-absorbed solution was calculated using Equations (5)–(10) below [44–48].

$$\mathrm{EC(S/m)} = k_o\gamma^2, \tag{5}$$

$$k_o = \sum z_i\lambda_i c_i, \tag{6}$$

$$IS = 500\sum c_i z_i^2, \tag{7}$$

$$\log \gamma = -Az_+z_-\sqrt{IS} \quad (\text{IS} < 0.01), \tag{8}$$

$$\log \gamma = -Az_+z_-\frac{\sqrt{IS}}{1+\sqrt{IS}} \quad (0.01 \leq \text{IS} \leq 0.1), \tag{9}$$

$$\log \gamma = -Az_+z_-\left(\frac{\sqrt{IS}}{1+\sqrt{IS}} - 0.2IS\right) \quad (0.1 < \text{IS} \leq 0.5), \tag{10}$$

The EC of a CO_2-absorbed solution is the product of the EC of an infinitely diluted solution (k_o: (S/m)) and the square of the mean ionic activity coefficient (γ), as described in Equation (5) [44,45]. First, the value of k_o is calculated via Equation (6), using the absolute charge (z), ionic conductivity (λ: $S{\cdot}m^2/mol{\cdot}z$), and concentration (c: mol/m^3) of ion i in solution. Secondly, the value of γ can be calculated selecting one of Equations (8)–(10) depending on the ionic strength (IS: mol/L) value of the solution, which can be calculated from Equation (7) [46–48]. Here, A is a Debye–Huckel constant—0.509 $(kg/mol)^{0.5}$ at 25 °C. Therefore, three factors are required to calculate the EC of a CO_2-absorbed solution—the concentration, charge value, and IC of the ions in the solution.

3.3. Calculation of Ionic Conductivity of DMEAH+

As CO_2 is absorbed into the DMEA solution, the ions produced or consumed are OH^-, HCO_3^-, and $DMEAH^+$, and the charge value of these ions is 1. Of these, the ICs of OH^- and HCO_3^- have previously been reported in the literature as 198.6 and 44.50 $S{\cdot}cm^2/(mol{\cdot}z)$, respectively [49,50]. However, the IC value of $DMEAH^+$ is not yet known; therefore, it is essential to know in order to calculate the EC value of the solution in accordance with the absorption reaction. The variation in the concentration of ions present in the solution is also necessary to calculate the EC value of the solution during the absorption reaction. Since the concentration of ions in the solution can be estimated based on the amount of CO_2 chemical absorption measured in real time using Equations (2) and (3), the EC variation in the solution during the absorption can be calculated after determining the IC of $DMEAH^+$. To find the IC of $DMEAH^+$, the initial guess value was assumed to be 1, and it was increased by increments of 0.01 until reaching the IC of H^+, which is the largest value (349.8 $S{\cdot}cm^2/(mol{\cdot}z)$) among all the ions [51]. Then, the EC was repeatedly calculated via Equation (5) (referred to as EC_C). Finally, the EC_C value was determined as the value entered at the time when the mean absolute percentage error (MAPE) between the EC_C value and the measured EC (EC_M) was minimized. The MAPE between the EC_M and the EC_C value during the absorption in 0.1–0.5 M DMEA solution was calculated using Equation (11) [52].

$$\text{MAPE}(\%) = \frac{100}{n}\sum_{i=1}^{n}\frac{|\text{EC}_\text{M} - \text{EC}_\text{C}|}{\text{EC}_\text{M}}, \tag{11}$$

The calculation was performed using MATLAB 2024a.

4. Results and Discussion

4.1. CO_2 Absorption Performance of 0.1–0.5 M DMEA Aqueous Solutions

The CAC and EC_M of the 0.1–0.5 M DMEA solutions according to time are shown in Figure 3a and Figure 3b, respectively.

The CAC of the solution varied with the shape of the square root function; the slope here refers to the CO_2 absorption rate. The initial absorption rates of all solutions were steep. However, as time progressed, the slope gradually decreased until absorption was completed. This is because the amine concentration in the solution is high at the beginning of absorption, but as the absorption progresses, the concentration decreases as the amine is consumed, as expressed in Equation (4). The trend of EC_M variation according to time, shown in Figure 3b, is also similar to that of CAC, which indicates that the EC_M value of the solution is directly affected by the concentration of ions generated by the CO_2 absorption reaction.

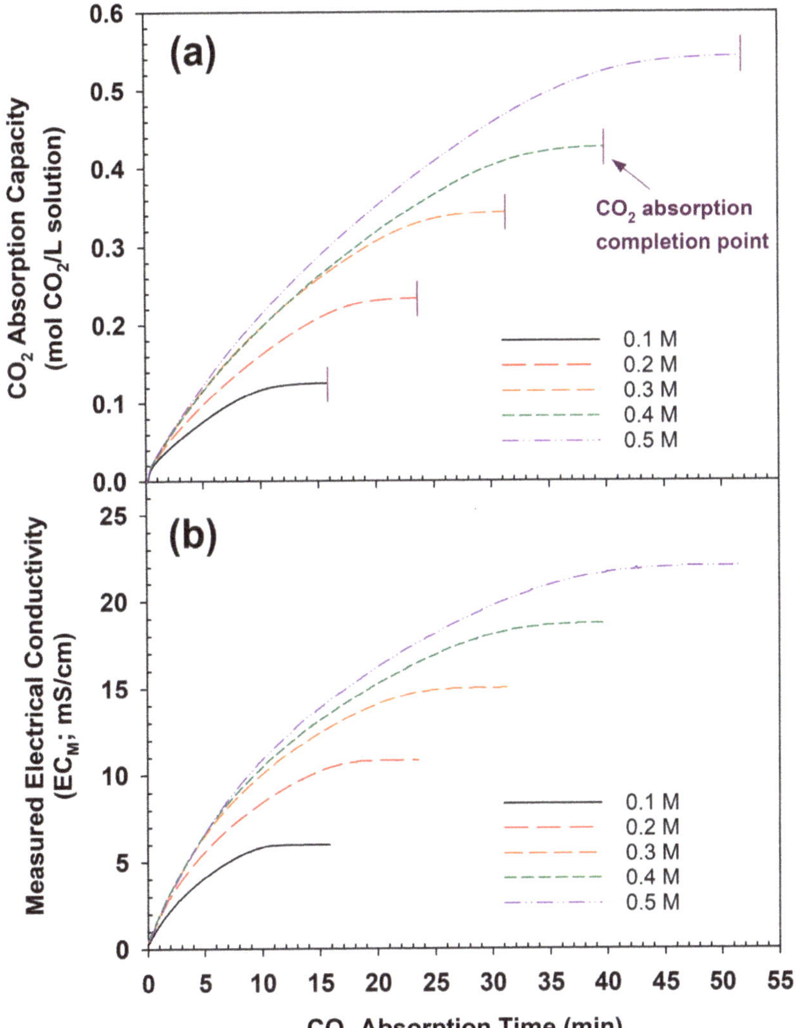

Figure 3. (**a**) CO_2 absorption capacity (CAC) and (**b**) measured electrical conductivity (EC_M) as a function of CO_2 absorption time in 0.1–0.5 M DMEA solutions.

Table 1 summarizes the total CAC of 0.1–0.5 M DMEA solutions at the absorption completion point, the amount of chemical and physical absorption, the total absorption time, and the average slope of Figure 3a, i.e., the overall CO_2 absorption rate of the solutions.

Here, the total CAC is the sum of the CO_2 chemical absorption, which is stoichiometrically calculated based on the DMEA concentration using Equations (2) and (3), and the CO_2 physical absorption given in Equation (4), which is calculated by subtracting the amount of chemical absorption from the total CAC. The amount of CO_2 physical absorption was at its lowest at 0.032 mol CO_2/L in a 0.1 M solution, whereas it was at its highest at 0.050 mol CO_2/L at 0.5 M. This amount is slightly higher than that of pure distilled water (0.023 mol/L), as referenced in Section 2, and it increases proportionally to the amine concentration. This is because the HCO_3^- produced by the chemical absorption of CO_2 with DMEA in the solution shifts the equilibrium point of the carbonate ions in the solution, which can produce more H_2CO_3 or CO_3^{2-} in the solution than in pure distilled water; consequently, additional trace amounts of CO_2 are absorbed.

Table 1. Total, chemical, and physical CO_2 absorption capacities (CACs), total absorption time, and overall CO_2 absorption rate in 0.1–0.5 M DMEA solutions.

	Initial Amine Concentration of DMEA Solution (mol/L)				
	0.1	**0.2**	**0.3**	**0.4**	**0.5**
Total CAC (mol CO_2/L)	0.135	0.238	0.345	0.448	0.550
Chemical CAC (mol CO_2/L)	0.100	0.200	0.300	0.400	0.500
Physical CAC (mol CO_2/L)	0.035	0.038	0.045	0.048	0.050
CO_2 absorption time (min)	15.0	23.8	35.0	44.9	52.8
Overall CO_2 absorption rate (mmol CO_2/(L·min))	9.0	10.0	9.9	10.0	10.4

The average standard deviation of CAC and CO_2 absorption time is ±0.012 and ±0.5, respectively.

Although the overall CO_2 absorption rate was slightly lower at 9.0 mmol CO_2/(L·min) in the 0.1 M solution, the overall absorption rate converged to about 10.0 mmol CO_2/(L·min), regardless of the initial amine concentration in the solution. This is consistent with previous studies on the kinetics of tertiary amines, which have shown similar behavior [53–56]. For example, W. Jiang et al. reported a pseudo-first-order rate constant of 1.5–3.0 s^{-1} in 0.075–0.175 M DMEA solutions, corresponding to a second-order rate constant of 22.19 m^3/(kmol·s) [53]. This indicates that the reaction is more dependent on CO_2 concentration than on DMEA concentration. This is attributed to the fact that DMEA, as a tertiary amine, follows a base-catalyzed reaction mechanism that facilitates the reaction of H_2O with CO_2. Therefore, the absorption rate is more dominated by the CO_2 concentration than the DMEA concentration in the solution. Thus, at relatively low concentrations of 0.5 M or less, the absorption rate was not found to be insensitive to the concentration of DMEA.

4.2. *Ionic Conductivity of Protonated DMEA (DMEAH⁺)*

The IC and charge values of the ions produced from the CO_2 absorption reaction (Equations (2) and (3)), including the dissociation reaction (Equation (1)) of the DMEA solution, are presented in Table 2.

Table 2. Ionic conductivity, absolute value of electric charge, and molar mass of ions generated from chemical CO_2 absorption in DMEA solutions with other tertiary amine solutions.

Ions	**Ionic Conductivity ($S·cm^2$/(mol·z))**	**Absolute Value of Electric Charge (z)**	**Molar Mass (g/mol)**	**Reference**
OH^-	198.6	1	17.0	[49]
HCO_3^-	44.5	1	61.0	[50]
$DMEAH^+$	53.1	1	90.1	This work
$MDEAH^+$	46.5	1	120.1	[41]
$TEAH^+$	37.6	1	150.1	[41]

As mentioned in Section 3.3, the IC of the OH^- and HCO_3^- ions have previously been shown to be 198.6 and 44.5 $S·cm^2$/(mol·z), respectively [49,50]. However, the IC of $DMEAH^+$ was not identified in the published literature. Consequently, in this study, the IC of $DMEAH^+$ was calculated using the trial-and-error method described in Section 3.3. This calculation was based on the ionic concentrations calculated from the amount of CO_2 chemical absorption (as determined using Equations (2) and (3)), as well as the EC_M. The calculated IC of $DMEAH^+$, as determined in this study, is 53.1 $S·cm^2$/(mol·z) with an MAPE of 13.06%. The IC of $DMEAH^+$, as determined by the present study, is larger than the values that have been reported for the other two of the three kinds of tertiary amines, namely $TEAH^+$ (37.6 S·cm^2/mol·z) and $MDEAH^+$ (46.5 S·cm^2/mol·z), in prior studies [41]. These values are inversely proportional to the molar masses of the three ions, which is likely attributable to the observation that the smaller and lighter ion with the same charge generally exhibits increased mobility in solution. This is consistent with the results of previous research, which has indicated that an increase in ionic mobility leads to an increase in ionic conductivity [57]. Moreover, as the range of absolute values obtained for IC in the case of $DMEAH^+$ is similar to that of the other two tertiary protonated amines, the result of the IC of $DMEAH^+$ obtained herein may be considered to be reliable.

4.3. EC_C and EC_M Variation According to CO_2 Absorption in 0.1–0.5 M DMEA Solutions

When using the IC of $DMEAH^+$ (53.1 S·cm^2/mol·z), the theoretical EC values of 0.1–0.5 M DMEA solutions, as calculated using Equation (5), are shown in Figure 4, along with the variation in EC_M (Figure 3b).

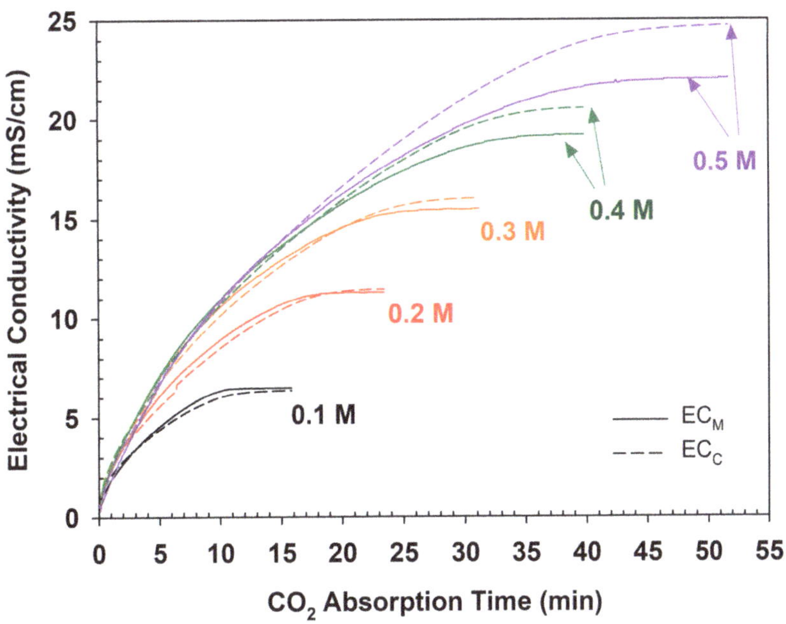

Figure 4. Comparison of measured electrical conductivity (EC_M) and calculated electrical conductivity (EC_C) during CO_2 absorption in 0.1–0.5 M DMEA solutions.

As can be seen in Figure 4, the variations in EC_M and EC_C were similar for all solutions. However, as CO_2 absorption neared completion, the discrepancy between the two EC values increased with the initial amine concentration. For example, at the absorption completion point of 0.1 M solution, the EC_M and EC_C values were 6.00 and 5.88 mS/cm, respectively, with an MAPE of 2.1%. In contrast, for a 0.5 M solution, the EC_M and EC_C values were 22.07 and 24.70 mS/cm, respectively, with an associated MAPE of 11.9%. These results are attributed to the tertiary amine nature of the $DMEAH^+$, the molecular structure of which is large and complex, resulting in a reduced mean ionic activity coefficient in the solution compared to that predicted using Equation (10). Sami-ullah et al. investigated the relationship between molecular weight and the activity coefficient in ionic liquid systems, revealing an inverse proportionality between the two [58]. H.-N Jeon et al. developed an artificial neural network-based model to predict the infinite dilution activity coefficients of organic molecules, utilizing parameters such as molecular structure, polarizability, and hydrogen bonding [59]. The results of our research also exhibited a similar trend to those of previous studies. Thus, as the concentration of amines increased, the mean ionic activity coefficient decreased, and this was accompanied by an increase in the difference between the EC_M and EC_C values, with the EC_M being measured at a lower value than the EC_C. However, in contrast to the other solutions, the EC_M was slightly higher than the EC_C in the 0.1 M solution, and it also had the smallest MAPE. This is due to the relatively very low concentration, as well as the fact that EC_M is augmented by the EC derived from the trace amount of CO_3^{2-} or HCO_3^- generated through the physical absorption process.

4.4. Correlation Between CO_2 Absorption Capacity and Electrical Conductivity

The trend in variations in CAC and EC_M according to absorption time was highly similar for the 0.1–0.5 M DMEA solutions, as illustrated in Figure 3a,b. The correlation between the CAC and the EC_M of the solution measured at the same time is depicted in Figure 5a, while the results of their regression analysis are presented in Figure 5b.

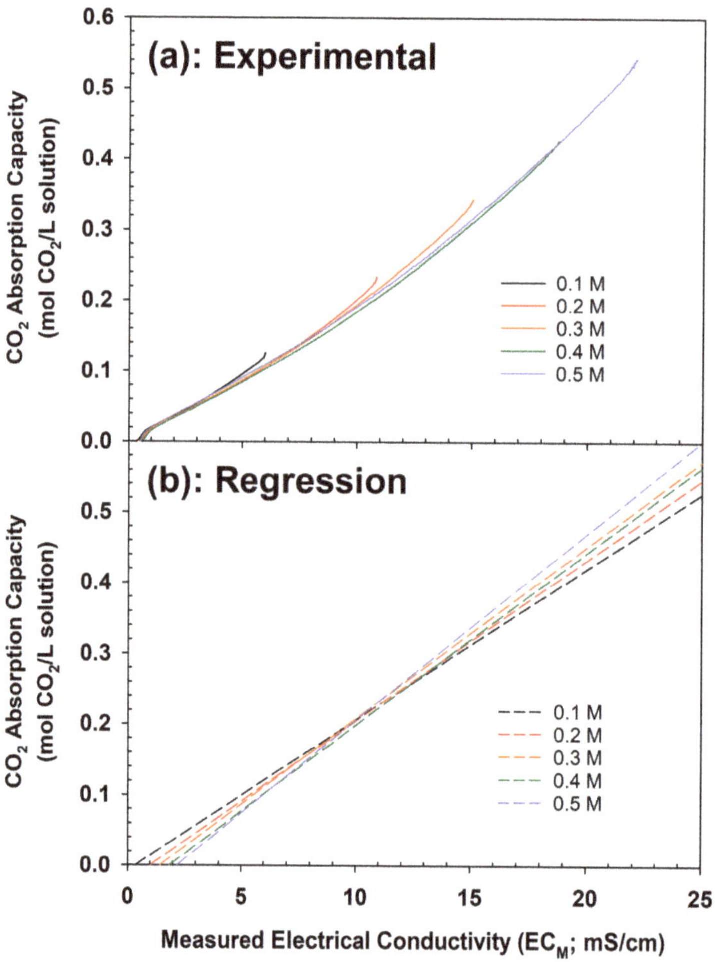

Figure 5. (**a**) Relationship between CO_2 absorption capacity (CAC) and measured electrical conductivity (EC_M) during CO_2 absorption and (**b**) linear regression analysis for 0.1–0.5 M DMEA solutions.

The correlation between CAC and EC_M in all solutions was linear, which indicates that EC_M was linearly proportional to CAC, and that both were directly related to the ionic (or molecular) concentration in the solution. This is valid for all CO_2-absorbed DMEA solutions with concentrations of 0.1–0.5 M, which showed linearity in the CAC and EC_M of the solution. As a result, the correlation equation between the two variables was derived through regression analysis (Figure 5b), and can be expressed as shown in Equation (12) below.

$$CO_2\text{absorption capacity}(\text{CAC}; \text{molCO}_2/\text{L}) = a \cdot EC_M + b, \quad (12)$$

The slope (a) and y-intercept (b) of Equation (12) were calculated for all solutions and summarized according to the initial amine concentration (C_{DMEA}) of the solution; these values are presented in Figure 6a,b.

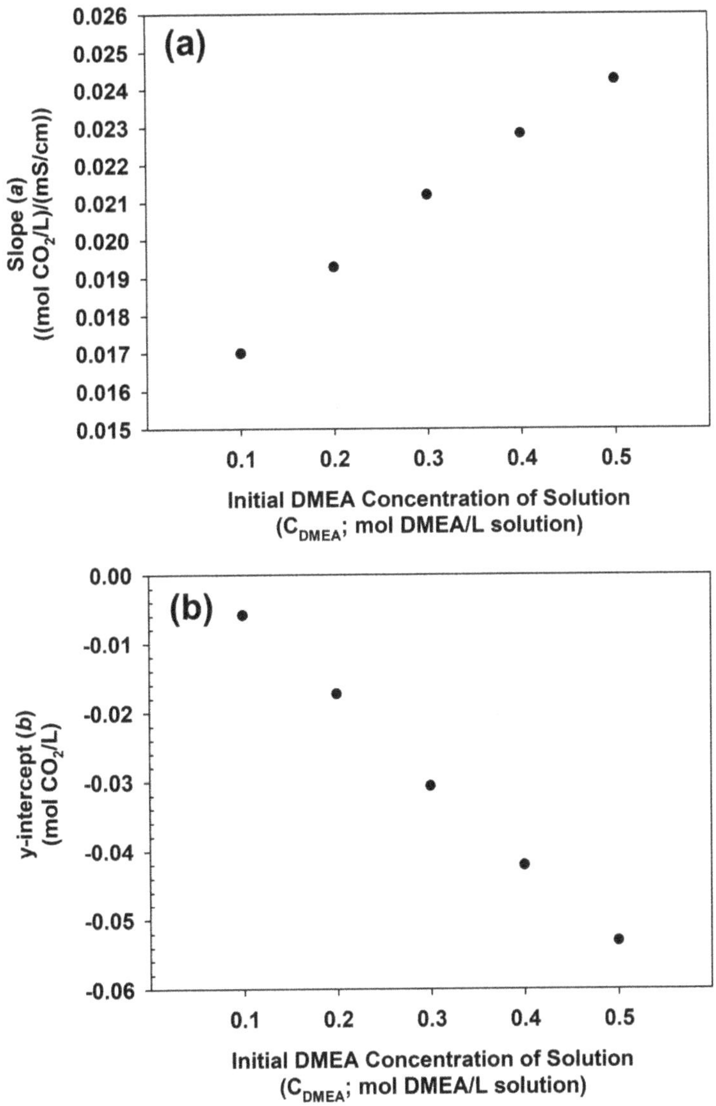

Figure 6. (**a**) Slope and (**b**) y-intercept values from the linear regression analysis between measured electrical conductivity (EC_M) and CO_2 absorption capacity (CAC) in 0.1−0.5 M DMEA solutions.

The slope (a) and y-intercept (b) exhibited linear proportionality to the initial amine concentration in solution. Here, the slope represents the requisite CO_2 absorption capacity (mol CO_2/L) in solution to achieve a 1 mS/cm increase in the EC of the solution, which exhibits a proportionality to the initial amine concentration of solutions (Figure 6a), as evidenced by the relation expressed in Equation (13).

$$\text{slope}(a; (\text{molCO}_2/\text{L})/(\text{mS/cm})) = 0.0181 \cdot C_{DMEA} + 0.0155, \quad (13)$$

The y-intercept (b) value of Equation (12) shown in Figure 6b decreased proportionally to the initial amine concentration, and the correlation coefficient was found to be 0.998 as a result of a linear regression, thus indicating a highly linear relationship with respect to the initial amine concentration. The correlation between the two variables was expressed as shown in Equation (14).

$$\text{y} - \text{intercept}(b; \text{molCO}_2/\text{L}) = -0.1193 \cdot C_{DMEA} + 6.0 \cdot 10^{-3}, \quad (14)$$

The y-intercept value represents the amount of CO_2 absorbed when the EC of the solution is zero. However, even before CO_2 absorption, a constant EC value for the DMEA solutions could be observed due to the presence of $DMEAH^+$ and OH^- ions in the solution, with a certain concentration, as described in Equation (1). Consequently, the y-intercept value is merely a mathematical representation of the correlation between EC and CAC.

Therefore, the slope and y-intercept of Equation (12) could, respectively, be generalized from Equations (13) and (14) for 0.1–0.5 M DMEA solutions to derive an empirical equation that estimates the CAC of each solution as a function of the EC_M and initial amine concentration, as shown in Equation (15).

$$\begin{aligned} \text{CO}_2 \text{ absorption capacity}(\text{mol CO}_2/\text{L}) &= 0.0181 \cdot C_{DMEA} \cdot EC_M + 0.0155 \cdot \\ &EC_M - 0.1193 \cdot C_{DMEA} + 6.0 \cdot 10^{-3}, \end{aligned} \quad (15)$$

Shown above, Equation (15) may be used for the estimation of CAC in real time based on the variations in the EC of DMEA solutions of the given initial amine concentration. Finally, Figure 7 illustrates the relative behavior of the CAC (y-axis) of DMEA solutions in terms of the C_{DMEA} ranging from 0.1 to 0.5 M (z-axis), as well as the EC_M value of the solution (x-axis). Moreover, this Figure presents variations in the amount of saturated CO_2 absorbed and the maximum EC_M value of each solution (parameters), and the data presented in Figure 7 provide information in relation to the relative behavior and interactions of the three variables in this system.

The correlation equation among CAC, EC_M, and DMEA concentration demonstrates that EC_M can serve as a reliable indicator for estimating CAC in real systems in real time, providing practical insights for CCS technologies. Recent studies emphasize the significance of such empirical models in enhancing the efficiency of CO_2 capture systems through real-time monitoring and control capabilities [13,60,61]. This approach has been confirmed to be applicable to other amine systems, including MEA, AMP, and TEA [41,42]. Consequently, this research offers a methodology for optimizing CCS processes by utilizing the variation in EC_M of DMEA solutions, establishing a reliable empirical model, and investigating their electrical properties.

However, the results are limited in two main respects. First, the research focused on low-concentration (0.1–0.5 M) DMEA solutions, which are considerably lower than those typically employed in industrial contexts. Second, the research was conducted under a restricted range of conditions, considering a single pressure and temperature to determine the ionic conductivity of $DMEAH^+$, which limits the scope for providing more detailed data on CO_2 absorption. Therefore, future research focuses on extending the experimental conditions to cover a broader range of amine concentrations, temperatures, and pressures. Furthermore, it is essential to incorporate a numerical model based on vapor–liquid equilibrium (VLE) simulations for the amine–CO_2–H_2O system to enhance the predictive capabilities of the electrical conductivity in the electrolyte model. Additionally, future research should focus on the application of more comprehensive models, such as the extended Non-Random Two-Liquid (NRTL) or Universal Quasi-Chemical (UNIQUAC) models that consider both long- and short-range interactions, and the traditional Debye–Huckel model that mainly considers long-range interactions.

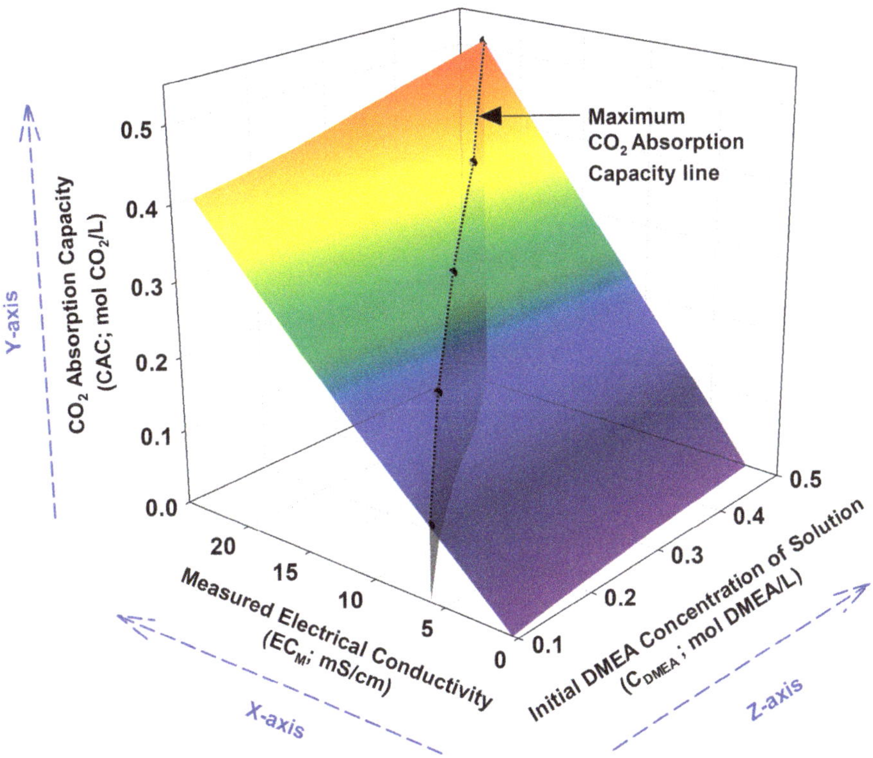

Figure 7. Variation in CO_2 absorption capacity (CAC; y-axis) as a function of initial DMEA concentration (C_{DMEA}; z-axis, ranging from 0.1 to 0.5 M) and measured electrical conductivity (EC_M; x-axis) in CO_2-absorbed DMEA solutions.

5. Conclusions

The present study investigated the CO_2 absorption characteristics of low-concentration (0.1–0.5 M) DMEA solutions, aiming to develop an empirical equation for the real-time estimation of CO_2 absorption capacity (CAC) based on the electrical conductivity (EC) of these solutions. A laboratory-scale semi-batch reactor was employed to evaluate the CO_2 absorption performance and the variation in EC of DMEA solutions during CO_2 absorption.

As CO_2 was absorbed into the DMEA solution, the measured electrical conductivity (EC_M) of the solution and CAC varied similarly over time, providing a basis for predicting CAC through changes in the EC_M. The overall CO_2 absorption rate for 0.1–0.5 M DMEA solutions was approximately 10.0 mmol CO_2/(L·min), independent of DMEA concentration. This consistency arises because the amine concentration is very low, while the absorption reaction follows a base-catalyzed mechanism predominantly influenced by CO_2 concentration. Moreover, the ionic conductivity (IC) of protonated DMEA ($DMEAH^+$) was determined to be 53.1 $S{\cdot}cm^2/(mol{\cdot}z)$, which was larger than that of other tertiary amines such as $TEAH^+$ (37.6) and $MDEAH^+$ (46.5). This outcome can be attributed to $DMEAH^+$

having the smallest molar mass among them. In the case of the 0.1 M DMEA solution saturated with CO_2, the EC_M was greater than the theoretically calculated EC (EC_C). This discrepancy is due to the low amine concentration and the presence of trace amount of HCO_3^- and CO_3^{2-} ions formed from physically absorbed CO_2. However, for 0.2–0.5 M DMEA solutions, the EC_M is lower than the EC_C. This discrepancy increased with higher DMEA concentrations. This can be explained by the larger ionic size of $DMEAH^+$, resulting in the actual mean ionic activity coefficient being smaller than the theoretical value as the concentration increases. Furthermore, the CAC required to increase the EC_M of the DMEA solution by 1 mS/cm increases proportionally with the initial amine concentration. This is due to the decreased activity of ions at higher concentrations, necessitating the production of a larger amount (concentration) of ions through absorption for a given change in EC.

Finally, the empirical equation derived from this study offers a practical approach for estimating CAC from EC measurements in DMEA solutions, enhancing the efficiency of CO_2 absorption processes through the real-time monitoring of CAC. However, limitations include the low concentrations of DMEA used in experiments and a restricted range of operational conditions such as temperatures and pressures. Further research is required to extend the study to higher DMEA concentrations and varied conditions to enhance the model's applicability and consider a more advanced activity coefficient model.

Author Contributions: Conceptualization, S.-J.H. and J.-H.W.; methodology, S.-J.H. and J.-H.W.; software, S.-J.H.; validation, S.-J.H.; formal analysis, S.-J.H. and J.Y.H.; investigation, S.-J.H. and J.Y.H.; resources, J.-H.W.; data curation, S.-J.H. and J.Y.H.; writing—original draft preparation, S.-J.H. and J.Y.H.; writing—review and editing, S.-J.H. and J.-H.W.; visualization, S.-J.H.; supervision, J.-H.W.; project administration, J.-H.W.; funding acquisition, S.-J.H. and J.-H.W. All authors have read and agreed to the published version of the manuscript.

Funding: This research was funded by the National Research Foundation of Korea (NRF) grant funded by the Korea government (MSIT) (2021R1C1C2093637 and 2023R1A2C1003698), as well as being supported by the Catholic University of Korea Research Fund, 2024.

Data Availability Statement: Dataset available on request from the authors.

Conflicts of Interest: The authors declare no conflicts of interest.

References

1. Global Carbon Budget. *Global Carbon Budget 2023*; Copernicus Publications: Göttingen, Germany, 2023.
2. Hockstad, L.; Hanel, L. Inventory of US Greenhouse Gas Emissions and Sinks. 2018, Environmental System Science Data Infrastructure for a Virtual Ecosystem. Available online: https://data.ess-dive.lbl.gov/datasets/doi:10.15485/1464240 (accessed on 1 November 2018).
3. 'The Era of Global Boiling Has Arrived,' Says UN Boss, as White House Announces Provisions to Protect Workers from Extreme Heat. Available online: https://cnb.cx/44J1a8k (accessed on 2 September 2024).
4. Patel, S.K.S.; Jeon, M.S.; Gupta, R.K.; Jeon, Y.; Kalia, V.C.; Kim, S.C.; Cho, B.K.; Kim, D.R.; Lee, J.-K. Hierarchical macroporous particles for efficient whole-cell immobilization: Application in bioconversion of greenhouse gases to methanol. *ACS Appl. Mater. Interfaces* **2019**, *11*, 18968–18977. [CrossRef] [PubMed]
5. Fu, S.; Angelidaki, I.; Zhang, Y. In situ biogas upgrading by CO_2-to-CH_4 bioconversion. *Trends Biotechnol.* **2021**, *39*, 336–347. [CrossRef] [PubMed]
6. Song, C.; Liu, Q.; Qi, Y.; Chen, G.; Song, Y.; Kansha, Y.; Kitamura, Y. Absorption-microalgae hybrid CO_2 capture and biotransformation strategy—A review. *Int. J. Greenh. Gas Control.* **2019**, *88*, 109–117. [CrossRef]
7. de Maria, P.D.; Kara, S.; Gallou, F. Biocatalysis in water or in non-conventional media? Adding the CO_2 production for the debate. *Molecules* **2023**, *28*, 6452. [CrossRef] [PubMed]
8. Santamaría, L.; Korili, S.A.; Gil, A. Layered double hydroxides for CO_2 adsorption at moderate temperatures: Synthesis and amelioration strategies. *Chem. Eng. J.* **2023**, *455*, 140551. [CrossRef]
9. Zhu, Z.; Shi, X.; Rao, Y.; Huang, Y. Recent progress of MgO-based materials in CO_2 adsorption and conversion: Modification methods, reaction condition, and CO_2 hydrogenation. *Chin. Chem. Lett.* **2024**, *35*, 108954. [CrossRef]
10. Moon, H.-S.; Moon, J.-H.; Chun, D.H.; Park, Y.C.; Yun, Y.N.; Sohail, M.; Baek, K.; Kim, H. Synthesis of [Mg_2(DOBDC)$(DMF)_2$]@ polystyrene composite and its carbon dioxide adsorption. *Microporous Mesoporous Mater.* **2016**, *232*, 161–166. [CrossRef]
11. Chouikhi, N.; Cecilia, J.A.; Vilarrasa-Garcia, E.; Besghaier, S.; Chlendi, M.; Duro, F.I.F.; Castellon, E.R.; Bagane, M. CO_2 adsorption of materials synthesized from clay minerals: A review. *Minerals* **2019**, *9*, 514. [CrossRef]

12. Miricioiu, M.G.; Zaharioiu, A.; Oancea, S.; Bucura, F.; Raboaca, M.S.; Filote, C.; Ionete, R.E.; Niculescu, V.C.; Constantinescu, M. Sewage sludge derived materials for CO_2 adsorption. *Appl. Sci.* **2021**, *11*, 7139. [CrossRef]
13. Guo, R.-T.; Li, G.-Y.; Liu, Y.; Pan, W.-G. Recent Progress on CO_2 Capture Based on Sterically Hindered Amines: A Review. *Energy Fuels* **2023**, *37*, 15429–15452. [CrossRef]
14. Tatarczuk, A.; Tańczyk, M.; Więcław-Solny, L.; Zdeb, J. Pilot plant results of amine-based carbon capture with heat integrated stripper. *Appl. Energy* **2024**, *367*, 123416. [CrossRef]
15. He, W.; Zhang, S.; Zhu, C.; Fu, T.; Ma, Y. CO_2 chemical absorption into AMP aqueous solution and mass transfer intensification in cascade sudden expansion microchannels. *Chem. Eng. Process.* **2022**, *181*, 109142. [CrossRef]
16. Ji, C.; Yuan, S.; Huffman, M.; El-Halwagi, M.M.; Wang, Q. Post-combustion carbon capture for tank to propeller via process modeling and simulation. *J. CO2 Util.* **2021**, *51*, 101655. [CrossRef]
17. Li, S.; Li, H.; Yu, Y.; Chen, J. Simulation and performance comparison for CO_2 capture by aqueous solvents of n-(2-hydroxyethyl) piperazine and another five single amines. *Processes* **2021**, *9*, 2184. [CrossRef]
18. Raznahan, M.M.; Riahi, S.; Mousavi, S.H. A simple, robust and efficient structural model to predict CO_2 absorption for different amine solutions: Concern to design new amine compounds. *J. Environ. Chem. Eng.* **2020**, *8*, 104572. [CrossRef]
19. Kayahan, E.; Caprio, U.D.; den Bogaert, A.V.; Khan, M.N.; Bulut, M.; Braeken, L.; Gerven, T.V.; Leblebici, M.E. A new look to the old solvent: Mass transfer performance and mechanism of CO_2 absorption into pure monoethanolamine in a spray column. *Chem. Eng. Process. Process Intensif.* **2023**, *184*, 109285. [CrossRef]
20. Kim, J.; Kim, K.; Lim, H.; Kang, J.H.; Park, H.S.; Park, J.; Song, H. Structural investigation of aqueous amine solutions for CO_2 capture: CO_2 loading, cyclic capacity, absorption–desorption rate, and pKa. *J. Environ. Chem. Eng.* **2024**, *12*, 112664. [CrossRef]
21. Li, Q.; Zhang, W.; Qin, Y.; An, A. Model predictive control for the process of mea absorption of CO_2 based on the data identification model. *Processes* **2021**, *9*, 183. [CrossRef]
22. Chen, Y.-H.; Shen, M.-T.; Chang, H.; Ho, C.-D. Control of solvent-based post-combustion carbon capture process with optimal operation conditions. *Processes* **2019**, *7*, 366. [CrossRef]
23. López, A.; La Rubia, M.; Pacheco, R.; Sánchez, S.; Navaza, J.; Gómez-Díaz, D. Carbon dioxide absorption by aqueous mixtures of diisopropanolamine and triethanolamine. *Chem. Eng. Process.* **2016**, *110*, 73–79. [CrossRef]
24. BChoi, K.; Kim, S.-M.; Kim, K.-M.; Lee, U.; Choi, J.H.; Lee, J.-S.; Baek, I.H.; Nam, S.C.; Moon, J.-H. Amine blending optimization for maximizing CO_2 absorption capacity in a diisopropanolamine–methyldiethanolamine–H_2O system using the electrolyte UNIQUAC model. *Chem. Eng. J.* **2021**, *419*, 129517. [CrossRef]
25. Akan, A.P.; Chau, J.; Gullu, G.; Sirkar, K.K. Life Cycle Assessment of Post-Combustion CO_2 Capture and Recovery by Hydrophobic Polypropylene Cross-Flow Hollow Fiber Membrane Contactors with Activated Methyldiethanolamine. *Atmosphere* **2023**, *14*, 490. [CrossRef]
26. Xiao, L.; Qiu, Z.; Feng, S.; Duan, X.; Zhao, Z.; Liu, Y.; Ma, L. Carbon dioxide absorption and desorption experiments based on MDEA. *Chem. Eng. Process.* **2024**, *204*, 109931. [CrossRef]
27. Pinto, D.D.; Zahraee, Z.; Buvik, V.; Hartono, A.; Knuutila, H.K. Vapor liquid equilibrium measurements of two promising tertiary amines for CO_2 capture. *Processes* **2019**, *7*, 951. [CrossRef]
28. Shiva Nagendra, S.M.; Yasa, P.R.; Narayana, M.V.; Khadirnaikar, S.; Rani, P. Mobile monitoring of air pollution using low cost sensors to visualize spatio-temporal variation of pollutants at urban hotspots. *Sustain. Cities Soc.* **2019**, *44*, 520. [CrossRef]
29. Sofia, D.; Giuliano, A.; Gioiella, F.; Barletta, D.; Poletto, M. Modeling of an air quality monitoring network with high space-time resolution. Comput. *Aided Chem. Eng.* **2018**, *43*, 193. [CrossRef]
30. Alam, A.U.; Clyne, D.; Jin, H.; Hu, N.-X.; Deen, M.J. Fully integrated, simple, and low-cost electrochemical sensor array for in situ water quality monitoring. *ACS Sens.* **2020**, *5*, 412. [CrossRef]
31. Lambrou, T.P.; Anastasiou, C.C.; Panayiotou, C.G.; Polycarpou, M.M. A low-cost sensor network for real-time monitoring and contamination detection in drinking water distribution systems. *IEEE Sens. J.* **2014**, *14*, 2765–2772. [CrossRef]
32. Fulton, S.G.; Stegen, J.C.; Kaufman, M.H.; Dowd, J.; Thompson, A. Laboratory evaluation of open source and commercial electrical conductivity sensor precision and accuracy: How do they compare? *PLoS ONE* **2023**, *18*, e0285092. [CrossRef]
33. Thirstrup, C.; Deleebeeck, L. Review on electrolytic conductivity sensors. *IEEE Trans. Instrum. Meas.* **2021**, *70*, 1008222. [CrossRef]
34. Liu, S.; Gao, H.; Luo, X.; Liang, Z. Kinetics and new mechanism study of CO_2 absorption i nto water and tertiary amine solutions by stopped-Flow technique. *AIChE J.* **2019**, *65*, 652–661. [CrossRef]
35. Xiang, J.; Wei, D.; Mao, W.; Liu, T.; Luo, Q.; Huang, Y.; Liang, Z.; Luo, X. Comprehensive kinetic study of carbon dioxide absorption in blended tertiary/secondary amine solutions: Experiments and simulations. *Sep. Purif. Technol.* **2024**, *330*, 125310. [CrossRef]
36. Littel, R.J.; Bos, M.; Knoop, G.J. Dissociation constants of some alkanolamines at 293, 303, 318, and 333 K. *J. Chem. Eng. Data* **1990**, *35*, 276–277. [CrossRef]
37. Kim, S.; Shi, H.; Lee, Y.Y. CO_2 absorption mechanism in amine solvents and enhancement of CO_2 capture capability in blended amine solvent. *Int. J. Greenh. Gas Control.* **2016**, *45*, 181–188. [CrossRef]
38. Shi, H.; Huang, M.; Wu, Q.; Zheng, L.; Cui, L.; Zhang, S.; Tontiwachwuthikul, P. Study of catalytic CO_2 absorption and desorption with tertiary amine DEEA and 1DMA-2p with the aid of solid acid and solid alkaline chemicals. *Molecules* **2019**, *24*, 1009. [CrossRef]

39. Krebs, H.A.; Roughton, F.J.W. Carbonic anhydrase as a tool in studying the mechanism of reactions involving H_2CO_3, CO_2 or HCO_3'. *Biochem. J.* **1948**, *43*, 550. [CrossRef]
40. Kladkaew, N.; Idem, R.; Tontiwachwuthikul, P.; Saiwan, C. Studies on corrosion and corrosion inhibitors for amine based solvents for CO_2 absorption from power plant flue gases containing CO_2, O_2 and SO_2. *Energy Procedia* **2011**, *4*, 1761–1768. [CrossRef]
41. Han, S.J.; Wee, J.H. Comparison of CO_2 absorption performance between methyl-di-ethanolamine and tri-ethanolamine solution systems and its analysis in terms of amine molecules. Greenh. *Gases* **2021**, *11*, 445–460. [CrossRef]
42. Han, S.J.; Wee, J.H. CO_2 absorption performance and electrical properties of 2-amino-2-methyl-1-propanol compared to monoethanolamine solutions as primary amine-based absorbents. *Energy Fuels* **2021**, *35*, 3197–3207. [CrossRef]
43. Han, S.J.; Wee, J.H. Estimation of the Amount of CO_2 Chemically Absorbed in Real Time by Measuring the Electrical Conductivity Variation of Monoethanol-Amine Aqueous Solutions. *Energy Fuels* **2023**, *37*, 19715–19725. [CrossRef]
44. Prini, R.F.; Harvey, A.H.; Palmer, D.A. *Aqueous Systems at Elevated Temperatures and Pressures: Physical Chemistry in Water, Steam and Hydrothermal Solutions*; Elsevier: Amsterdam, The Netherlands, 2004.
45. Rice, E.W.; Bridgewater, L.; American Public Health Association. *Standard Methods for the Examination of Water and Wastewater*; American Public Health Association: Washington, DC, USA, 2012; Volume 10.
46. Debye, P.; Hückel, E. De la theorie des electrolytes. I. abaissement du point de congelation et phenomenes associes. *Phys. Z.* **1923**, *24*, 185–206.
47. Davies, C.; Malpass, V. Ion association and the viscosity of dilute electrolyte solutions. Part 1.—Aqueous inorganic salt solutions. *Trans. Faraday Soc.* **1964**, *60*, 2075–2084. [CrossRef]
48. Stumm, W.; Morgan, J.J. *Aquatic Chemistry: Chemical Equilibria and Rates in Natural Waters*; John Wiley & Sons: Hoboken, NJ, USA, 2013.
49. Felippe, A.C.; Bellettini, I.C.; Eising, R.; Minatti, E.; Giacomelli, F.C. Supramolecular complexes formed by the association of poly (ethyleneimine)(PEI), sodium cholate (NaC) and sodium dodecyl sulfate (SDS). *J. Braz. Chem.* **2011**, *22*, 1539–1548. [CrossRef]
50. Corti, H.R.; Trevani, L.N.; Anderko, A. *Transport Properties in High Temperature and Pressure Ionic Solutions, in Aqueous Systems at Elevated Temperatures and Pressures*; Elsevier: Amsterdam, The Netherlands, 2004; pp. 321–375.
51. Artemov, V.; Volkov, A.; Sysoev, N.; Volkov, A. On autoionization and pH of liquid water. *Dokl. Phys.* **2016**, *61*, 1–4. [CrossRef]
52. de Myttenaere, A.; Golden, B.; Grand, B.L.; Rossi, F. Mean absolute percentage error for regression models. *Neurocomputing* **2016**, *192*, 38–48. [CrossRef]
53. Jiang, W.; Gao, H.; Liang, Z.; Liu, B.; Tontiwachwuthikul, P.; Hu, X. A comparative kinetics study of CO_2 absorption into aqueous DEEA/MEA and DMEA/MEA blended solutions. *AIChE J.* **2018**, *64*, 1350–1358. [CrossRef]
54. Henni, A.; Li, J.; Tontiwachwuthikul, P. Reaction kinetics of CO_2 in aqueous 1-amino-2-propanol, 3-amino-1-propanol, and dimethylmonoethanolamine solutions in the temperature range of 298–313 K using the stopped-flow technique. *Ind. Eng. Chem. Res.* **2008**, *47*, 2213–2220. [CrossRef]
55. Tong, C.; Perez, C.C.; Chen, J.; Marcos, J.-C.V.; Neveux, T.; Moullec, Y.L. Measurement and calculation for CO_2 solubility and kinetic rate in aqueous solutions of two tertiary amines. *Energy Procedia* **2013**, *37*, 2084–2093. [CrossRef]
56. Liu, B.; Luo, X.; Liang, Z.; Olson, W.; Liu, H.; Idem, R.; Tontiwachwuthikul, P. The development of kinetics model for CO_2 absorption into tertiary amines containing carbonic anhydrase. *AIChE J.* **2017**, *63*, 4933–4943. [CrossRef]
57. Kaur, D.P.; Yamada, K.; Park, J.-S.; Sekhon, S.S. Correlation between ion diffusional motion and ionic conductivity for different electrolytes based on ionic liquid. *J. Phys. Chem. B* **2009**, *113*, 5381–5390. [CrossRef]
58. Zhang, A.; Yang, X.; Yang, F.; Zhang, C.; Zhang, Q.; Duan, G.; Jiang, S. Research progress of the ion activity coefficient of polyelectrolytes: A review. *Molecules* **2023**, *28*, 2042. [CrossRef] [PubMed]
59. Jeon, H.-N.; Shin, H.K.; Hwang, S.; No, K.T. Development of an Infinite Dilution Activity Coefficient Prediction Model for Organic Solutes in Ionic Liquids with Modified Partial Equalization Orbital Electronegativity Method Derived Descriptors. *ACS Omega* **2021**, *6*, 15361–15373. [CrossRef] [PubMed]
60. Zhu, L.; Schade, G.W.; Nielsen, C.J. Real-time monitoring of emissions from monoethanolamine-based industrial scale carbon capture facilities. *Environ. Sci. Technol.* **2013**, *47*, 14306–14314. [CrossRef] [PubMed]
61. Chan, C.W.; Zhou, Q.; Tontiwachiwuthikul, P. Part 4a: Applications of knowledge-based system technology for the CO_2 capture process system. *Carbon Manag.* **2012**, *3*, 69–79. [CrossRef]

processes

MDPI

Review

CO_2 to Methanol Conversion: A Bibliometric Analysis with Insights into Reaction Mechanisms, and Recent Advances in Catalytic Conversion

Shahdev Sajnani [1,†], Mazhar Ahmed Memon [2,*,†], Shabir Ahmed Memon [3], Akash Kumar [4], Darakhshan Mehvish [5], Somavia Ameen [6], Mukarama [7], Wei Zhou [2] and Yuan Liu [2,*]

1 School of Environmental Science and Engineering, Tianjin University, Tianjin 300072, China; shahdevsajnani@tju.edu.cn
2 Tianjin Key Laboratory of Applied Catalysis Science and Technology, School of Chemical Engineering and Technology, Tianjin University, Tianjin 300072, China; weizhou1234@tju.edu.cn
3 Key Laboratory of Coast Civil Structure Safety of Ministry of Education, Tianjin University, Tianjin 300072, China; shabirmemon@tju.edu.cn
4 School of Civil Engineering, Guangzhou University, Guangzhou 510006, China; akash.kumar@tju.edu.cn
5 School of Chemistry and Chemical Engineering and Technology, Beijing Institute of Technology, Beijing 100081, China; 3820242064@bit.edu.cn
6 Tianjin Key Laboratory of Molecular Optoelectronic Science, Department of Chemistry, School of Science, Tianjin University, Tianjin 300072, China; somaviaqamer297@tju.edu.cn
7 Department of Chemistry, University of Punjab, Lahore P.O. Box 54590, Pakistan; mk0tma098@gmail.com
* Correspondence: mazharahmed@tju.edu.cn (M.A.M.); yuanliu@tju.edu.cn (Y.L.)
† These authors contributed equally to this work.

Abstract: The rising levels of atmospheric carbon dioxide (CO_2) necessitate urgent and effective strategies for its capture and utilization. Among the various CO_2 valorization pathways, the conversion of CO_2 into methanol has gained considerable attention due to its dual role in reducing greenhouse gas emissions and serving as a renewable fuel and chemical feedstock. This review uniquely combines bibliometric analysis of 13,289 peer-reviewed publications (2012–2023) with an evaluation of Cu-based catalyst advancements, addressing critical gaps in the literature. A bibliometric analysis highlights the key trends, collaborations, and research gaps in the field. Among the catalytic systems, noble metals, though highly active, are uneconomical for large-scale applications, while non-noble metals, such as nickel, exhibit limited activity due to undesired reaction pathways. In comparison, Cu-based catalysts overcome these challenges by offering a balance of activity, selectivity, and cost-effectiveness. Special emphasis is placed on the CO_2 to methanol conversion pathways, with insights into thermodynamic constraints, emerging solutions, and potential directions for future research. By consolidating the current state of knowledge, this review identifies significant opportunities for advancing CO_2 conversion technologies, particularly in methanol synthesis, positioning it as a promising strategy for sustainable carbon management and energy production.

Keywords: global warming; carbon dioxide; CO_2 conversion; methanol; catalysis

1. Introduction

Climate change poses a formidable environmental challenge, principally driven by increased atmospheric concentrations of carbon dioxide (CO_2). This elevation in CO_2, primarily resulting from the combustion of fossil fuels, has seen levels rise from 381 ppm in 2006 to 423 ppm by 2024, with annual increases ranging from 1.5 to 3 ppm [1]. Addressing this issue demands sustainable approaches to mitigating CO_2 emissions while also advancing renewable fuel production. Among the strategies employed, Carbon Capture and Storage (CCS) offers a pathway by collecting and sequestering CO_2 through methods like direct air capture, saline aquifers, and depleted oil reservoirs. However, the scalability of CCS is limited by substantial spatial and financial demands. Alternatively, Carbon Capture and Utilization (CCU) proposes a solution by converting captured CO_2 into less harmful, value-added products [2,3].

In the quest for sustainable industrial solutions, the thermo-catalytic conversion of CO_2 has shown considerable promise. Specifically, CO_2 hydrogenation is a critical process where hydrogen (H_2), produced sustainably via the electrolysis of water using renewable energy sources such as wind and solar power, serves as a vital reagent [4,5]. This process can yield several crucial chemical products, including methane [2,6], alkanes [7], carbon monoxide [8], olefins [9], and alcohols [10]. Among these, methanol is particularly significant due to its extensive industrial demand exceeding 110 million tons annually. Methanol is utilized as a precursor in the production of formaldehyde, acetic acid, methyl tert-butyl ether (MTBE), and gasoline [11,12]. It also offers potential as a fuel alternative in modified diesel engines due to its high-octane number and is utilized in direct methanol fuel cells (DMFC) to generate electricity [13]. The conversion of CO_2 to methanol not only has environmental benefits but also presents economic advantages. This process is highly selective, generating fewer byproducts and operating under milder reaction conditions compared to traditional methods. It offers a sustainable solution for CO_2 recycling, supporting global efforts to reduce greenhouse gas emissions and address climate change. Methanol synthesized from CO_2 is a clean fuel and a versatile raw material for producing various chemicals, benefiting from an ideal stoichiometric ratio that enhances conversion efficiency. Furthermore, it requires less energy, avoids the need for high-pressure storage like hydrogen, and can be safely utilized as a fuel without requiring modifications to engines.

The transformation of CO_2 into methanol necessitates catalysts due to the molecule's stability. Advances in catalyst development have significantly enhanced the efficiency of this conversion process. Various metal-based catalytic systems have been explored, including noble metals (e.g., Pd, Pt) and transition metals (e.g., Fe, Ni, Co). While noble metals demonstrate excellent catalytic activity and selectivity, their high cost and limited availability hinder widespread application [14]. Transition metal catalysts, on the other hand, are more abundant and cost-effective but often favor competing reactions such as methane or carbon monoxide formation, reducing methanol selectivity [1,15]. Additionally, these systems tend to lack the thermal stability and tunable active sites necessary for efficient CO_2 hydrogenation under industrially relevant conditions [16]. In contrast, Cu-based catalysts provide an optimal balance of cost, availability, and catalytic efficiency [17]. Their ability to generate and stabilize active Cu^0/Cu^+ sites facilitates the activation of CO_2 and selective methanol synthesis under milder conditions [18,19].

The scholarly work surrounding CO_2 hydrogenation to methanol has been growing annually, indicating an active field of research with substantial progress [17,19,20]. Although several reviews have explored the catalytic conversion of CO_2 to methanol, most focus on general overviews of catalytic systems, reaction mechanisms, and thermodynamic aspects [21–24]. However, they often lack detailed bibliometric analyses along with highlighting the emerging role of Cu-based catalysts in this field. Additionally, many existing studies do not comprehensively examine the recent technological advancements and trends driving progress in CO_2 to-methanol conversion. This review addresses these gaps by combining a bibliometric analysis of publication trends with an in-depth evaluation of thermodynamic and reaction pathways, with a particular focus on the recent developments in Cu-based catalysts. By doing so, this manuscript provides insights into both academic and industrial advancements in CO_2 conversion technologies. This assessment will critically explore these advancements and propose directions for future research. By consolidating existing knowledge, this review aims to illuminate pathways for future enhancements in CO_2 conversion technologies, positioning methanol synthesis as an effective strategy for sustainable carbon management and energy production.

2. Search Strategy and Bibliometric Evaluation Technique

The bibliometric analysis of CO_2 conversion to methanol was conducted using a robust search strategy to capture the breadth of relevant research publications. Data for this analysis was extracted from the Scopus database, a widely recognized platform for high-quality, peer-reviewed academic publications (See Figure 1). The initial data set comprised 14,517 documents, spanning multiple disciplines and types of publications. To ensure a focused and high-quality analysis, a systematic filtering process was employed, narrowing the data set to 13,289 relevant publications. This filtering excluded non-peer-reviewed content and other less relevant document types, ensuring the data set consisted of high-impact research outputs.

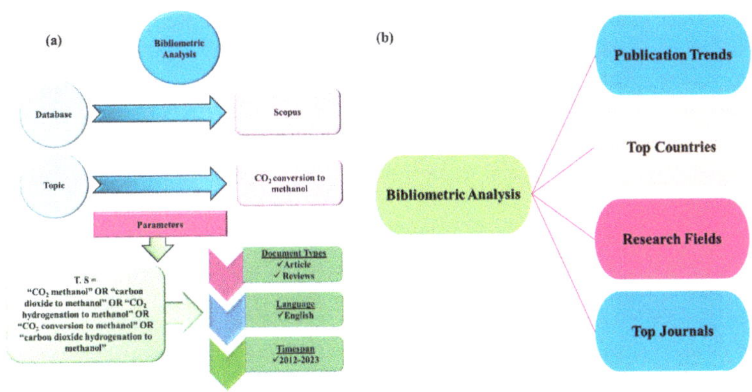

Figure 1. (**a**) Search Strategy Technique, (**b**) Bibliometric Evaluation Trends.

The search string used for data retrieval included combinations of key terms relevant to CO_2 conversion to methanol, such as "CO_2 methanol", "carbon dioxide to methanol", "CO_2 hydrogenation to methanol", "CO_2 conversion to methanol", and "carbon dioxide hydrogenation to methanol". The search was limited to documents published between 2012 and 2023, ensuring that only the most recent research on CO_2 methanol conversion was included. Furthermore, the search was restricted to English language publications, although a few significant contributions in Chinese, Polish, Portuguese, and Russian were also considered to reflect the global nature of research in this field.

To further understand the thematic focus within the CO_2 to-methanol research community, a keyword analysis was conducted. A total of 9363 distinct keywords were identified across the selected articles, providing insights into the prevalent themes and areas of focus within the literature. To process and visualize the collected data, VOSviewer (version 1.6.20) was employed to construct co-authorship networks, co-citation analyses, and keyword co-occurrence maps. This software allowed for the identification of collaborative networks among researchers, institutions, and countries, while also illuminating the key research topics that have emerged over the past decade. Microsoft Excel was used for supplementary data management and statistical analysis, particularly in generating publication trends and other performance indicators.

This comprehensive bibliometric approach provides a detailed view of the global research landscape surrounding CO_2 conversion to methanol. The findings not only highlight the rapid growth of research activity in this field but also reveal key collaborations, influential contributors, and emerging research themes. These insights are invaluable for identifying research gaps, fostering collaboration, and guiding future research efforts in the pursuit of efficient and scalable CO_2 methanol conversion technologies.

2.1. *Publication Trends and Top Research Fields*

A comprehensive analysis of the research landscape on CO_2 conversion to methanol revealed a total of 14,517 documents indexed in the Scopus database. After applying filters to ensure data quality, a final selection of 13,289 documents was made for further analysis. These filtered results comprised 11,058 original research articles, 2584 review articles, 200 conference proceedings, and 803 books, indicating a robust and diverse body of work addressing various aspects of CO_2 to methanol conversion.

Most of the publications were in English, accounting for 14,135 documents, underscoring the dominance of English as the primary language for academic discourse in this field. There was also a notable, albeit smaller, contribution from non-English languages, with 59 publications in Chinese, seven each in Polish and Portuguese, and five in Russian. This diversity in language reflects the global interest in research on CO_2 conversion, though English language publications dominate the field.

Additionally, the publication was analyzed with the inclusion of the query term "Cu-Based Catalyst". A total of 9361 papers were identified that featured Cu-based catalysts, representing a subset of the initial 13,283 publications. These included 2063 review articles and 7298 research articles.

2.1.1. Publication Trend Analysis

The trend of publications over time indicates a consistent and significant rise in research output, particularly in recent years. From 2013 to 2023, the number of published works has increased dramatically (Figure 2a). In 2013, the field saw only 219 publications, while in 2023, this number surged to 2833, a more than tenfold increase over a decade. This upward trend underscores the growing importance of CO_2 conversion to methanol in the context of global efforts to mitigate climate change and develop sustainable energy solutions.

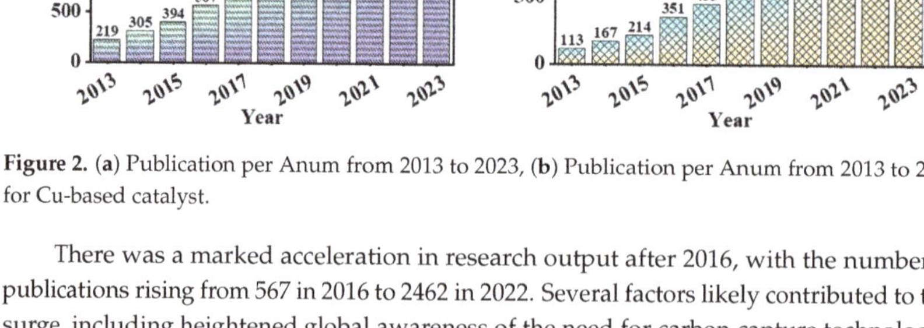

Figure 2. (**a**) Publication per Anum from 2013 to 2023, (**b**) Publication per Anum from 2013 to 2023 for Cu-based catalyst.

There was a marked acceleration in research output after 2016, with the number of publications rising from 567 in 2016 to 2462 in 2022. Several factors likely contributed to this surge, including heightened global awareness of the need for carbon capture technologies and sustainable fuel-production methods. Moreover, the year 2020 saw 1623 publications, reflecting a continued growth in research activity despite global disruptions caused by the COVID-19 pandemic. The largest yearly increase in research output occurred between 2019 and 2020, where the number of publications rose from 1129 to 1623, an increase of approximately 44%. This spike is possibly attributable to increased investments in sustainable energy research and the ongoing development of catalytic processes and new technologies for CO_2 to methanol conversion. The upward trend continued post-pandemic, with 2078 articles published in 2021 and 2462 in 2022, reflecting a sustained interest in this research area.

Interestingly, a similar publication trend is observed specifically for Cu-based catalysts, as shown in Figure 2b. From 2013 to 2023, research activity in this area increased substantially, rising from 113 publications in 2013 to 2182 in 2023. This exponential growth demonstrates the intensifying global focus on Cu-based catalysts, particularly after 2020, likely due to their crucial role in advancing catalytic efficiency.

2.1.2. Research Areas

The field of CO_2 conversion to methanol spans a broad range of scientific disciplines, reflecting its inherently interdisciplinary nature. The distribution of research areas demonstrates the multifaceted approach required to address the technical and scientific challenges associated with converting CO_2 into value-added products like methanol.

Figure 3 exhibits that the most prominent research domain is Chemistry, with 7722 publications, representing the central role that chemical processes, catalysis, and reaction mechanisms play in CO_2 methanol conversion. Chemistry is essential for developing and optimizing catalytic systems, understanding reaction pathways, and improving efficiency in the conversion process. Closely following is Chemical Engineering, with 7073 publications. Chemical engineering research contributes significantly to the practical implementation and scaling of CO_2 to methanol conversion technologies. This includes the design of reactors, process intensification, and the integration of CO_2 conversion systems into existing industrial frameworks. The near parity in the number of publications between chemistry and chemical engineering highlights the close collaboration between these disciplines in tackling CO_2 conversion challenges.

The Energy sector, with 3258 publications, underscores the importance of this research in the broader context of renewable energy and sustainable fuel production. Research in this area focuses on developing energy-efficient processes, integrating renewable H_2 sources, and scaling up methanol production for use as an alternative fuel. Materials Science contributes 3102 publications, reflecting its critical role in the development of advanced catalysts and novel materials that enhance CO_2 conversion efficiency. The design and synthesis of materials with high surface area, stability, and catalytic activity are key to improving the conversion rates of CO_2 to methanol. Environmental Science, with 3013 publications, represents the growing recognition of CO_2 conversion as a crucial strategy for reducing greenhouse gas emissions and mitigating climate change. Research in this area focuses on the environmental impact of CO_2 utilization technologies and their potential role in circular carbon economies. Engineering more broadly, excluding chemical engineering, also contributes significantly, with 1913 publications. This research area includes mechanical and industrial engineering approaches to process optimization, system integration, and the practical deployment of CO_2 conversion technologies on a commercial scale. Physics and Astronomy contribute 1565 publications, primarily in theoretical studies, modeling, and simulations of reaction mechanisms at the atomic and molecular levels. Understanding the fundamental physical principles of CO_2 conversion is essential for improving reaction efficiency and developing new catalytic processes. The field of Biochemistry, Genetics, and Molecular Biology has 876 publications, highlighting the potential for biological approaches to CO_2 conversion, such as using enzymes or microorganisms for methanol production. These studies often explore the intersection of biocatalysis and chemical processes for CO_2 utilization. Other contributing fields include Multidisciplinary research (317 publications), which brings together insights from various scientific and engineering domains to address the complex challenges of CO_2 methanol conversion.

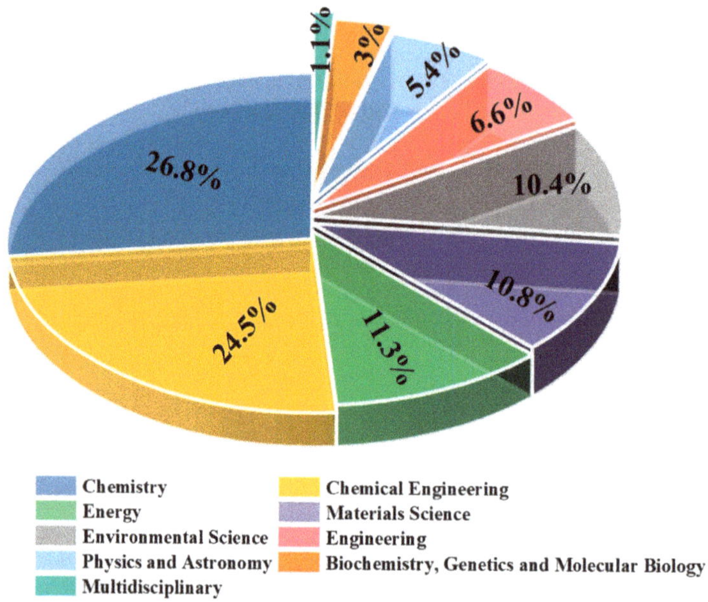

Figure 3. Leading research areas for CO_2 methanol conversion.

This diverse distribution of research areas demonstrates that the successful conversion of CO_2 to methanol requires collaboration across multiple scientific disciplines. Chemistry, chemical engineering, and energy research form the core of this effort, while contributions from materials science, environmental science, and engineering ensure that these technologies can be effectively scaled and integrated into broader energy systems.

2.2. Top Countries

A country-wise analysis of the research output in the field of CO_2 conversion to methanol provides insights into the global distribution of expertise and contributions (Figure 4a). The analysis reveals that research in this domain is concentrated in a few key regions, with several countries emerging as global leaders in the field.

China leads the world by a significant margin, with 6247 publications, making it the dominant force in CO_2 conversion to methanol research. This prominence reflects China's significant investment in sustainable energy technologies and its emphasis on reducing greenhouse gas emissions through innovative chemical conversion processes. China's research output in this area accounts for a substantial portion of the global literature, underscoring its pivotal role in advancing CO_2 utilization technologies.

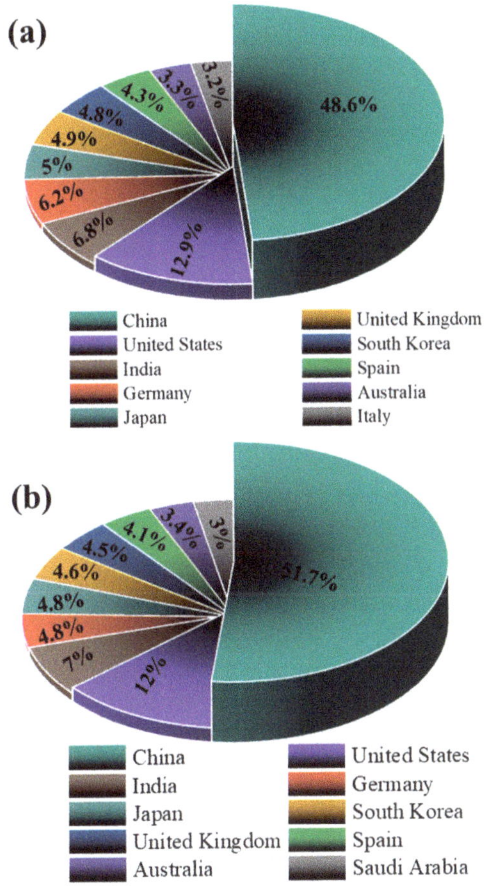

Figure 4. (**a**) Top countries for CO_2-to-methanol conversion, (**b**) Top countries for CO_2-to-methanol conversion for Cu-based catalyst.

The United States follows with 1657 publications, maintaining its position as a key player in this domain. The U.S. has a well-established research infrastructure and a history of innovation in chemical engineering and catalysis, which are critical to the development of CO_2 to methanol technologies. While the number of publications is considerably lower than China's, the U.S. remains a major contributor to both fundamental and applied research in this field. India ranks third with 878 publications, highlighting the country's growing focus on sustainable energy and CO_2-reduction technologies. India's increasing research output is indicative of its rising prominence in global efforts to address climate change and develop alternative energy sources. Germany and Japan are also significant contributors, with 797 and 645 publications, respectively. Both countries are known for their advancements in chemical engineering, catalysis, and sustainable technologies, making them key players in the CO_2 conversion research landscape. Germany's strong research focus on catalysis and renewable energy aligns well with the goals of CO_2 to methanol conversion, while Japan's contributions reflect its innovation in material sciences and energy conversion processes. The United Kingdom follows closely with 624 publications, showcasing its contributions to both fundamental research and industrial applications of CO_2 conversion. South Korea (611 publications) and Spain (549 publications) have also established themselves as important contributors to this field, with research focusing on catalysis, renewable energy integration, and innovative conversion technologies. Rounding out the top 10 are Australia, with 426 publications, and Italy, with 423 publications. Both countries have increasingly invested in renewable energy research, with a particular focus on CO_2 utilization and the development of sustainable chemical processes.

Upon comparing the country-wise research outputs in CO_2 conversion to methanol and CO_2 conversion to methanol specified with Cu-based catalysts, a notable correlation emerges (see Figure 4b). China leads both, with 6247 publications in CO_2 to methanol research and 4774 in Cu-based catalyst studies, underscoring its significant advancements in CO_2 conversion as sustainable energy technologies. Similarly, the United States and India maintain prominent positions in both fields, reflecting their commitment to advancing catalytic processes for CO_2 utilization. This parallel trend suggests that nations prioritizing CO_2 conversion research also focus on developing Cu-based catalysts, highlighting the integral role of such catalysts in efficient CO_2-to-methanol conversion processes.

In total, these top 10 countries account for a significant majority of the global research output in CO_2 conversion to methanol, reflecting the strategic importance of this field in the context of global efforts to mitigate climate change. This analysis underscores the collaborative nature of this research, with multiple countries contributing to the development of scalable and efficient CO_2 utilization technologies.

2.3. Top Journals

To assess the influence and contribution of various journals to the field of CO_2 conversion to methanol, an in-depth journal analysis was conducted. A total of 10 prominent journals emerged as key platforms for publishing research in this area, as shown in Figure 5. The data reflects a broad distribution across multidisciplinary and specialized journals, highlighting the interdisciplinary nature of this research, which spans catalysis, energy technologies, and material sciences. The most prolific journals, ranked by the number of articles published, are presented below.

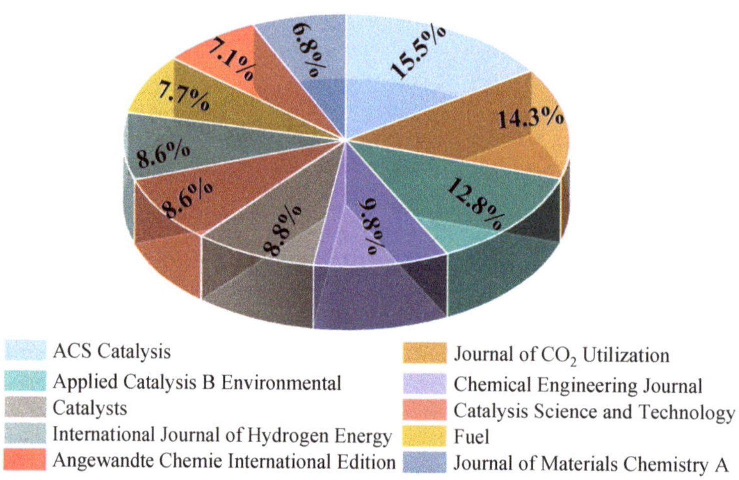

Figure 5. Leading journals for CO_2-to-methanol conversion.

ACS Catalysis leads with 509 publications, positioning itself as the most significant contributor to research on CO_2-to-methanol conversion. As one of the premier journals in the field of catalysis, its focus on both theoretical and applied research makes it a preferred outlet for cutting-edge studies in this domain. Following closely is the *Journal of CO_2 Utilization*, with 469 publications, underscoring its pivotal role in disseminating research specifically targeting CO_2 conversion and valorization technologies. These journals collectively represent the forefront of innovation in catalysis and CO_2 utilization. *Applied Catalysis B: Environmental* ranks third, with 420 publications, highlighting its contribution to sustainable catalysis and environmental applications of CO_2 conversion. *Chemical Engineering Journal* and *Catalysts* follow, with 321 and 290 articles, respectively, reflecting their importance in both theoretical advancements and practical implementations of catalytic processes. Other notable contributors include *Catalysis Science and Technology* and the *International Journal of Hydrogen Energy*, each publishing 284 articles. Their focus spans across catalysis, energy conversion, and H_2 production, reflecting the growing interconnection between CO_2-reduction technologies and the H_2 economy. *Fuel* contributed 253 articles, reinforcing its relevance in the study of alternative fuels and energy systems. Meanwhile, *Angewandte Chemie International Edition*, a prestigious journal known for publishing pioneering chemical research, produced 235 publications on CO_2-to-methanol conversion, reflecting its influence in high-impact, foundational research. Finally, *Journal of Materials Chemistry A* contributed 225 articles, indicating the significant role of advanced materials in catalysis and CO_2 utilization.

This analysis demonstrates that CO_2 conversion to methanol research is disseminated through a variety of high-impact journals, spanning catalysis, chemical engineering, energy systems, and materials science. These journals not only shape the current research landscape but also drive the future directions of CO_2 conversion technologies.

2.4. Top Keywords

The analysis of top keywords in the field of CO_2 conversion to methanol highlights the prominent research areas and emerging trends in this domain. From a total of 9363 keywords identified across the selected publications, 102 keywords were chosen for detailed analysis based on their frequency of occurrence (Figure 6). These keywords were further divided into five distinct clusters, each representing key themes within the research landscape.

The most frequently used keyword, CO_2 hydrogenation, appeared 460 times and falls under Cluster 1, indicating its central role in research focusing on the catalytic conversion processes necessary for producing methanol from CO_2. This keyword also reflects the high level of interest in using hydrogenation as a method to convert CO_2 into valuable chemicals. The total link strength of this term, 481, highlights its significance in the ongoing discussions surrounding CO_2-reduction technologies and efficient methanol synthesis. Another critical term, CO_2 reduction, appeared 339 times and is part of Cluster 3, emphasizing the growing attention toward reducing CO_2 emissions through chemical and catalytic processes. This keyword is closely linked with the study of advanced catalysts and reaction mechanisms, particularly in the context of sustainable and energy-efficient solutions. As CO_2 reduction plays a pivotal role in mitigating climate change, its frequent occurrence underscores its relevance to carbon capture and utilization strategies.

The keyword methanol was cited 290 times, positioning it within Cluster 5. Its high frequency reflects the emphasis on methanol as the primary product of interest in CO_2 conversion research. Methanol synthesis, a key component of carbon utilization efforts, has been the focus of extensive research aiming to optimize production processes and improve catalyst efficiency. Similarly, carbon dioxide, mentioned 244 times in the same cluster, illustrates the ongoing research efforts to utilize CO_2 as a feedstock, highlighting the environmental focus of this field. In addition to these dominant keywords, terms such as photocatalysis (226 occurrences in Cluster 3) and electrocatalysis (146 occurrences in Cluster 2) demonstrate the importance of exploring alternative, renewable energy-driven methods for CO_2 conversion. Photocatalysis represents a promising approach using light to drive CO_2-reduction reactions, while electrocatalysis focuses on electrochemical methods powered by sustainable energy sources. These areas represent innovative pathways for improving the sustainability of methanol production.

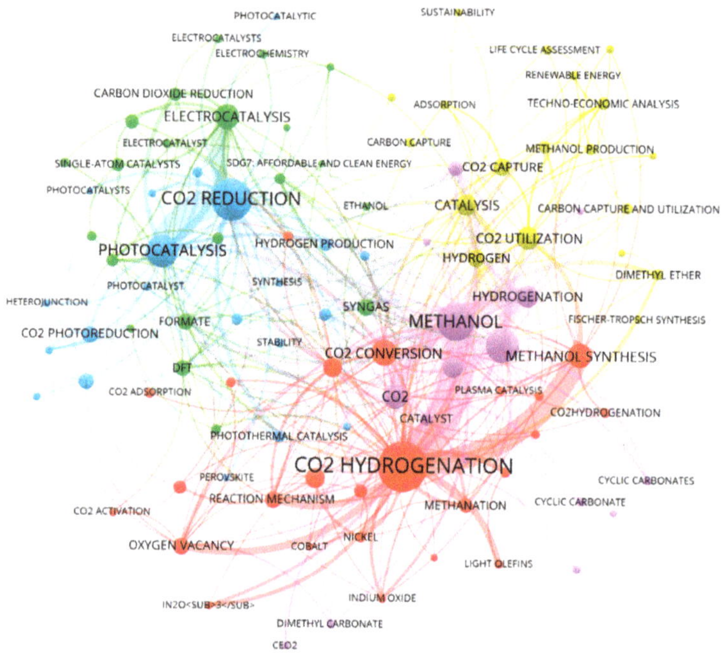

Figure 6. The network visualization of the most frequent keywords.

Other noteworthy keywords include methanol synthesis (135 occurrences in Cluster 1), reflecting the focus on improving the efficiency of converting CO_2 into methanol, and CO_2 conversion (154 occurrences in Cluster 1), which encompasses a broad spectrum of research dedicated to the transformation of CO_2 into value-added products. Together, these terms illustrate a diverse and multifaceted research landscape, with efforts ranging from fundamental catalytic studies to applied engineering solutions. The keyword clusters reveal the diversity of research approaches: Cluster 1 emphasizes hydrogenation and catalytic conversion, Cluster 2 focuses on electrocatalysis and sustainable energy, Cluster 3 delves into photocatalysis, Cluster 4 explores chemical engineering aspects, and Cluster 5 involves materials science and catalytic system advancements. This comprehensive keyword analysis offers a clear understanding of the primary research directions, showcasing the field's focus on catalysis, sustainable processes, and innovative chemical engineering solutions in CO_2-to-methanol conversion.

3. CO_2 Conversion to Methanol

The transformation of CO_2 into alcohols, particularly methanol, is a topic of significant interest both in the chemical industry and the scientific community. As a vital C1 alcohol, methanol serves as a precursor to a diverse array of chemical compounds. The concept of a "methanol economy", introduced by Nobel Laureate George Olah, centers around the hydrogenation of CO_2 into methanol and its derivatives. This approach encompasses the capture of CO_2 from various sources, including natural, human-made, or industrial, followed by its effective catalytic conversion into methanol, illustrated in Figure 7.

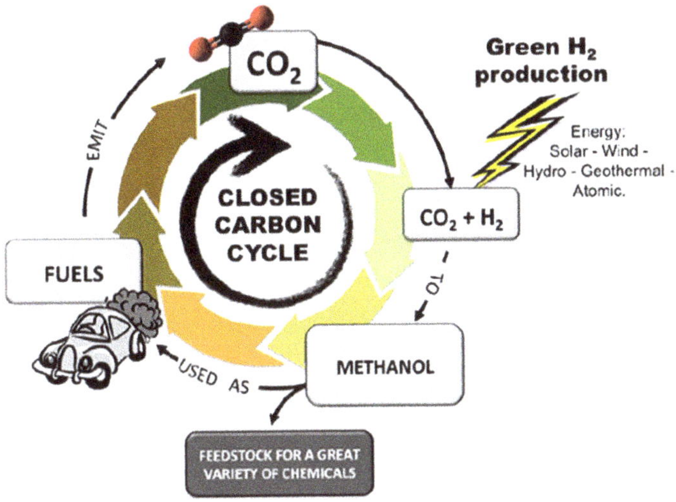

Figure 7. Schematic illustration for CO_2 conversion to methanol [25].

In recent developments, research and development (R&D) in methanol synthesis have been increasingly geared towards more sustainable methodologies. This includes the reduction of CO_2 using H_2 derived from technologies that utilize natural and renewable energy sources. Prominent companies in this field, such as Mitsui Chemicals and Carbon Recycling International (CRI) Inc., have pioneered such methods for methanol production. Specifically, CRI's facility in Iceland boasts a production capacity of approximately 5 million liters of methanol annually (4 kta). Remarkably, the H_2 required for this process is generated through water electrolysis, utilizing energy sourced primarily from natural elements like geothermal, hydroelectric, and wind power [26].

3.1. Thermodynamics and Reaction Mechanism of CO_2 Conversion to Methanol

The commercial production of methanol predominantly relies on syngas (a mixture of CO and H_2), which is derived from fossil resources such as coal and natural gas through processes like coal gasification and natural gas steam reforming [27]. Typically, small amounts of CO_2 (2–8%) are introduced to the CO/H_2 stream to adjust the H/C ratio to achieve the required stoichiometry and enhance the reaction rate.

$$CO + 2H_2 = CH_3OH \quad \Delta H_{298K} = -90.6 \text{ kJ mol}^{-1} \tag{1}$$

$$CO_2 + 3H_2 = CH_3OH + H_2O \quad \Delta H_{298K} = -49.5 \text{ kJ mol}^{-1} \tag{2}$$

Reverse water-gas-shift reaction (RWGS)

$$CO_2 + H_2 = COH + H_2O \quad \Delta H_{298K} = -41.2 \text{ kJ mol}^{-1} \tag{3}$$

Compared to syngas-based methanol production (Equation (1)), the hydrogenation of CO_2 to methanol demands an additional amount of H_2 due to the need to remove one oxygen atom from CO_2 via water formation as a byproduct (Equation (2)). This results in thermodynamically less favorable conditions for methanol production from CO_2 compared to CO, resulting in lower one-pass methanol yields (Figure 8). For instance, at 200 °C, the equilibrium methanol yield from CO exceeds 80%, while from CO_2, it is less than 40% [28]. Additionally, the formation of side products like ethers, ketones, higher alcohols, or hydrocarbons is thermodynamically more favorable under typical synthesis conditions [5,29].

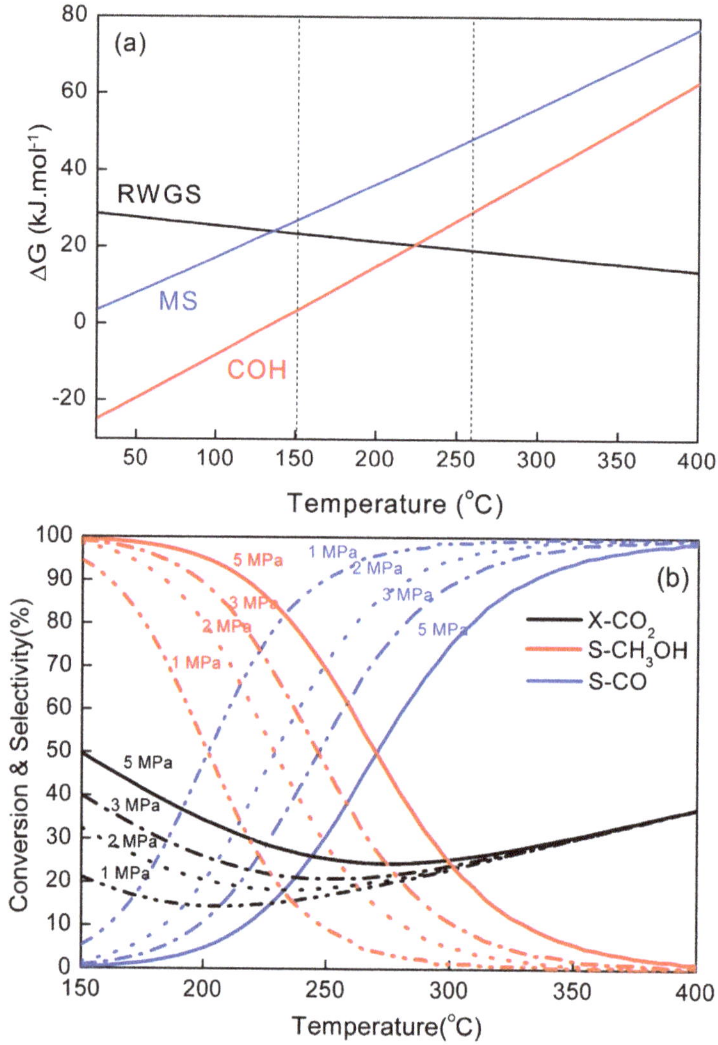

Figure 8. (**a**) Free energy from CO_2, RWGS, and syngas hydrogenation. (**b**) Equilibrium conversion-selectivity values of the CO_2 hydrogenation reaction at various pressures [5].

The synthesis of methanol from CO_2 and H_2 is an exothermic reaction that reduces the number of gas-phase molecules. According to Le Chatelier's principle, high pressure and low temperature should thermodynamically favor methanol production. However, due to the inert nature of CO_2, a higher reaction temperature (typically above 240 °C) is needed to increase the rate of CO_2 activation and methanol formation [19,30,31]. At these temperatures, the reverse water-gas shift (RWGS) reaction (Equation (3)) becomes thermodynamically favorable, reducing methanol yields and wasting H_2 [31,32].

The RWGS reaction will not only wasteH_2 supplies but also reduce the yield of methanol.

Although the thermodynamic equilibrium limits the maximum methanol yield, this limitation can be mitigated through optimization of reaction conditions, reactor design, and innovations such as recycling unconverted feed gas after product separation or in situ product removal (e.g., water removal via distillation or membranes) [28].

Methanol production from CO_2 is highly exothermic and limited by kinetics, resulting in only 15–25% conversion under typical conditions [33]. The endothermic RWGS reaction competes with methanol synthesis, further complicating the process. Efficient reactor design is essential to remove the heat generated during methanol synthesis and maintain isothermal conditions [34]. Water, a byproduct of this reaction, can deactivate conventional copper-based catalysts, highlighting the importance of catalyst design and kinetic studies [35,36].

Methanol synthesis from CO_2 can proceed through various reaction pathways, with the formate, RWGS, and trans-COOH* mechanisms being the most prominent (Figure 9a). In the classical formate pathway, CO_2 reacts with adsorbed atomic H, forming a formate intermediate (HCOO*), which subsequently transforms into intermediates such as dioxomethylene (H_2COO*), formaldehyde (H_2CO*), methoxy (CH_3O*), and finally, methanol (CH_3OH) [5,36]. Alternatively, formate can hydrogenate into formic acid (HCOOH*) and follow a similar sequence leading to methanol through the methoxy intermediate.

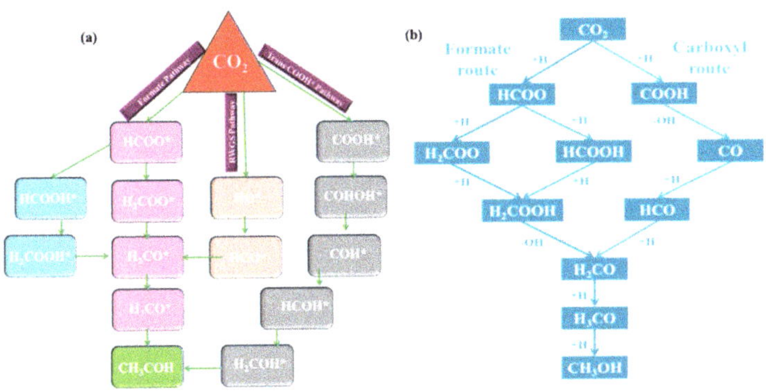

Figure 9. (**a**) Schematic representation of reaction mechanisms for CO_2 methanol [37], (**b**) reaction mechanism on Cu-based catalysts. * Indicates intermediates that are adsorbed on the catalyst surface.

The RWGS pathway involves the initial transformation of CO_2 into CO*, which is then hydrogenated to form methanol via intermediates such as formyl (HCO*), formaldehyde (H_2CO*), and methoxy (CH_3O*). A variation of this mechanism involves the formation of hydrocarboxyl (COOH*) intermediates, which can convert into CO* and proceed through a similar route to methanol. In the trans-COOH mechanism, COOH* can also produce dihydroxycarbene (COHOH*), which dissociates into hydroxymethylidyne (COH*) and subsequently into hydroxymethylene (HCOH*), leading to methanol formation [5,36].

Methanol is an attractive product due to its versatility as a clean, biodegradable fuel that can be converted into other valuable chemicals such as dimethyl ether, olefins, and longer-chain hydrocarbons [38–40]. However, two major thermodynamic challenges hinder methanol generation: high-pressure requirements (up to 200 bar) and low reaction kinetics at lower temperatures [41], which reduce methanol yield. Moreover, to maintain high methanol selectivity and minimize CO production, the RWGS reaction must be suppressed [10,41–44].

The methanol formation mechanism via CO_2 hydrogenation is complex and not yet fully understood [31]. Effective catalysts are required to reduce operating pressure while ensuring high methanol yields. Copper (Cu) is the most widely used catalyst due to its high activity at low temperatures, affordability, diverse oxidation states, and stable interactions with oxygen, which prevent surface poisoning and unstable intermediates [45–47]. Cu also exhibits strong interactions with other materials (e.g., Zn), improving catalyst stability and selectivity while lowering operational pressures [28,48]. However, Cu-based catalysts face challenges such as deactivation from impurities, a high activation barrier for H_2, and limited CO_2 activation [49,50]. To address these issues, supports, promoters, and bimetallic catalyst are often added to modulate the active sites and surface properties of Cu catalysts, improving their performance. For example, adding these materials enhances Cu's ability to dissociate H_2 and provides smaller particle sizes that increase the surface area available for reactions [51,52].

Kinetic models, such as the Eley–Rideal (ER) and Langmuir–Hinshelwood (LH) mechanisms, have been proposed to describe the methanol synthesis process. The formate mechanism is widely believed to dominate on Cu-based catalysts [53]. For instance, DFT calculations on $Cu/ZnO/Al_2O_3$ catalysts, particularly on the Cu (111) surface, showed that CO_2 hydrogenation occurs via the formate pathway ($CO_2^* \rightarrow HCOO^* \rightarrow HCOOH^* \rightarrow CH_3O_2^* \rightarrow CH_2O^* \rightarrow CH_3O^* \rightarrow CH_3OH^*$). Despite the thermodynamic preference for methanol production from CO, studies indicate that CO_2 hydrogenation contributes to approximately two-thirds of the methanol yield, contrary to the predictions of potential energy surface (PES) analysis, which suggest CO hydrogenation is more favorable. The promotion of CO_2 hydrogenation by CO* may involve its role as an H-donor, facilitating the formation of HCO* and enhancing CO_2 conversion [53].

Most recently, Cao et al. [54] investigated CO_2 hydrogenation to methanol on Cu-based catalysts using DFT calculations and kinetic modeling. Their work revealed that the formate (HCOO*) pathway is the dominant reaction mechanism, with a reaction rate four to six orders of magnitude higher than the carboxyl (COOH*) pathway (see Figure 9b). The rate-determining step involves the hydrogenation of HCOOH* to H_2COOH^*, leading to methanol via CH_3O^* intermediates. CO* acts primarily as a byproduct due to its high hydrogenation barrier.

3.2. Recent Advances in Cu-Based Catalyst for CO_2 Hydrogenation to Methanol

The conversion of CO_2 to methanol is a highly promising route, offering dual benefits of mitigating greenhouse gas emissions and reducing dependence on fossil fuels. The hydrogenation of CO_2 to methanol ($CO_2 + 3H_2 \rightleftharpoons CH_3OH + H_2O$) presents several challenges, including the formation of water vapor as a byproduct, which can negatively impact catalyst performance [44,55]. This is a significant difference compared to methanol synthesis from CO hydrogenation, where no water is produced. Therefore, developing catalysts that maintain stability under high-water partial pressure and resist the reverse water–gas shift (RWGS) reaction remains a primary research focus [44]. Among the transition metal, Noble metals are limited and unfavorable for industrial applications, and non-noble metals like Ni or Fe are not effective due to reaction limitations. Based on our bibliometric analysis and effective catalytic activity of Cu-based catalyst, this review further explores the recent advancement in the Cu-based catalyst. The Cu-based catalysts for CO_2 to methanol, along with the recently reported reaction promoters, are summarized in Table 1.

Historically, the first industrial process for methanol synthesis was developed in the 1920s by Badische Anilin Soda and Fabrik (BASF), using $ZnO–Cr_2O_3$ catalysts under harsh reaction conditions (300–400 °C, 25–30 MPa). In the 1960s, Imperial Chemical Industries (ICI) introduced the $Cu/ZnO/Al_2O_3$ catalyst, which enabled methanol synthesis under milder conditions (220–300 °C, 5–10 MPa). This catalyst quickly became the industry standard due to its higher activity and stability [44,56,57]. However, despite its success, $Cu/ZnO/Al_2O_3$ catalysts are not without limitations, as the water produced during the reaction can facilitate the agglomeration of ZnO and the oxidation of Cu^0 to Cu^{2+}, leading to catalyst deactivation [58].

Recent research has focused on developing highly effective Cu-based catalysts that demonstrate better activity, selectivity, water tolerance, and stability. Cu and ZnO remain the primary active components in these catalysts, often supported by various modifiers and metal oxides such as Al_2O_3, ZrO_2, and CeO_2 [1,3,59–61]. These catalysts typically operate at 220–300 °C and 5–10 MPa, with varying degrees of conversion, selectivity, and space–time yield (STY) depending on the composition and preparation method [59,62]. Cu-based catalysts are the most extensively studied systems for CO_2 hydrogenation to methanol. They are widely recognized for their high activity, operational longevity, and resistance to poisoning under industrial conditions. The primary mechanism for methanol synthesis involves the direct hydrogenation of CO_2, although the RWGS reaction ($CO_2 + H_2 \rightleftharpoons CO + H_2O$) often competes with methanol formation, which can lower methanol yield [63,64]. Over the years, scientists have developed a wide range of catalysts to mitigate this issue, focusing primarily on Cu-based systems supported by metal oxides [65].

The primary strategy for improving Cu-based catalysts has been the incorporation of metal oxides such as ZnO, Al_2O_3, and ZrO_2, which act as carriers and promoters, optimizing the interaction between Cu, the support and bimetallic/alloy [66]. The addition of ZrO_2, for instance, has been shown to enhance methanol selectivity by stabilizing reaction intermediates like formate and methoxide [63,67]. Additionally, CeO_2, with its unique redox properties, has proven to be an excellent promoter, increasing the number of oxygen vacancies and improving CO_2 activation [68]. Research trends over the past decade show that Cu-based catalysts continue to dominate, with approximately 79% of publications on CO_2 hydrogenation to methanol focusing on Cu–ZnO systems [69,70]. Among these, the addition of Al_2O_3 and ZrO_2 remains the most popular approach for enhancing catalyst performance.

3.2.1. Role of Supports or Support Effect in Cu-Based Catalysts

Selecting an appropriate support is essential for optimizing catalytic activity, stability, and selectivity for CO_2 hydrogenation to methanol. Originally, supports were used primarily to ensure well-dispersed metal particles and reduce sintering [71,72]. However, it has become clear that the SMSI effect can significantly influence catalyst morphology, chemical states, and activity levels [73,74].

ZnO is one of the most studied supports due to its dual function as a structural and electronic promoter in Cu-based catalysts. ZnO acts as a geometric spacer between Cu NPs, enhancing Cu dispersion and increasing the exposed Cu surface area [75,76]. Additionally, ZnO modulates electronic properties through SMSI with Cu, impacting the electron density at the Cu–ZnO interface. Tsang et al. investigated these morphology-dependent electronic interactions, finding that the polar (002) facet in plate-like ZnO, with higher oxygen defect density, increases electron transfer from ZnO to Cu, resulting in methanol selectivity above 70% in Cu/plate–ZnO/Al_2O_3 catalysts [48]. Further enhancements in electron density were demonstrated with the encapsulation of CdSe quantum dots in ZnO rods, leading to an impressive 75% methanol selectivity [77].

Most recently, Xuan et al. [78] explored the impact of solvent choice on the preparation and catalytic performance of CuO–ZnO–ZrO_2–Al_2O_3 (CZZA) catalysts using the citrate complexing method for CO_2 hydrogenation to methanol. The improved results are attributed to the better dispersion of active Cu species and the increased oxygen vacancy content, which are critical for CO_2 activation and methanol synthesis. The CZZA catalyst achieved a CO_2 conversion rate of 21.8% and a methanol selectivity of 51.0% under optimal reaction conditions (240 °C, 3 MPa). The study also underscores the formate pathway as the reaction mechanism and highlights hydrogen spillover as a pivotal factor in the reaction.

ZrO_2 is recognized for its high thermal stability, mechanical strength, and specific surface area, making it a preferred support for Cu-based catalysts in CO_2 hydrogenation. Comparative studies have shown that Cu/ZrO_2-based catalysts outperform Cu/Al_2O_3 counterparts due to ZrO_2's lower hydrophilicity and its ability to facilitate a dual-site reaction mechanism for CO_2 adsorption and hydrogenation. Grabowski et al. observed that catalysts with a higher proportion of tetragonal ZrO_2 (t-ZrO_2) phase, stabilized by oxygen vacancies, demonstrated increased methanol production rates by enhancing the availability of Cu^+ ions at the surface [79]. Similarly, nanocrystalline ZrO_2, with more surface edges and defects, was shown to strengthen Cu–ZrO_2 interactions, enhancing CO_2 adsorption and promoting catalyst reduction [70].

Spinel-structured supports, such as $ZnFe_2O_4$, enable fine control over Cu nanoparticle size and distribution, influencing CO adsorption and selectivity. Recent studies have demonstrated that Cu/$ZnFe_2O_4$ spinel catalysts with a 33Cu/ZnFe-0.5 configuration achieved a methanol selectivity of 71.6% and CO_2 conversion of 9.4% at 260 °C and 4.5 MPa, attributed to the SMSI effect and the stabilization of ZnO adjacent to surface Cu nanoparticles [80–82]. Despite their effectiveness, spinel catalysts typically have lower surface areas due to high calcination temperatures. However, microwave-hydrothermal synthesis methods have improved their surface area and thermal stability, allowing the preparation of Cu-based catalysts with stable microstructures and smaller particle sizes [80,83].

Metal-organic frameworks (MOFs) offer a novel approach by stabilizing metal NPs within their porous structures, providing unique catalytic properties. For instance, Cu-based catalysts encapsulated within Zr-based MOF UiO-66 achieve 100% methanol selectivity and an eightfold increase in catalytic activity compared to traditional Cu/ZnO/Al_2O_3 catalysts with the help of SMSI effects between Zr oxide secondary building units and Cu NPs [84]. This structure enables efficient H_2 activation by metallic Cu and stabilizes intermediates, enhancing both reaction rates and stability [85,86].

Commonly used supports like SiO_2 provide large surface areas but suffer from thermal instability at elevated temperatures, limiting their long-term applicability in methanol synthesis due to their transformation into less stable forms in the presence of water vapor [54]. Despite these limitations, advanced support materials such as carbon-based carriers, MOFs, and structured metal oxides continue to expand the potential for durable and efficient methanol synthesis from CO_2.

The role of supports in Cu-based catalysts for CO_2 hydrogenation to methanol is multifaceted, encompassing structural, electronic, and chemical effects. By optimizing the choice of support material, researchers can significantly enhance the catalytic performance, selectivity, and stability of Cu-based systems. The use of advanced supports, such as mixed oxides, spinel structures, and rare earth oxides, has opened new avenues for improving methanol synthesis, offering better resistance to deactivation and higher methanol yields. Moving forward, continued research into the design of tailored support materials will be key to further advancements in this field.

3.2.2. Role of Promoters in Cu-Based Catalysts

Promoters in Cu-based catalysts for CO_2 hydrogenation enhance both structural and electronic properties, optimizing catalytic activity, stability, and methanol selectivity. By influencing factors such as metal dispersion, redox properties, and acid-base interactions, promoters enable more efficient pathways for CO_2 reduction to methanol [87].

Alkali metals, particularly potassium (K), act as promoters in Cu–Zn–Al catalysts by modifying the Cu^+/u^0 ratio, which is critical for enhanced CO_2 hydrogenation activity.

For instance, potassium-promoted Cu–Zn–Al (CZA-K) catalysts show a higher Cu^+/u^0 ratio than sodium-promoted systems, which correlates with improved methanol selectivity. The increased Cu^+ concentration is associated with the formation of K-O-(CO)-O surface species, which limits the reverse water-gas shift (RWGS) reaction and increases methanol production [88]. This effect arises because the K-O species inhibit CO_2 dissociation, thereby promoting methanol over CO formation.

Ga_2O_3 is extensively studied as a promoter in Cu–ZnO-based catalysts due to its ability to increase specific surface area, reduce particle size, and stabilize Cu^+ species. These structural and electronic modifications contribute to a higher methanol yield. Studies have shown that Ga_2O_3 in Cu–ZnO/HZSM-5 catalysts enhance the dispersion of Cu and stabilizes intermediate Cu^+ states, which are essential for CO_2 hydrogenation [32,55,89,90]. For instance, Schumann et al. observed that Ga^{3+} incorporation increases ZnO conductivity and generates redox-active defect sites, resulting in improved methanol synthesis rates [91]. Furthermore, Ga_2O_3 facilitates the formation of a $ZnGa_2O_4$ spinel phase, enabling type II heterojunctions in ZnO–Ga_2O_3 mixtures, which drive the reduction of ZnO and the formation of CuZn alloys, both of which are highly active for methanol synthesis [48].

Promotion of Indium (In) has demonstrated significant effects in Cu/CeO_2 catalysts, enhancing methanol yields by promoting smaller Cu particle sizes and improving dispersion and stability. Loading only 1 wt% indium in Cu/CeO_2 resulted in marked improvements in the methanol yield rate due to the enhanced stability and distribution of Cu on the catalyst surface [92]. The reduction in particle size and improved distribution likely results from In's role in preventing Cu agglomeration, which stabilizes the catalyst's active sites during CO_2 hydrogenation.

Rare earth metals, such as lanthanum (La), calcium (Ca), and cerium (Ce), have shown substantial promotional effects on $Cu/ZnO/ZrO_2$ catalysts. La_2O_3, for example, not only increases Cu dispersion but also introduces oxygen vacancies by partially substituting Zr^{4+} with La^{4+}, thereby enhancing CO_2 adsorption and methanol selectivity up to 72% [93]. Additionally, the enhanced basicity from La_2O_3 increases the number of basic sites, favoring formate adsorption and hydrogenation, which are crucial for methanol formation. Recent studies using La-modified CZA catalysts demonstrated up to a 30% increase in CH_3OH production, attributed to the improved porosity and surface area, which facilitate intermediate adsorption and CO_2 hydrogenation steps [94,95].

Recent literature [14,18,19] has explored the significant influence of catalyst design strategies and support interactions, as evidenced by the varied results across systems such as Cu/TiO_2 and Cu-$Ca_{0.8}La_{0.2}TiO_3$. Cu/TiO_2-600, prepared using a lattice confinement strategy, exhibited superior performance with a methanol selectivity of 55.5% and a CO_2 conversion of 45.2% under reaction conditions of 240 °C, 3 MPa, and a feed ratio of $CO_2/H_2/N_2 = 1/3/1$. This high efficiency is attributed to the enhanced dispersion of Cu species within the TiO_2 lattice, which not only improves sintering resistance but also maintains a high density of active sites for CO_2 activation and methanol synthesis. In comparison, Cu/TiO_2-500, operated under harsher conditions of 300 °C and 4 MPa, showed lower methanol selectivity (43.3%) and CO_2 conversion (12.5%), suggesting that the catalytic efficiency is sensitive to both the reaction environment and the interplay between Cu and its support. In this system, the role of oxygen vacancies in TiO_2 was found to be crucial for activating CO_2, but excessive metal-support interaction (SMSI) likely hindered catalytic activity by partially covering Cu active sites, underscoring the importance of balancing metal-support interactions for optimal performance. On the other hand, Cu-$Ca_{0.8}La_{0.2}TiO_3$, a perovskite-structured catalyst, displayed a well-balanced performance with a methanol selectivity of 58.5% and a CO_2 conversion of 22.5% at 300 °C and 3 MPa, using a higher hydrogen-rich feed ratio of $CO_2/H_2/N_2 = 6/18/1$. The incorporation of

La into the perovskite lattice enhanced the dispersion of Cu species and optimized the ratio of Cu^0/Cu^+, with Cu^0 playing a pivotal role in H_2 activation and Cu^+ facilitating methanol synthesis. This dual functionality, combined with the stability of the perovskite structure, enabled efficient CO_2 activation and selectivity toward methanol. Comparatively, the results indicate that while lattice confinement strategies in Cu/TiO_2-600 maximize methanol selectivity and thermal stability, the Cu–$Ca_{0.8}La_{0.2}TiO_3$ system strikes a balance between selectivity and conversion by leveraging perovskite properties and active site tuning. These findings collectively highlight the need for precise control over catalyst composition and support interactions to achieve high-efficiency methanol production in CO_2 hydrogenation processes.

Noble metals, including palladium (Pd) and gold (Au), have been used to boost hydrogenation rates in Cu-based catalysts by facilitating H_2 spillover. This mechanism enhances reducibility, improves surface hydrogenation reactions, and stabilizes Cu sites, resulting in increased methanol production [20,96–98]. For instance, Martin et al. reported that Au acts as an electron-withdrawing agent, promoting electronic transfer from ZnO to Cu and stabilizing Cu^0 species against reoxidation, thus enhancing methanol yield by increasing resistance to CO_2 or H_2O-induced oxidation [99].

Interestingly, methanol itself may act as a promoter in Cu/ZnO/MgO catalysts. The presence of a small amount of methanol in the reaction feed has been shown to reduce the apparent activation energy from 117.9 kJ/mol to 67.9 kJ/mol, suggesting that methanol interacts with the catalyst in a way that lowers the energy barrier for CO_2 hydrogenation. Although the precise mechanism of this effect remains unclear, it may relate to methanol's interaction with active sites, enhancing reaction kinetics under certain conditions.

The study by He et al. [100] investigates the catalytic performance of Mg-modified Cu–ZnO–ZrO_2 catalysts with a co-precipitation method for CO_2 hydrogenation to methanol, the incorporation of MgO into the Cu–ZnO–ZrO_2 system optimized the catalyst's microstructure, enhanced the dispersion of active species, and introduced appropriate basic sites, which significantly improved the adsorption and activation of CO_2. Among the catalysts tested, the optimized Cu–ZnO–ZrO_2/MgO ($CZZ_{0.8}M_{0.2}$) achieved a CO_2 conversion of 7.3% and methanol selectivity of 71.8% under reaction conditions of 220 °C, 3 MPa. This performance represents nearly double the conversion rate compared to unmodified Cu–ZnO–ZrO_2 catalysts. The MgO as a promoter has showed significant potential for improving CO_2 hydrogenation catalysts, making $CZZ_{0.8}M_{0.2}$ a highly promising candidate for efficient methanol production.

In summary, the incorporation of promoters, whether alkali metals, metal oxides, rare earth elements, or noble metals, significantly impacts the performance of Cu-based catalysts in CO_2 hydrogenation to methanol. Each promoter provides unique benefits, such as improved Cu dispersion, stabilization of reactive intermediates, and enhanced electronic properties, all of which contribute to greater methanol yields and selectivity. The strategic use of these promoters thus forms a critical approach in designing more effective and sustainable catalysts for methanol synthesis from CO_2.

Table 1. Cu-based catalysts for CO_2 to methanol.

Catalysts	T (°C)	P (MPa)	Gas Composition	Space Velocity ($mL \cdot g^{-1} h^{-1}$)	CO_2 Conversion (%)	CH_3OH Selectivity (%)	Ref.
Cu-ZrO_2	230	1	$CO_2/H_2/N_2$ = 1/3/1	50,000	1.6	72.2	[101]
Cu/ZrO_2	260	3	$CO_2/H_2/N_2$ = 23/69/8	6000	17.1	58.5	[102]
Cu-Zn-Zr/$CuBr_2$	250	5	$CO_2/H_2/N_2$ = 23/69/8	3000	10.7	97.1	[103]
Cu/TiO_2-600	240	3	$CO_2/H_2/N_2$ = 1/3/1	3600	45.2	55.5	[18]
Cu/TiO_2-500	300	4	$CO_2/H_2/N_2$ = 1/3/1	10,000	12.5	43.3	[14]

Table 1. *Cont.*

Catalysts	T (°C)	P (MPa)	Gas Composition	Space Velocity ($mL{\cdot}g^{-1}\ h^{-1}$)	CO_2 Conversion (%)	CH_3OH Selectivity (%)	Ref.
Cu-$Ca_{0\cdot8}La_{0\cdot2}TiO_3$	300	3	$CO_2/H_2/N_2$ = 6/18/1	3000	22.5	58.5	[19]
0.5% Cu-Zn-Zr	290	4.5	CO_2/H_2 = 1/3	10,800	9.5	76	[104]
Cu-Zn-Al-Ce	250	3	CO_2/H_2 = 1/3	12,000	14.2	37.8	[105]
20% ZnO-ZrO_2	320	2	CO_2/H_2 = 24/72	24,000	6.4	78.5	[63]
inverse-ZrO_2/Cu	220	3	CO_2/H_2 = 1/3	48,000	<5	~70	[65]
Cu@ZnO-1.0	240	3	CO_2/H_2 = 1/3	12,000	19.6	76.9	[61]
CuO/$Ce_{0.4}$ $Zr_{0.6}$ O_2	220	3	CO_2/H_2 = 1/3	10,000	7	96.4	[66]
Cu-Zn-Al-K	240	3	CO_2/H_2 = 1/4	2400	14	96	[106]
Cu-ZnO-$A1_2O_3$	260	3.6	CO_2/H_2 = 1/10	18,000	65.8	77.3	[107]
Cu-ZnO-$A1_2O_3$	280	4	CO_2/H_2 = 1/3	10,000	65.3	91.9	[108]
Cu-ZnO	250	3	CO_2/H_2 = 1/3	18,000	2.3	100	[109]
Cu-ZnO-ZrO_2	200	1	$CO_2/H_2/N_2$ = 3/9/1	4000	5.8	55.2	[110]
Cu-ZnO-ZrO_2	230	3	-	3000	15.2	35.1	[111]
Cu-ZnO-ZrO_2/MgO	220	3	$CO_2/H_2/N_2$ = 23/69/8	18,000	7.3	71.8	[100]
Cu-ZnO-ZrO_2	240	3	CO_2/H_2 = 1/3	3600	12.1	54.1	[112]
Cu/ZrO_2	280	3	CO_2/H_2 = 1/3	7200	12	32	[113]
30Cu-ZnO-ZrO_2	280	5	-	10,000	23	33	[114]
Fe-Cu-ZnO-ZrO_2	250	3	CO_2/H_2 = 1/4	60,000	18.7	53.8	[115]
Cu-Ga_2O_3-ZrO_2	250	2	CO_2/H_2 = 1/3	2500	13.7	75.6	[116]
7.7 Ga_2O_3/IE/Cu/ZrO_2	250	3	CO_2/H_2 = 22/75	20,000	1.3	74	[117]
Cu/Al_2O_3	280	9.5	CO_2/H_2 = 1/4	12,000	30	80	[118]
$LaO_{0.8}Zr_{0.2}Cu_{0.7}Zn_{0.3}Ox$	250	5	CO_2/H_2 = 1/3	3600	12.6	52.5	[119]
30Cu-ZnO-ZrO_2	280	5	-	10,000	21	34	[120]
Cu-ZnO-ZrO_{20}-Al_2O_3	240	3	$CO_2/H_2/N_2$ = 23.5/64.5/12	2400	21.8	51	[78]

3.2.3. Role of Alloy in Cu-Based Catalysts

The Cu/ZnO/Al_2O_3 (CZA) system has been commercially employed for over 60 years, but the exact nature of its active sites remains a topic of debate. The active sites for CO_2 to methanol conversion in these catalysts are believed to be located at the metal-oxide interfaces (Cu–ZnO) and surface Cu–Zn alloys (Alloys). Nakamura and Fujitani first provided evidence using XRD that Cu–Zn alloys form in Cu/ZnO catalysts, which significantly improve the specific activity and indicate that these alloys are the active sites for methanol synthesis [121]. Other researchers, such as Malte Behrens, observed through HRTEM that Cu nanoparticles in these systems are covered by disordered ZnOx layers, forming a surface Cu–Zn alloy. DFT calculations further confirmed that these surface alloying sites can enhance methanol synthesis [122].

Further support comes from Topsøe researchers, who used H_2-TPD and XPS to provide definitive evidence of surface Cu–Zn alloy formation [123]. Their findings suggest that increased H_2 partial pressure leads to more Zn in the metallic state, while DFT calculations indicated that Cu–Zn alloying weakens H_2 adsorption, promoting methanol production [124]. The group also proposed that ZnO particle size plays a key role in controlling Zn coverage on Cu, affecting the overall catalytic activity [59].

In contrast, a team at Brookhaven National Laboratory provided evidence that the metal-oxide interface (ZnO–Cu) is the primary active site. They found that Zn/Cu catalysts initially exhibited low activity but were gradually oxidized to form ZnO/Cu interfaces during the reaction, achieving high activity levels comparable to directly prepared ZnO/Cu systems [125].

Palladium (Pd) has been introduced into Cu–ZnO catalysts to increase methanol yield, not by creating new active sites, but by promoting H_2 spillover and stabilizing active Cu sites, preventing their oxidation by CO_2 [50]. Pd and Cu can form stable alloys after reduction, which has been demonstrated to provide significant synergistic effects for methanol synthesis [126]. For instance, Chunshan Song's team developed a series of SiO_2 supported Cu–Pd bimetallic catalysts, showing that an optimized atomic ratio of Pd/(Pd + Cu) between 0.25 and 0.34 doubled the methanol yield compared to monometallic Cu and Pd catalysts [126,127]. Their results indicated that Cu–Pd alloy nanoparticles increase the surface coverage of formate species, a key intermediate in methanol synthesis. Furthermore, PdCu-enriched catalysts outperformed PdCu3-enriched catalysts in methanol production [128].

The effect of support materials on the performance of Cu–Pd catalysts has also been investigated. Supports such as TiO_2, ZrO_2, and Al_2O_3 were found to enhance methanol production by creating moderate MSI. In contrast, CeO_2, which created stronger MSI, hindered the formation of Cu–Pd alloys and led to competitive adsorption of H_2 and CO_2, resulting in poorer performance [129].

Although CZA catalysts perform well, their efficiency in CO_2 hydrogenation can be negatively affected by water formation and the hydrophilic nature of alumina [110,126]. As a result, ZrO_2 has been introduced as a less hydrophilic promoter to improve catalytic performance [130]. Studies have shown that adding ZrO_2 increases Cu dispersion, enhancing methanol selectivity [111]. Various synthesis methods have been developed to create Cu–ZnO–ZrO_2 catalysts, including surfactant-assisted coprecipitation, which enhances the interaction between Cu and Zr, resulting in higher methanol selectivity [131]. Other studies have focused on optimizing the crystallographic structure of ZrO_2 to increase catalytic activity [79]. For example, tetrahedral ZrO_2 has been found to be more beneficial than amorphous forms for methanol synthesis [113].

In addition to ZrO_2, Ga_2O_3 and CeO_2 have also been used as bimetallic promoters. Ga_2O_3 has been shown to improve catalytic performance by increasing SACu and enhancing the Cu–ZnO interaction [114], while CeO_2 provides new sites for CO_2 activation, significantly increasing methanol production [132]. Furthermore, ternary Cu/Ga_2O_3/ZrO_2 systems have outperformed binary Cu–ZrO_2 and Cu–Ga_2O_3 catalysts, particularly when specific preparation conditions are followed [117].

4. Conclusions

The conversion of CO_2 to methanol is a pivotal strategy for addressing global carbon emissions while enabling the production of a versatile chemical and energy carrier. This review integrates a comprehensive bibliometric analysis with a detailed examination of thermodynamic constraints, reaction mechanisms, and advances in catalyst development. The bibliometric analysis, encompassing over 13,000 publications from 2012 to 2023, highlights the significant growth in research on CO_2-to-methanol technologies, particularly the central role of Cu-based catalysts. Notably, 9361 of these publications focus specifically on Cu-based catalysts, underscoring the growing interest in their potential for CO_2 conversion. The analysis underscores China's leadership in the field, along with substantial contributions from the United States, Germany, and other nations, reflecting the global urgency to address climate change.

Cu-based catalysts, enhanced through doping with elements such as ZrO_2, CeO_2, and Ga_2O_3, and innovations in composite supports like Cu–ZnO–ZrO_2–Al_2O_3, have emerged as the cornerstone for industrial-scale CO_2 hydrogenation. These systems balance activity, stability, and cost-effectiveness, addressing the challenges posed by the reverse water-gas shift reaction, water-induced deactivation, and the thermodynamic stability of CO_2. The review also explores advanced catalytic strategies like perovskite-type oxides, highlighting their potential to overcome existing limitations and unlock new avenues for catalyst design. Emerging materials, including metal-organic frameworks and bimetallic alloys, offer further opportunities for optimizing reaction pathways and improving selectivity under industrial conditions.

Further challenges and outlook for industrial applications are discussed in the next section. Addressing these challenges requires an interdisciplinary approach that integrates renewable energy sources, innovative reactor designs, and policy-driven incentives to enable sustainable industrial adoption. By consolidating current knowledge and identifying critical gaps, this review not only provides a roadmap for future research but also emphasizes the transformative potential of CO_2-to-methanol conversion in realizing a circular carbon economy. Continued innovation and collaboration across disciplines will be essential in transitioning this technology from the laboratory to large-scale implementation, making it a cornerstone of sustainable energy and carbon management.

5. Challenges and Outlook

While significant strides have been made in advancing CO_2 hydrogenation to methanol, several technical and economic challenges remain unresolved, limiting the scalability and industrial adoption of this promising technology. The activation of CO_2 remains a primary obstacle, driven by the molecule's inherent thermodynamic stability, which necessitates substantial energy input. In Cu-based catalytic systems, this challenge is compounded by the weak interaction between CO_2 and the catalyst surface, leading to a reliance on high operational pressures—up to 200 bar—to drive reactions forward. This reliance significantly increases energy consumption and operational costs, forming a barrier to large-scale commercialization. Addressing this requires innovative approaches to catalyst design. Emerging strategies, such as the integration of machine learning and quantum simulations, offer the potential to predict and engineer advanced catalytic materials with enhanced CO_2 affinity, enabling activation at lower pressures and temperatures. By optimizing the electronic and structural properties of catalysts, such as doping Cu-based systems with promoters that create active surface sites for CO_2 adsorption, future research can directly target these bottlenecks.

In addition to activation challenges, the reverse water-gas shift (RWGS) reaction poses a persistent issue, particularly at elevated temperatures where it competes with methanol production. The RWGS reaction not only reduces methanol selectivity but also wastes valuable hydrogen, undermining process efficiency. To overcome this, catalyst systems must be designed to suppress RWGS while favoring methanol formation under milder conditions. Recent advances in nanostructured and multifunctional catalysts have shown promise in achieving this balance, particularly through the stabilization of intermediates that favor methanol synthesis pathways. Further research should explore the synergy between catalyst structure and reaction engineering, such as optimizing reactor configurations to enable selective intermediate stabilization. For example, reactors that facilitate in situ water removal could minimize the impact of water as a byproduct, which otherwise deactivates Cu-based catalysts by promoting agglomeration and oxidation of active sites. Enhancing the resistance of catalysts to water-induced deactivation through surface modifications or the incorporation of hydrophobic supports is another critical avenue for improving process longevity and efficiency.

Scaling up CO_2 to methanol conversion remains a key challenge due to the economic and logistical barriers associated with CO_2 capture and hydrogen production. While small-scale demonstration plants, such as Carbon Recycling International's facility in Iceland, have validated the feasibility of this technology, achieving economic viability at an industrial scale demands substantial advancements. Current CO_2 capture methods, whether from industrial emissions or direct air capture (DAC) systems, are energy-intensive and costly, with DAC technologies exhibiting low technological readiness levels. Innovations in capture materials and processes, such as adsorption-based systems with higher efficiency or bioengineered solutions leveraging microorganisms for CO_2 fixation, could significantly lower costs. Coupling these advancements with decentralized production hubs that integrate CO_2 capture and methanol synthesis at emission sources offers a practical pathway to scalability.

Hydrogen production remains another critical bottleneck. Conventional methods like steam methane reforming and biomass gasification generate carbon byproducts, undercutting the environmental benefits of methanol synthesis. Electrolysis, powered by renewable energy sources such as wind or geothermal energy, provides a more sustainable alternative, as demonstrated by the geothermal-powered production model in Iceland. However, the high costs and relatively low efficiencies of current electrolysis systems, typically ranging between 40–60%, limit their widespread adoption. Increasing the efficiency of electrolysis through advances in membrane technologies, such as solid oxide electrolysis cells, and scaling renewable energy sources to supply hydrogen at competitive costs, are essential for the long-term viability of CO_2-to-methanol conversion. Policy incentives and investment in renewable energy infrastructure will also play a pivotal role in reducing the economic barriers to green hydrogen production.

Catalyst development continues to be a central focus in advancing CO_2 hydrogenation to methanol, addressing the limitations of Cu-based systems that, while effective and affordable, are prone to deactivation under industrial conditions, particularly in the presence of water. Recent progress has demonstrated the potential of composite supports, bimetallic systems, and structural modifications to overcome these challenges and enhance catalyst performance. Composite supports such as Cu–ZnO–ZrO_2–Al_2O_3 have shown improved stability by optimizing the dispersion of active Cu species and resisting sintering and deactivation, critical for sustained industrial applications. Bimetallic catalysts, such as Pd–Cu or Ga–Cu systems, exploit synergistic interactions between metals to stabilize intermediates, improve reaction kinetics, and suppress competing reactions like the reverse water-gas shift (RWGS). Structural innovations, including perovskite-type oxides like $Ca_{0.8}La_{0.8}TiO_3$, introduce dynamic lattice oxygen and defect sites that facilitate CO_2 activation and enhance hydrogenation efficiency, while layered double hydroxides (LDHs) offer a high density of tunable active sites and robust thermal stability, creating a supportive environment for Cu nanoparticles. These advances collectively aim to enhance catalyst durability, activity, and selectivity under practical operating conditions, addressing key industrial requirements. Moving forward, hybrid systems that integrate the complementary benefits of these approaches hold significant promise, and their validation through pilot-scale studies will be essential to establish their feasibility and performance under real-world conditions, paving the way for resilient and efficient catalytic technologies for CO_2 hydrogenation.

Finally, the economic viability of methanol production from CO_2 remains a critical hurdle. Although market demand for methanol is growing, the cost of production through CO_2 hydrogenation currently exceeds that of conventional methods due to the expenses associated with CO_2 capture and hydrogen generation. Integrating methanol synthesis plants with renewable energy sources and industries emitting high CO_2 volumes could reduce costs and improve economic feasibility. Additionally, exploring the valorization of methanol into higher-value products, such as olefins or specialty chemicals, could provide supplementary revenue streams. Participation in carbon credit markets through Certified Emissions Reduction (CER) programs may also offset production costs, further incentivizing industrial adoption.

The successful commercialization of CO_2-to-methanol technologies will ultimately depend on a confluence of scientific innovation, strategic integration of renewable energy, and supportive policy frameworks. By setting measurable performance targets—such as achieving methanol selectivity above 90% at pressures below 50 bar—and fostering interdisciplinary collaborations, the field can overcome its current limitations. As catalyst systems evolve and production processes are optimized, CO_2 hydrogenation to methanol has the potential to transition from a laboratory-scale innovation to a cornerstone of sustainable carbon management and green energy production.

Author Contributions: Conceptualization, S.S. and M.A.M.; Methodology, S.S., M.A.M. and S.A.M.; Writing—original draft preparation, S.S. and M.A.M.; Writing—review and editing, S.S., M.A.M., A.K., S.A.M., D.M., S.A., M., W.Z. and Y.L.; Visualization, S.S. and M.A.M.; Supervision, M.A.M. and Y.L. All authors have read and agreed to the published version of the manuscript.

Funding: This research received no external funding.

Data Availability Statement: No new data were created or analyzed in this study.

Conflicts of Interest: The authors declare no conflicts of interest.

References

1. Memon, M.A.; Jiang, Y.; Hassan, M.A.; Ajmal, M.; Wang, H.; Liu, Y. Heterogeneous Catalysts for Carbon Dioxide Methanation: A View on Catalytic Performance. *Catalysts* **2023**, *13*, 1514. [CrossRef]
2. Memon, M.A.; Zhou, W.; Ajmal, M.; Afzal; Jiang, Y.; Zhang, C.; Zhang, J.; Liu, Y. Ni–$CaZrO_3$ with perovskite phase loaded on ZrO_2 for CO_2 methanation. *Int. J. Hydrogen Energy* **2024**, *92*, 1202–1213. [CrossRef]
3. Liu, R.; Song, P.; Memon, M.A.; Liu, P.; Liu, Y. Manganese Oxide Decorated Co Nanoparticles on $CaTiO_3$ for Higher Alcohols Synthesis from Syngas. *Ind. Eng. Chem. Res.* **2024**, *63*, 16111–16123. [CrossRef]
4. Gao, P.; Zhang, L.; Li, S.; Zhou, Z.; Sun, Y. Novel heterogeneous catalysts for CO_2 hydrogenation to liquid fuels. *ACS Cent. Sci.* **2020**, *6*, 1657–1670. [CrossRef] [PubMed]
5. Zhong, J.; Yang, X.; Wu, Z.; Liang, B.; Huang, Y.; Zhang, T. State of the art and perspectives in heterogeneous catalysis of CO_2 hydrogenation to methanol. *Chem. Soc. Rev.* **2020**, *49*, 1385–1413. [CrossRef]
6. Durga Devi, A.; Pushpavanam, S.; Singh, N.; Verma, J.; Kaur, M.P.; Roy, S.C. Enhanced methane yield by photoreduction of CO_2 at moderate temperature and pressure using Pt coated, graphene oxide wrapped TiO_2 nanotubes. *Results Eng.* **2022**, *14*, 100441. [CrossRef]
7. Cubeiro, M.L.; Morales, H.; Goldwasser, M.R.; Pérez-Zurita, M.J.; González-Jiménez, F.; Urbina De N, C. Hydrogenation of carbon oxides over Fe/Al_2O_3 catalysts. *Appl. Catal. A Gen.* **1999**, *189*, 87–97. [CrossRef]
8. Sakurai, H.; Ueda, A.; Kobayashi, T.; Haruta, M. Low-temperature water-gas shift reaction over gold deposited on TiO_2. *Chem. Commun.* **1997**, 271–272. [CrossRef]
9. Gao, P.; Dang, S.; Li, S.; Bu, X.; Liu, Z.; Qiu, M.; Yang, C.; Wang, H.; Zhong, L.; Han, Y.; et al. Direct Production of Lower Olefins from CO_2 Conversion via Bifunctional Catalysis. *ACS Catal.* **2018**, *8*, 571–578. [CrossRef]
10. Studt, F.; Sharafutdinov, I.; Abild-Pedersen, F.; Elkjær, C.F.; Hummelshøj, J.S.; Dahl, S.; Chorkendorff, I.; Nørskov, J.K. Discovery of a Ni-Ga catalyst for carbon dioxide reduction to methanol. *Nat. Chem.* **2014**, *6*, 320–324. [CrossRef] [PubMed]
11. Ali, K.A.; Abdullah, A.Z.; Mohamed, A.R. Recent development in catalytic technologies for methanol synthesis from renewable sources: A critical review. *Renew. Sustain. Energy Rev.* **2015**, *44*, 508–518. [CrossRef]
12. Jadhav, S.G.; Vaidya, P.D.; Bhanage, B.M.; Joshi, J.B. Catalytic carbon dioxide hydrogenation to methanol: A review of recent studies. *Chem. Eng. Res. Des.* **2014**, *92*, 2557–2567. [CrossRef]
13. Olah, G.A. Towards oil independence through renewable methanol chemistry. *Angew. Chemie—Int. Ed.* **2013**, *52*, 104–107. [CrossRef]
14. Zhang, C.; Wang, L.; Etim, U.J.; Song, Y.; Gazit, O.M.; Zhong, Z. Oxygen vacancies in Cu/TiO_2 boost strong metal-support interaction and CO_2 hydrogenation to methanol. *J. Catal.* **2022**, *413*, 284–296. [CrossRef]
15. Yang, X.; Li, S.; Wang, L.; Liu, Z. Transition metal catalysts for CO_2 hydrogenation: Mechanistic insights and performance trends. *Catal. Today* **2023**, *423*, 101–116. [CrossRef]
16. Sun, J.; Fujimoto, K. Economic feasibility of Cu-based catalysts for industrial-scale CO_2-to-methanol synthesis. *Ind. Eng. Chem. Res.* **2017**, *56*, 6060–6069. [CrossRef]

17. Gao, B.; Wen, Z.; Wang, Y.; Chen, D.; Yang, B.; Ishihara, T.; Guo, L. Recent Advances in Alloy Catalysts for CO_2 Hydrogenation to Methanol. *ChemCatChem* **2024**, *16*, e202400814. [CrossRef]
18. Chen, H.; Li, S.; Ma, P.; Chang, K.; Zhao, Z.; Xie, Z. Lattice-confined Cu-TiO_2 catalysts with improved thermal stability for CO_2 hydrogenation. *ACS Sustain. Chem. Eng.* **2023**, *11*, 18112–18122. [CrossRef]
19. Zhou, W.; Memon, M.A.; Niu, M.; Fan, X.; Jiang, Y.; Zhang, X.; Liu, Y. Active pairs of Cu0/Cu+ for CO_2 hydrogenation. *Int. J. Hydrogen Energy* **2024**. [CrossRef]
20. Xie, X.; Li, C.; Lu, Z.; Wang, Y.; Yang, W.; Chen, M.; Li, W. Noble metal modified copper-exchanged mordenite zeolite (Cu-ex-MOR) catalysts for catalyzing the methane efficient gas-phase synthesis methanol. *Energy* **2024**, *300*, 131595. [CrossRef]
21. Wu, C.; Cheng, D.; Wang, M.; Ma, D. Understanding and Application of Strong Metal-Support Interactions in Conversion of CO_2 to Methanol: A Review. *Energy Fuels* **2021**, *35*, 19012–19023. [CrossRef]
22. Guil-López, R.; Mota, N.; Llorente, J.; Millán, E.; Pawelec, B.; Fierro, J.L.G.; Navarro, R.M. Methanol Synthesis from CO_2: A Review of the Latest Developments in Heterogeneous Catalysis. *Materials* **2019**, *12*, 3902. [CrossRef]
23. Rodriguez, J.A.; Liu, P.; Stacchiola, D.J.; Senanayake, S.D.; White, M.G.; Chen, J.G. Hydrogenation of CO_2 to Methanol: Importance of Metal-Oxide and Metal-Carbide Interfaces in the Activation of CO_2. *Acs Catal.* **2015**, *5*, 6696–6706. [CrossRef]
24. Bowker, M. Methanol Synthesis from CO_2 Hydrogenation. *ChemCatChem* **2019**, *11*, 4238–4246. [CrossRef] [PubMed]
25. Álvarez, A.; Bansode, A.; Urakawa, A.; Bavykina, A.V.; Wezendonk, T.A.; Makkee, M.; Gascon, J.; Kapteijn, F. Challenges in the Greener Production of Formates/Formic Acid, Methanol, and DME by Heterogeneously Catalyzed CO_2 Hydrogenation Processes. *Chem. Rev.* **2017**, *117*, 9804–9838. [CrossRef] [PubMed]
26. Ganesh, I. Conversion of carbon dioxide into methanol—A potential liquid fuel: Fundamental challenges and opportunities (a review). *Renew. Sustain. Energy Rev.* **2014**, *31*, 221–257. [CrossRef]
27. Bart, J.C.J.; Sneeden, R.P.A. Copper-zinc oxide-alumina methanol catalysts revisited. *Catal. Today* **1987**, *2*, 1–124. [CrossRef]
28. Studt, F.; Behrens, M.; Kunkes, E.L.; Thomas, N.; Zander, S.; Tarasov, A.; Schumann, J.; Frei, E.; Varley, J.B.; Abild-Pedersen, F.; et al. The Mechanism of CO and CO_2 Hydrogenation to Methanol over Cu-Based Catalysts. *ChemCatChem* **2015**, *7*, 1105–1111. [CrossRef]
29. Murthy, P.S.; Liang, W.; Jiang, Y.; Huang, J. Cu-Based Nanocatalysts for CO_2 Hydrogenation to Methanol. *Energy and Fuels* **2021**, *35*, 8558–8584. [CrossRef]
30. Aresta, M.; Dibenedetto, A.; Quaranta, E. State of the art and perspectives in catalytic processes for CO_2 conversion into chemicals and fuels: The distinctive contribution of chemical catalysis and biotechnology. *J. Catal.* **2016**, *343*, 2–45. [CrossRef]
31. Wang, W.; Wang, S.; Ma, X.; Gong, J. Recent advances in catalytic hydrogenation of carbon dioxide. *Chem. Soc. Rev.* **2011**, *40*, 3703–3727. [CrossRef] [PubMed]
32. Cai, W.; Ramirez, P.; Piscina, D.; Toyir, J.; Homs, N. CO_2 hydrogenation to methanol over CuZnGa catalysts prepared using microwave-assisted methods. *Catal. Today* **2015**, *242*, 193–199. [CrossRef]
33. Yang, Y.; White, M.G.; Liu, P. Theoretical Study of Methanol Synthesis from CO_2 Hydrogenation on Metal-Doped Cu(111) Surfaces. *J. Phys. Chem. C* **2011**, *116*, 248–256. [CrossRef]
34. Park, N.; Park, M.-J.; Lee, Y.-J.; Ha, K.-S.; Jun, K.-W. Kinetic modeling of methanol synthesis over commercial catalysts based on three-site adsorption. *Fuel Process. Technol.* **2014**, *125*, 139–147. [CrossRef]
35. Cui, X.; Kær, S.K. A comparative study on three reactor types for methanol synthesis from syngas and CO_2. *Chem. Eng. J.* **2020**, *393*, 124632. [CrossRef]
36. Zhang, S.; Wu, Z.; Liu, X.; Hua, K.; Shao, Z.; Wei, B.; Huang, C.; Wang, H.; Sun, Y. A Short Review of Recent Advances in Direct CO_2 Hydrogenation to Alcohols. *Top. Catal.* **2021**, *64*, 371–394. [CrossRef]
37. Azhari, N.J.; Erika, D.; Mardiana, S.; Ilmi, T.; Gunawan, M.L.; Makertihartha, I.G.B.N.; Kadja, G.T.M. Methanol synthesis from CO_2: A mechanistic overview. *Results Eng.* **2022**, *16*, 100711. [CrossRef]
38. Yarulina, I.; Chowdhury, A.D.; Meirer, F.; Weckhuysen, B.M.; Gascon, J. Recent trends and fundamental insights in the methanol-to-hydrocarbons process. *Nat. Catal.* **2018**, *1*, 398–411. [CrossRef]
39. Ilias, S.; Bhan, A. Mechanism of the Catalytic Conversion of Methanol to Hydrocarbons. *ACS Catal.* **2012**, *3*, 18–31. [CrossRef]
40. Sehested, J. Industrial and scientific directions of methanol catalyst development. *J. Catal.* **2019**, *371*, 368–375. [CrossRef]
41. Behrens, M.; Armbrüster, M. Methanol Steam Reforming. In *Catalysis for Alternative Energy Generation*; Springer: New York, NY, USA, 2012; pp. 175–235. ISBN 9781461403449.
42. Mahyuddin, M.H.; Shiota, Y.; Yoshizawa, K. Methane selective oxidation to methanol by metal-exchanged zeolites: A review of active sites and their reactivity. *Catal. Sci. Technol.* **2019**, *9*, 1744–1768. [CrossRef]
43. Porosoff, M.D.; Yan, B.; Chen, J.G. Catalytic reduction of CO_2 by H_2 for synthesis of CO, methanol and hydrocarbons: Challenges and opportunities. *Energy Environ. Sci.* **2016**, *9*, 62–73. [CrossRef]
44. Yang, H.; Zhang, C.; Gao, P.; Wang, H.; Li, X.; Zhong, L.; Wei, W.; Sun, Y. A review of the catalytic hydrogenation of carbon dioxide into value-added hydrocarbons. *Catal. Sci. Technol.* **2017**, *7*, 4580–4598. [CrossRef]

45. Rodemerck, U.; Holeňa, M.; Wagner, E.; Smejkal, Q.; Barkschat, A.; Baerns, M. Catalyst Development for CO_2 Hydrogenation to Fuels. *ChemCatChem* **2013**, *5*, 1948–1955. [CrossRef]
46. Koh, M.K.; Khavarian, M.; Chai, S.P.; Mohamed, A.R. The morphological impact of siliceous porous carriers on copper-catalysts for selective direct CO_2 hydrogenation to methanol. *Int. J. Hydrogen Energy* **2018**, *43*, 9334–9342. [CrossRef]
47. Koh, M.K.; Wong, Y.J.; Chai, S.P.; Mohamed, A.R. Carbon dioxide hydrogenation to methanol over multi-functional catalyst: Effects of reactants adsorption and metal-oxide(s) interfacial area. *J. Ind. Eng. Chem.* **2018**, *62*, 156–165. [CrossRef]
48. Li, M.M.; Zeng, Z.; Liao, F.; Hong, X.; Chi, S.; Tsang, E. Enhanced CO_2 hydrogenation to methanol over CuZn nanoalloy in Ga modified Cu / ZnO catalysts. *J. Catal.* **2016**, *343*, 157–167. [CrossRef]
49. Lucci, F.R.; Marcinkowski, M.D.; Lawton, T.J.; Sykes, E.C.H. H_2 Activation and Spillover on Catalytically Relevant Pt–Cu Single Atom Alloys. *J. Phys. Chem. C* **2015**, *119*, 24351–24357. [CrossRef]
50. Hu, B.; Yin, Y.; Liu, G.; Chen, S.; Hong, X.; Tsang, S.C.E. Hydrogen spillover enabled active Cu sites for methanol synthesis from CO_2 hydrogenation over Pd doped CuZn catalysts. *J. Catal.* **2018**, *359*, 17–26. [CrossRef]
51. Tursunov, O.; Kustov, L.; Kustov, A. A Brief Review of Carbon Dioxide Hydrogenation to Methanol Over Copper and Iron Based Catalysts. *Oil Gas Sci. Technol.—Rev. d'IFP Energies Nouv.* **2017**, *72*, 30. [CrossRef]
52. Gesmanee, S.; Koo-amornpattana, W. Catalytic hydrogenation of CO_2 for methanol production in fixed-bed reactor using Cu-Zn supported on gamma-Al 2 O 3. *Energy Procedia* **2017**, *138*, 739–744. [CrossRef]
53. Grabow, L.C.; Mavrikakis, M. Mechanism of Methanol Synthesis on Cu through CO_2 and CO Hydrogenation. *ACS Catal.* **2011**, *1*, 365–384. [CrossRef]
54. Cao, H.S.; Li, S.; Pan, Y.X.; Zhang, X.B.; Luo, Z.H. Exploring the reaction mechanism and kinetic properties of CO_2 hydrogenation to methanol on Cu/CeO_2. *Chem. Eng. Sci.* **2025**, *304*, 120990. [CrossRef]
55. Toyir, J.; Ramírez De La Piscina, P.; Fierro, J.L.G.; Homs, N. Highly effective conversion of CO_2 to methanol over supported and promoted copper-based catalysts: Influence of support and promoter. *Appl. Catal. B Environ.* **2001**, *29*, 207–215. [CrossRef]
56. Hu, J.; Yu, L.; Deng, J.; Wang, Y.; Cheng, K.; Ma, C.; Zhang, Q.; Wen, W.; Yu, S.; Pan, Y.; et al. Sulfur vacancy-rich MoS2 as a catalyst for the hydrogenation of CO_2 to methanol. *Nat. Catal.* **2021**, *4*, 242–250. [CrossRef]
57. Chen, Y.; Wei, J.; Duyar, M.S.; Ordomsky, V.V.; Khodakov, A.Y.; Liu, J. Carbon-based catalysts for Fischer-Tropsch synthesis. *Chem. Soc. Rev.* **2021**, *50*, 2337–2366. [CrossRef] [PubMed]
58. Xu, X.; Tong, Y.; Huang, J.; Zhu, J.; Fang, X.; Xu, J.; Wang, X. Insights into CO_2 methanation mechanism on cubic ZrO_2 supported Ni catalyst via a combination of experiments and DFT calculations. *Fuel* **2021**, *283*, 118867. [CrossRef]
59. Kuld, S.; Thorhauge, M.; Falsig, H.; Elkjær, C.F.; Helveg, S.; Chorkendorff, I.; Sehested, J. Quantifying the promotion of Cu catalysts by ZnO for methanol synthesis. *Science* **2016**, *352*, 969–974. [CrossRef]
60. Posada-Pérez, S.; Ramírez, P.J.; Gutiérrez, R.A.; Stacchiola, D.J.; Viñes, F.; Liu, P.; Illas, F.; Rodriguez, J.A. The conversion of CO_2 to methanol on orthorhombic β-Mo2C and Cu/β-Mo2C catalysts: Mechanism for admetal induced change in the selectivity and activity. *Catal. Sci. Technol.* **2016**, *6*, 6766–6777. [CrossRef]
61. Kondrat, S.A.; Smith, P.J.; Wells, P.P.; Chater, P.A.; Carter, J.H.; Morgan, D.J.; Fiordaliso, E.M.; Wagner, J.B.; Davies, T.E.; Lu, L.; et al. Stable amorphous georgeite as a precursor to a high-activity catalyst. *Nature* **2016**, *531*, 83–87. [CrossRef]
62. Behrens, M.; Girgsdies, F. Structural Effects of Cu/Zn Substitution in the Malachite–Rosasite System. *Zeitschrift für Anorg. und Allg. Chemie* **2010**, *636*, 919–927. [CrossRef]
63. Han, Z.; Tang, C.; Sha, F.; Tang, S.; Wang, J.; Li, C. CO_2 hydrogenation to methanol on ZnO-ZrO_2 solid solution catalysts with ordered mesoporous structure. *J. Catal.* **2021**, *396*, 242–250. [CrossRef]
64. Guo, X.; Mao, D.; Lu, G.; Wang, S.; Wu, G. CO_2 hydrogenation to methanol over Cu/ZnO/ZrO_2 catalysts prepared via a route of solid-state reaction. *Catal. Commun.* **2011**, *12*, 1095–1098. [CrossRef]
65. Wu, C.; Lin, L.; Liu, J.; Zhang, J.; Zhang, F.; Zhou, T.; Rui, N.; Yao, S.; Deng, Y.; Yang, F.; et al. Inverse ZrO_2/Cu as a highly efficient methanol synthesis catalyst from CO_2 hydrogenation. *Nat. Commun.* **2020**, *11*, 5767. [CrossRef]
66. Wang, W.; Qu, Z.; Song, L.; Fu, Q. Effect of the nature of copper species on methanol synthesis from CO_2 hydrogenation reaction over CuO/Ce0.4Zr0.6O_2 catalyst. *Mol. Catal.* **2020**, *493*, 111105. [CrossRef]
67. Singh, R.; Tripathi, K.; Pant, K.K. Investigating the role of oxygen vacancies and basic site density in tuning methanol selectivity over Cu/CeO_2 catalyst during CO_2 hydrogenation. *Fuel* **2021**, *303*, 121289. [CrossRef]
68. Gothe, M.L.; Pérez-Sanz, F.J.; Braga, A.H.; Borges, L.R.; Abreu, T.F.; Bazito, R.C.; Gonçalves, R.V.; Rossi, L.M.; Vidinha, P. Selective CO_2 hydrogenation into methanol in a supercritical flow process. *J. CO_2 Util.* **2020**, *40*, 101195. [CrossRef]
69. Jiang, F.; Yang, Y.; Wang, L.; Li, Y.; Fang, Z.; Xu, Y.; Liu, B.; Liu, X. Dependence of copper particle size and interface on methanol and CO formation in CO_2 hydrogenation over Cu@ZnO catalysts. *Catal. Sci. Technol.* **2022**, *12*, 551–564. [CrossRef]
70. Lam, E.; Corral-Pérez, J.J.; Larmier, K.; Noh, G.; Wolf, P.; Comas-Vives, A.; Urakawa, A.; Copéret, C. CO_2 Hydrogenation on Cu/Al_2O_3: Role of the Metal/Support Interface in Driving Activity and Selectivity of a Bifunctional Catalyst. *Angew. Chemie Int. Ed.* **2019**, *58*, 13989–13996. [CrossRef] [PubMed]

71. Klier, K. Methanol Synthesis. In *Advances in Catalysis*; Elsevier: Amsterdam, The Netherlands, 1982; pp. 243–313, ISBN 9780120078318.
72. Sankar, M.; He, Q.; Engel, R.V.; Sainna, M.A.; Logsdail, A.J.; Roldan, A.; Willock, D.J.; Agarwal, N.; Kiely, C.J.; Hutchings, G.J. Role of the Support in Gold-Containing Nanoparticles as Heterogeneous Catalysts. *Chem. Rev.* **2020**, *120*, 3890–3938. [CrossRef]
73. Tauster, S.J.; Fung, S.C.; Baker, R.T.K.; Horsley, J.A. Strong Interactions in Supported-Metal Catalysts. *Science* **1981**, *211*, 1121–1125. [CrossRef] [PubMed]
74. Wang, L.; Yi, Y.; Guo, H.; Tu, X. Atmospheric Pressure and Room Temperature Synthesis of Methanol through Plasma-Catalytic Hydrogenation of CO_2. *ACS Catal.* **2017**, *8*, 90–100. [CrossRef]
75. Liao, F.; Huang, Y.; Ge, J.; Zheng, W.; Tedsree, K.; Collier, P.; Hong, X.; Tsang, S.C. Morphology-Dependent Interactions of ZnO with Cu Nanoparticles at the Materials' Interface in Selective Hydrogenation of CO_2 to CH_3OH. *Angew. Chemie Int. Ed.* **2011**, *50*, 2162–2165. [CrossRef]
76. Meunier, F.C. Mixing Copper Nanoparticles and ZnO Nanocrystals: A Route towards Understanding the Hydrogenation of CO_2 to Methanol? *Angew. Chem. Int. Ed.* **2011**, *50*, 4053–4054. [CrossRef] [PubMed]
77. Li, M.M.-J.; Tsang, S.C.E. Bimetallic catalysts for green methanol production via CO_2 and renewable hydrogen: A mini-review and prospects. *Catal. Sci. Technol.* **2018**, *8*, 3450–3464. [CrossRef]
78. Ling, X.; Wang, G.; Han, J.; Wang, L.; Yu, J.; Mao, D. Solvent effects on the preparation of CuO-ZnO-ZrO_2-Al_2O_3 catalyst by citrate complexing method for CO_2 hydrogenation to methanol. *Fuel* **2025**, *382*, 133653. [CrossRef]
79. Samson, K.; Śliwa, M.; Socha, R.P.; Góra-Marek, K.; Mucha, D.; Rutkowska-Zbik, D.; Paul, J.-F.; Ruggiero-Mikołajczyk, M.; Grabowski, R.; Słoczyński, J. Influence of ZrO_2 Structure and Copper Electronic State on Activity of Cu/ZrO_2 Catalysts in Methanol Synthesis from CO_2. *ACS Catal.* **2014**, *4*, 3730–3741. [CrossRef]
80. Conrad, F.; Massue, C.; Kühl, S.; Kunkes, E.; Girgsdies, F.; Kasatkin, I.; Zhang, B.; Friedrich, M.; Luo, Y.; Armbrüster, M.; et al. Microwave-hydrothermal synthesis and characterization of nanostructured copper substituted ZnM_2O_4 (M = Al, Ga) spinels as precursors for thermally stable Cu catalysts. *Nanoscale* **2012**, *4*, 2018. [CrossRef]
81. Liu, T.; Xu, D.; Wu, D.; Liu, G.; Hong, X. Spinel $ZnFe_2O_4$ Regulates Copper Sites for CO_2 Hydrogenation to Methanol. *ACS Sustain. Chem. Eng.* **2021**, *9*, 4033–4041. [CrossRef]
82. Kühl, S.; Schumann, J.; Kasatkin, I.; Hävecker, M.; Schlögl, R.; Behrens, M. Ternary and quaternary Cr or Ga-containing ex-LDH catalysts—Influence of the additional oxides onto the microstructure and activity of Cu/$ZnAl_2O_4$ catalysts. *Catal. Today* **2015**, *246*, 92–100. [CrossRef]
83. Conrad, F.; Zhou, Y.; Yulikov, M.; Hametner, K.; Weyeneth, S.; Jeschke, G.; Günther, D.; Grunwaldt, J.; Patzke, G.R. Microwave-Hydrothermal Synthesis of Nanostructured Zinc-Copper Gallates. *Eur. J. Inorg. Chem.* **2010**, *2010*, 2036–2043. [CrossRef]
84. Rungtaweevoranit, B.; Baek, J.; Araujo, J.R.; Archanjo, B.S.; Choi, K.M.; Yaghi, O.M.; Somorjai, G.A. Copper Nanocrystals Encapsulated in Zr-based Metal-Organic Frameworks for Highly Selective CO_2 Hydrogenation to Methanol. *Nano Lett.* **2016**, *16*, 7645–7649. [CrossRef] [PubMed]
85. Yang, Q.; Xu, Q.; Jiang, H.L. Metal-organic frameworks meet metal nanoparticles: Synergistic effect for enhanced catalysis. *Chem. Soc. Rev.* **2017**, *46*, 4774–4808. [CrossRef] [PubMed]
86. Kobayashi, H.; Taylor, J.M.; Mitsuka, Y.; Ogiwara, N.; Yamamoto, T.; Toriyama, T.; Matsumura, S.; Kitagawa, H. Charge transfer dependence on CO_2 hydrogenation activity to methanol in Cu nanoparticles covered with metal-organic framework systems. *Chem. Sci.* **2019**, *10*, 3289–3294. [CrossRef]
87. Frontera, P.; Macario, A.; Ferraro, M.; Antonucci, P. Supported Catalysts for CO_2 Methanation: A Review. *Catalysts* **2017**, *7*, 59. [CrossRef]
88. Pasupulety, N.; Driss, H.; Abobakor, Y.; Ahmed, A.; Daous, M.A.; Petrov, L. Applied Catalysis A: General Studies on Au / Cu–Zn–Al catalyst for methanol synthesis from CO_2. *Applied Catal. A Gen.* **2015**, *504*, 308–318. [CrossRef]
89. Zheng, J.; Lebedev, K.; Wu, S.; Huang, C.; Ayvall, T.; Wu, T.S.; Li, Y.; Ho, P.L.; Soo, Y.L.; Kirkland, A.; et al. High Loading of Transition Metal Single Atoms on Chalcogenide Catalysts. *J. Am. Chem. Soc.* **2021**, *143*, 7979–7990. [CrossRef]
90. Du, J.; Zhang, Y.; Wang, K.; Ding, F.; Jia, S.; Liu, G.; Tan, L. Investigation on the promotional role of Ga_2O_3 on the CuO–ZnO/HZSM-5 catalyst for CO_2 hydrogenation. *RSC Adv.* **2021**, *11*, 14426–14433. [CrossRef]
91. Schumann, J.; Eichelbaum, M.; Lunkenbein, T.; Thomas, N.; Consuelo, M.; Galva, A.; Schlo, R.; Behrens, M. Promoting Strong Metal Support Interaction: Doping ZnO for Enhanced Activity of Cu/ZnO:M (M = Al, Ga, Mg) Catalysts. *ACS Catal.* **2015**, *5*, 3260–3270. [CrossRef]
92. Sharma, S.K.; Paul, B.; Pal, R.S.; Bhanja, P.; Banerjee, A.; Samanta, C.; Bal, R. Influence of Indium as a Promoter on the Stability and Selectivity of the Nanocrystalline Cu/CeO_2 Catalyst for CO_2 Hydrogenation to Methanol. *ACS Appl. Mater. Interfaces* **2021**, *13*, 28201–28213. [CrossRef] [PubMed]
93. Guo, X.; Mao, D.; Lu, G.; Wang, S.; Wu, G. The influence of la doping on the catalytic behavior of Cu/ZrO_2 for methanol synthesis from CO_2 hydrogenation. *J. Mol. Catal. A Chem.* **2011**, *345*, 60–68. [CrossRef]

94. Ban, H.; Li, C.; Asami, K.; Fujimoto, K. Influence of rare-earth elements (La, Ce, Nd and Pr) on the performance of Cu/Zn/Zr catalyst for CH_3OH synthesis from CO_2. *Catal. Commun.* **2014**, *54*, 50–54. [CrossRef]
95. Kourtelesis, M.; Kousi, K.; Kondarides, D.I. CO_2 Hydrogenation to Methanol over La_2O_3-Promoted $CuO/ZnO/Al_2O_3$ Catalysts: A Kinetic and Mechanistic Study. *Catalysts* **2020**, *10*, 183. [CrossRef]
96. Jeong, E.; Hee, Y.; Lee, D.; Moon, D.; Lee, K. Hydrogenation of CO_2 to methanol over Pd–Cu/CeO_2 catalysts. *Mol. Catal.* **2023**, *434*, 146–153. [CrossRef]
97. Luk, H.T.; Mondelli, C.; Ferré, D.C.; Stewart, J.A.; Pérez-Ramírez, J. Status and prospects in higher alcohols synthesis from syngas. *Chem. Soc. Rev.* **2017**, *46*, 1358–1426. [CrossRef] [PubMed]
98. Bowker, M.; Lawes, N.; Gow, I.; Hayward, J.; Esquius, J.R.; Richards, N.; Smith, L.R.; Slater, T.J.A.; Davies, T.E.; Dummer, N.F.; et al. The Critical Role of βpdZn Alloy in Pd/ZnO Catalysts for the Hydrogenation of Carbon Dioxide to Methanol. *ACS Catal.* **2022**, *12*, 5371–5379. [CrossRef]
99. Martin, O.; Mondelli, C.; Curulla-ferre, D.; Drouilly, C.; Hauert, R.; Pe, J. Zinc-Rich Copper Catalysts Promoted by Gold for Methanol Synthesis. *ACS Catal.* **2015**, *5*, 5607–5616. [CrossRef]
100. He, Q.; Li, Z.; Li, D.; Ning, F.; Wang, Q.; Liu, W.; Zhang, W.; Cui, Y.; Zhang, J.; Liu, C. Mg enhanced the performance of Cu/ZnO/ZrO_2 for CO_2 hydrogenation to methanol and the mechanism investigation. *Mol. Catal.* **2024**, *558*, 114008. [CrossRef]
101. Oshima, K.; Honma, Y.; Kinoshita, K.; Gao, Z.; Honma, T.; Tada, S.; Satokawa, S. Mechanochemical Effect in Mixing Sponge Copper with Amorphous ZrO_2 Creates Effective Active Sites for Methanol Synthesis by CO_2 Hydrogenation. *J. Phys. Chem. C* **2021**, *125*, 8155–8162. [CrossRef]
102. Dai, W.; Meng, X.; Xu, B.; Zhao, R.; Jin, D.; Xu, F.; Yang, D.; Xin, Z. Effect of Reflux Time on the Performance of the Cu/ZrO_2 Catalyst for CO_2 Hydrogenation to Methanol. *ACS Appl. Energy Mater.* **2023**, *6*, 9417–9426. [CrossRef]
103. Chen, S.; Zhang, J.; Song, F.; Zhang, Q.; Yang, G.; Zhang, M.; Wang, X.; Xie, H.; Tan, Y. Induced high selectivity methanol formation during CO_2 hydrogenation over a CuBr2-modified CuZnZr catalyst. *J. Catal.* **2020**, *389*, 47–59. [CrossRef]
104. Xu, D.; Hong, X.; Liu, G. Highly dispersed metal doping to ZnZr oxide catalyst for CO_2 hydrogenation to methanol: Insight into hydrogen spillover. *J. Catal.* **2021**, *393*, 207–214. [CrossRef]
105. Mureddu, M.; Lai, S.; Atzori, L.; Rombi, E.; Ferrara, F.; Pettinau, A.; Cutrufello, M.G. Ex-ldh-based catalysts for CO_2 conversion to methanol and dimethyl ether. *Catalysts* **2021**, *11*, 615. [CrossRef]
106. Pasupulety, N.; Al-Zahrani, A.A.; Daous, M.A.; Podila, S.; Driss, H. A study on highly active Cu-Zn-Al-K catalyst for CO_2 hydrogenation to methanol. *Arab. J. Chem.* **2021**, *14*, 102951. [CrossRef]
107. Bansode, A.; Urakawa, A. Towards full one-pass conversion of carbon dioxide to methanol and methanol-derived products. *J. Catal.* **2014**, *309*, 66–70. [CrossRef]
108. Gaikwad, R.; Bansode, A.; Urakawa, A. High-pressure advantages in stoichiometric hydrogenation of carbon dioxide to methanol. *J. Catal.* **2016**, *343*, 127–132. [CrossRef]
109. Le Valant, A.; Comminges, C.; Tisseraud, C.; Canaff, C.; Pinard, L.; Pouilloux, Y. The Cu–ZnO synergy in methanol synthesis from CO_2, Part 1: Origin of active site explained by experimental studies and a sphere contact quantification model on Cu + ZnO mechanical mixtures. *J. Catal.* **2015**, *324*, 41–49. [CrossRef]
110. Arena, F.; Barbera, K.; Italiano, G.; Bonura, G.; Spadaro, L.; Frusteri, F. Synthesis, characterization and activity pattern of Cu-ZnO/ZrO_2 catalysts in the hydrogenation of carbon dioxide to methanol. *J. Catal.* **2007**, *249*, 185–194. [CrossRef]
111. Raudaskoski, R.; Niemelä, M.V.; Keiski, R.L. The effect of ageing time on co-precipitated Cu/ZnO/ZrO_2 catalysts used in methanol synthesis from CO_2 and H_2. *Top. Catal.* **2007**, *45*, 57–60. [CrossRef]
112. Frei, E.; Schaadt, A.; Ludwig, T.; Hillebrecht, H.; Krossing, I. The Influence of the Precipitation/Ageing Temperature on a Cu/ZnO/ZrO_2 Catalyst for Methanol Synthesis from H_2 and CO_2. *ChemCatChem* **2014**, *6*, 1721–1730. [CrossRef]
113. Witoon, T.; Chalorngtham, J.; Dumrongbunditkul, P.; Chareonpanich, M.; Limtrakul, J. CO_2 hydrogenation to methanol over Cu/ZrO 2 catalysts: Effects of zirconia phases. *Chem. Eng. J.* **2016**, *293*, 327–336. [CrossRef]
114. Angelo, L.; Kobl, K.; Tejada, L.M.M.; Zimmermann, Y.; Parkhomenko, K.; Roger, A.-C. Study of CuZnMOx oxides (M = Al, Zr, Ce, CeZr) for the catalytic hydrogenation of CO_2 into methanol. *Comptes Rendus. Chim.* **2015**, *18*, 250–260. [CrossRef]
115. Kanuri, S.; Singh, S.A.; Dinda, S. Prominence of Fe on Cu/ZnO/ZrO_2 catalyst for methanol synthesis from CO_2: Material preparation, performance demonstration, and kinetic analysis. *Chem. Eng. Sci.* **2024**, *286*, 119661. [CrossRef]
116. Liu, X.M.; Yan, Z.F.; Lu, G.Q. Role of nanosized zirconia on the properties of Cu/Ga2O 3/ZrO_2 catalysts for methanol synthesis. *Chinese J. Chem.* **2006**, *24*, 172–176. [CrossRef]
117. Fornero, E.L.; Sanguineti, P.B.; Chiavassa, D.L.; Bonivardi, A.L.; Baltanás, M.A. Performance of ternary Cu–Ga_2O_3–ZrO_2 catalysts in the synthesis of methanol using CO_2-rich gas mixtures. *Catal. Today* **2013**, *213*, 163–170. [CrossRef]
118. Tidona, B.; Koppold, C.; Bansode, A.; Urakawa, A.; Rudolf Von Rohr, P. CO_2 hydrogenation to methanol at pressures up to 950 bar. *J. Supercrit. Fluids* **2013**, *78*, 70–77. [CrossRef]

119. Zhan, H.; Li, F.; Gao, P.; Zhao, N.; Xiao, F.; Wei, W.; Zhong, L.; Sun, Y. Methanol synthesis from CO_2 hydrogenation over La-M-Cu-Zn-O (M = Y, Ce, Mg, Zr) catalysts derived from perovskite-type precursors. *J. Power Sources* **2014**, *251*, 113–121. [CrossRef]
120. Angelo, L.; Girleanu, M.; Ersen, O.; Serra, C.; Parkhomenko, K.; Roger, A.C. Catalyst synthesis by continuous coprecipitation under micro-fluidic conditions: Application to the preparation of catalysts for methanol synthesis from CO_2/H_2. *Catal. Today* **2016**, *270*, 59–67. [CrossRef]
121. Fujitani, T.; Saito, M.; Kanai, Y.; Watanabe, T.; Nakamura, J.; Uchijima, T. Development of an active Ga_2O_3 supported palladium catalyst for the synthesis of methanol from carbon dioxide and hydrogen. *Appl. Catal. A Gen.* **1995**, *125*, L199–L202. [CrossRef]
122. Behrens, M.; Studt, F.; Kasatkin, I.; Kühl, S.; Hävecker, M.; Abild-Pedersen, F.; Zander, S.; Girgsdies, F.; Kurr, P.; Kniep, B.-L.; et al. The Active Site of Methanol Synthesis over Cu/ZnO/Al 2 O 3 Industrial Catalysts. *Science* **2012**, *336*, 893–897. [CrossRef]
123. Nakamura, J.; Fujitani, T.; Kuld, S.; Helveg, S.; Chorkendorff, I.; Sehested, J. Comment on "Active sites for CO_2 hydrogenation to methanol on Cu/ZnO catalysts". *Science* **2017**, *357*, eaan8074. [CrossRef] [PubMed]
124. Muhler, M.; Nielsen, L.P.; Törnqvist, E.; Clausen, B.S.; Topsøe, H. Temperature-programmed desorption of H_2 as a tool to determine metal surface areas of Cu catalysts. *Catal. Letters* **1992**, *14*, 241–249. [CrossRef]
125. Amann, P.; Klötzer, B.; Degerman, D.; Köpfle, N.; Götsch, T.; Lömker, P.; Rameshan, C.; Ploner, K.; Bikaljevic, D.; Wang, H.Y.; et al. The state of zinc in methanol synthesis over a Zn/ZnO/Cu(211) model catalyst. *Science* **2022**, *376*, 603–608. [CrossRef]
126. Jiang, X.; Koizumi, N.; Guo, X.; Song, C. Bimetallic Pd–Cu catalysts for selective CO_2 hydrogenation to methanol. *Appl. Catal. B Environ.* **2015**, *170–171*, 173–185. [CrossRef]
127. Nie, X.; Li, W.; Jiang, X.; Guo, X.; Song, C. Chapter Two—Recent advances in catalytic CO_2 hydrogenation to alcohols and hydrocarbons. In *Advances in Catalysis*; Song, C., Ed.; Academic Press: Cambridge, MA, USA, 2019; Volume 65, pp. 121–233.
128. Nie, X.; Jiang, X.; Wang, H.; Luo, W.; Janik, M.J.; Chen, Y.; Guo, X.; Song, C. Mechanistic Understanding of Alloy Effect and Water Promotion for Pd-Cu Bimetallic Catalysts in CO_2 Hydrogenation to Methanol. *ACS Catal.* **2018**, *8*, 4873–4892. [CrossRef]
129. Lin, F.; Jiang, X.; Boreriboon, N.; Wang, Z.; Song, C.; Cen, K. Effects of supports on bimetallic Pd-Cu catalysts for CO_2 hydrogenation to methanol. *Appl. Catal. A Gen.* **2019**, *585*, 117210. [CrossRef]
130. Wambach, J.; Baiker, A.; Wokaun, A. CO_2 hydrogenation over metal/zirconia catalysts. *Phys. Chem. Chem. Phys.* **1999**, *1*, 5071–5080. [CrossRef]
131. Li, L.; Mao, D.; Yu, J.; Guo, X. Highly selective hydrogenation of CO_2 to methanol over CuO–ZnO–ZrO_2 catalysts prepared by a surfactant-assisted co-precipitation method. *J. Power Sources* **2015**, *279*, 394–404. [CrossRef]
132. Graciani, J.; Mudiyanselage, K.; Xu, F.; Baber, A.E.; Evans, J.; Senanayake, S.D.; Stacchiola, D.J.; Liu, P.; Hrbek, J.; Sanz, J.F.; et al. Highly active copper-ceria and copper-ceria-titania catalysts for methanol synthesis from CO_2. *Science* **2014**, *345*, 546–550. [CrossRef] [PubMed]

processes

MDPI

Review

A Compact Review of Current Technologies for Carbon Capture as Well as Storing and Utilizing the Captured CO_2

Tim M. Thiedemann * and Michael Wark *

Institute of Chemistry, Chemical Technology 1, Carl von Ossietzky Universität, 26129 Oldenburg, Germany
* Correspondence: tim.thiedemann@uol.de (T.M.T.); michael.wark@uol.de (M.W.)

Abstract: With the consequences of climate change becoming more urgent, there has never been a more pressing need for technologies that can help to reduce the carbon dioxide (CO_2) emissions of the most polluting sectors, such as power generation, steel, cement, and the chemical industry. This review summarizes the state-of-the-art technologies for carbon capture, for instance, post-combustion, pre-combustion, oxy-fuel combustion, chemical looping, and direct air capture. Moreover, already established carbon capture technologies, such as absorption, adsorption, and membrane-based separation, and emerging technologies like calcium looping or cryogenic separation are presented. Beyond carbon capture technologies, this review also discusses how captured CO_2 can be securely stored (CCS) physically in deep saline aquifers or depleted gas and oil reservoirs, stored chemically via mineralization, or used in enhanced oil recovery. The concept of utilizing the captured CO_2 (CCU) for producing value-added products, including formic acid, methanol, urea, or methane, towards a circular carbon economy will also be shortly discussed. Real-life applications, e.g., already pilot-scale continuous methane (CH_4) production from flue gas CO_2, are shown. Actual deployment of the most crucial technologies for the future will be explored in real-life applications. This review aims to provide a compact view of the most crucial technologies that should be considered when choosing to capture, store, or convert CO_2, informing future researchers with efforts aimed at mitigating CO_2 emissions and tackling the climate crisis.

Keywords: CCSU; CCU; CCS; CO_2; BECCS; carbon capture; post-combustion; pre-combustion; oxy-fuel combustion; chemical looping; DAC

1. Introduction

From 2011 to 2020, average global temperatures increased by 1.1 °C compared to pre-industrial level in 1850–1900. By 2021–2040, it is likely that the 1.5 °C level will be reached or even exceeded if our annual emissions remain the same. A rise up to even 2 °C is possible if we do not take action on a larger scale [1]. With the rise in global temperatures, we will face an imbalance in our unique ecological balance. Impacts become visible through extreme weather, rising sea levels, and ecological disruptions [2]. Total greenhouse gas (GHG) emissions, typically from carbon dioxide (CO_2), methane (CH_4), fluorinated gases (FG), and nitrous oxide (N_2O), are the fundamental drivers of global temperature rise. The CO_2 emissions are, in fact, the largest of all these GHGs [3]. The Global Carbon Project estimates global CO_2 emissions for 2022 at 37.1 (±2) Gt. Emissions were expected to hit 37.5 Gt in 2023, a 1.1% increase compared to 2022 [3]. Reducing these emissions has become one of our greatest challenges. To tackle the rising temperatures, the European Union declared 2019 an ambitious goal to have net-zero emissions by 2050 and decoupled

https://doi.org/10.3390/pr13010283

economic growth from resource usage [4]. However, reducing emissions is difficult because the underlying causes are deeply embedded in complex infrastructure, which includes many sectors, such as energy, transportation, production, and agriculture. These sectors have been developed and optimized over decades. Thus, transitioning to something more sustainable is difficult, and long-term strategies can be costly. In addition, political and economic issues easily complicate GHG mitigation. The public interest and the industry's capabilities must be in harmony. Otherwise, this often causes prolonged negotiations and delayed implementation of policies and regulations. In addition, many developing nations, particularly China and India, are experiencing rapid economic development, leading to a large demand for coal as the main energy resource [5]. As more and more industrial output is produced to meet demand from such countries, a massive ramp-up in CO_2 emissions can be seen. For example, China produced 11.4 Gt CO_2 in 2022, 30.73% of all global CO_2 emissions. Industrial countries, such as the USA and the European Union, emitted 2022, 5.1, and 2.8 Gt CO_2, respectively. Yet, concerning the CO_2 emissions per capita, China released 6.9 Gt CO_2 less into the atmosphere than the USA last year [3].

The main goal of reducing GHG emissions, particularly CO_2 emissions, should be finding urgent solutions to curb emissions without disrupting economic growth and industrial practices in industrial and developing countries. We should invest in reliable, cost-effective, and already-established strategies. More specifically, the focus should be on technologies that can be easily scaled [6], have high and efficient CO_2 fixation capabilities [7], have a low energy penalty for each ton of CO_2 captured [8,9], and have low investment and operating costs [10,11]. Hereby, it heavily depends on the industry sector whether a technology is considered efficient [12]. For high-temperature processes, waste heat can be applied to carbon capture processes, lowering the additional energy demand [13]. In contrast, technologies with low energy consumption are interesting for power-generating industries such as power plants [14–16].

This work provides a compact overview of possible CO_2 capture technologies, and the primary techniques used, like pre- and post-combustion, oxy-fuel combustion, and chemical looping for capturing flue gas CO_2. The application of direct air capture to capture airborne CO_2 is also presented. The technologies will be discussed in terms of their advancements and limitations. In addition to carbon capture, carbon storage (CCS), utilization (CCU), and bio-based technologies (BECCSUs) will be assessed in terms of feasibility and today's application. The review will not discuss other important factors in CO_2 reduction, e.g., limiting deforestation processes or indirect carbon capture methods. Nevertheless, those factors play an important role and contribute to the big picture.

2. Current Challenges of Reducing CO_2 from Flue Gases

To reduce emissions, capturing CO_2 from flue gases, e.g., power plants or chemical production plants, plays an important role. It is even bigger if the power plant is fairly big and generates energy for many production sites or households. However, in addition to political challenges, there are a number of key challenges in CO_2 capture from flue gas, identified in the literature [17–21], which are summarized below. The main goal of making the process of carbon capture and storage more practical should be to address these challenges.

2.1. High Energy Consumption and Carbon Penalty

A big challenge in CO_2 capture from flue gases is the high energy demand associated with the capture process. Much of the technology employed today for CO_2 capture is based on solvent-based systems, which consume large amounts of energy in both the absorption and regeneration steps. The energy used in this process is generally produced by the combustion of extra fossil fuels, which increases the overall CO_2 emissions of the facility and causes what can be described as a "carbon penalty". In other words, the energy needed for CO_2 capture can diminish the benefit of capturing the gas if it raises the emissions. Thus, it becomes very significant to lower the energy intensity of CO_2 capture processes to make the technology more effective and efficient at net emission reduction.

2.2. High Initial Capital Costs or Operational Costs

The deployment and operation of CO_2 capture systems come with significant capital investments in the beginning and associated continuous operational costs. Infrastructure developments, including gas treatment systems, capturing units, and storage facilities for CO_2, require a huge financial investment. That is why the higher cost may greatly influence the feasibility of CO_2 capture projects and hinder large-scale applications. In addition to the initial capital costs, there are also continuing operating costs due to energy requirements, e.g., solvent regeneration, maintenance of complex systems, and monitoring of CO_2 emissions. These financial strains represent a strong disincentive to the widespread adoption of CO_2 capture technologies unless full economic incentives, such as governmental funding or carbon credit programs, are provided.

2.3. Low CO_2 Purity in Flue Gases and Inefficient Capture Rates

The quite low CO_2 concentration in flue gases poses a significantly greater challenge to the efficiency of CO_2 capture. The flue gas basically contains all sorts of impurities, like NO_x, SO_x, moisture, and soot, which complicate the process of capture. These components lower the rate of capture and lower the purity of the captured CO_2, thereby making it difficult to use or store it afterward. Therefore, it is of great importance to develop new technologies for the removal of these additional substances to improve the overall efficiency of the CO_2 capture systems. Advanced membrane or filtration technologies can be used to remove unwanted impurities from the gas stream before the actual capture of CO_2. This will not only produce high-purity CO_2 but also increase the efficiency of the whole process.

2.4. Solvent Degradation and Environmental Concerns

Another problem that exists in CO_2 capture is the degradation and environmental effects of solvents involved in the process. Solvents, like amines, undergo chemical changes during the process of absorbing CO_2 and should be replaced or renewed after certain periods of time. Partial degradation of these solvents can further cause the release of harmful byproducts, either inflicting environmental damage or requiring costly disposal measures. Many of the solvents used can contaminate soil and water if not disposed of properly. The development of less harmful and stable solvents that degrade less easily is an important step toward sustainable CO_2 capture.

2.5. Scalability and Infrastructure Requirements

Scalability of technologies designed for CO_2 capture is a big challenge. They must be flexible enough to be used in small-scale industrial facilities and large-scale power generation plants. While CO_2 capture has been demonstrated effectively in pilot plants, the question of how to scale up these technologies to an industrial level without losing efficiency or driving up costs remains very relevant.

In addition, infrastructure for the storage or processing of large amounts of captured CO_2 is required. The general transportation of CO_2 via pipelines to its place of storage or being used in other processes requires substantial infrastructure development. Moreover, the wide-scale deployment of CO_2 capture technologies will need massive financial investment in infrastructure and logistical solutions to ensure CO_2 is stored or utilized in a safe and efficient manner.

2.6. Uncertain Long-Term Storage Solutions

After capturing, the CO_2 has to be stored somewhere long-term. This is one of the uncertain aspects of CCS technologies. Geological storage in deep saline aquifers or depleted oil/gas fields seems to be promising, but many questions remain unanswered about the long-term safety of such storage and the possibility of CO_2 leakage over time. This poses some real risks, as CO_2 may escape from underground storage sites for decades or many centuries. In addition, the effectiveness and stability of CO_2 storage reservoirs are still under study. Establishing and maintaining public confidence in CO_2 storage will require much stricter regulation, continued monitoring, and transparent reporting to show that storage sites remain secure over time.

3. CO_2 from Flue Gases

Power plants burn fuel, e.g., coal, to produce primary electricity and some heat. Fuel combustion with air results in a flue gas with a low concentration of CO_2 because of a high proportion of nitrogen gas from the air. However, for coal, combustion can also produce other components that occur in the flue gas stream, such as nitrogen dioxide (NO_2), sulfur dioxide (SO_2), and soot particles. To capture the CO_2 from the resulting flue gas stream, a gas purification step can be required. This ensures that the subsequent carbon capture steps are neither damaged nor rendered inefficient. Typical flue gas streams of natural gas or coal-fired power plants consist mostly of about 8% and 14% CO_2 by volume, respectively. In the cement industry, the kiln off-gas can have a concentration of CO_2 of up to 33% by volume. For other applications, such as partial oxidation of natural gas or coal combustion with pure O_2, CO_2 concentration can reach up to 40% or >80% by volume, respectively [22].

3.1. Post-Combustion Route (PCCC)

Post-combustion carbon capture (PCCC) of CO_2 is a conventional approach that is mature for many industrial applications (Figure 1).

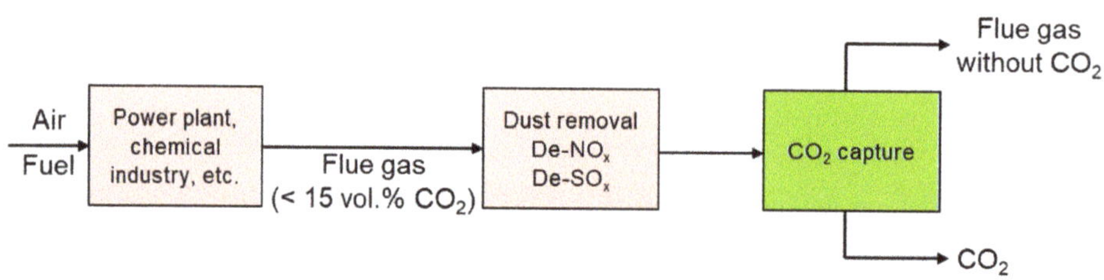

Figure 1. Simplified scheme for the post-combustion route for carbon capture (based on [17]).

In this method, CO_2 is captured from the flue gases produced after combustion, as typically found in facilities like power plants. Even though this is a very promising application, the concentration of CO_2 in flue gases can be very low (less than 15 vol.% for coal-based power plants). Thus, the economic capture of CO_2 remains very challenging from a technical perspective. Such a dilute concentration also diminishes the thermodynamic driving force of the capture process [23]. One of the greatest selling points of the post-combustion route is its ability to retrofit existing power plants, making it a stepping stone towards a greener path without new infrastructure. This flexibility makes post-combustion capture a preferable option for numerous existing facilities seeking a process to lower their CO_2 emissions. A current operational example of a retrofitted post-combustion coal-fired power plant is the Boundary Dam Power Station, owned by SaskPower in Canada. It is the world's first integrated and continuously used post-combustion CCS facility in a coal-fired power plant [24]. The plant consists of five units with a total power capacity of 531 MW.

In 2014, unit #3, with a power capacity of 120 MW, was retrofitted to capture around 1 Mt CO_2 per year. The captured CO_2 is then used for enhanced oil recovery (EOR) or stored underground in deep saline aquifers [24,25]. Future possibilities for CCS application include retrofitted coal-fired power plants in countries with high CO_2 emissions, highly installed coal-fired power plant capacities, and additional storage opportunities near such facilities. In 2023, these conditions were investigated for several Asian countries because of their high portion of the global CO_2 emissions. The study concludes that India and China have the highest CO_2 removal possibility using retrofitted power plants. Of the 520 power plants screened in all countries, 305 (59%) passed the screening. These coal-fired power plants have a total capacity of 378 GW; if 90% CO_2 capture is implemented, 1.55 Gt CO_2 per year can be captured and stored in designated areas around the power plants [26].

In addition to the already established methods, there are numerous other techniques for post-combustion CO_2 capture. Among these are emerging approaches that show promising potential for greater cost efficiency.

3.1.1. Absorption via Fluid Absorbents

The absorption technology is a widely used and common separation method for industry applications in the petroleum, coal, natural gas power plants, and chemical industries for CO_2 separation [20]. The process can be divided into two main types of absorption: chemical or physical absorption. First, the flue gas stream must be cooled down. Subsequently, it is inserted into an absorber column where the CO_2 is absorbed into the absorbent. The lean flue gas is then released, and the CO_2-rich absorbent liquid is transferred into a stripper column, where the solvent is regenerated by heat or pressure change. The released pure CO_2 is then captured and prepared for further storage or utilization. Conversely, the sorbent is transferred back into the absorber column for another cycle (Figure 2) [19].

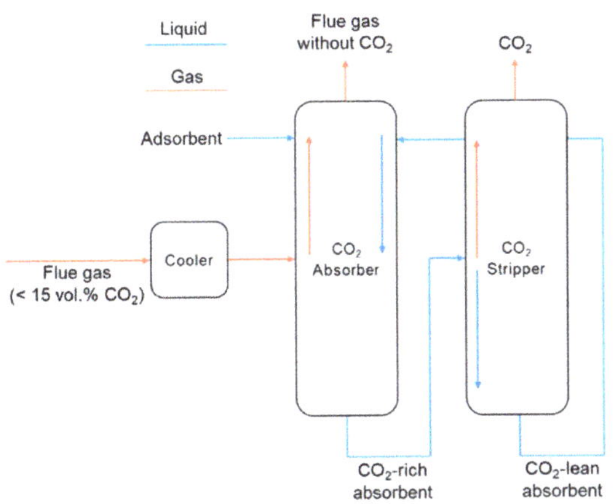

Figure 2. Simplified flowsheet of CO_2 absorption via fluid absorbents (based on [19]).

The most common method of capturing CO_2 from flue gases is chemical absorption. More specifically, the property of some solvents to selectively absorb and react with CO_2 and release it upon regeneration [27]. The process is well-established in the natural gas industry and is also used to generate food- and beverage-grade CO_2 [28]. Due to their ability to establish a strong interaction with CO_2, many aqueous amines, including monoethanolamine (MEA), methyl diethanolamine (MDEA), and diethanolamine (DEA), have emerged as the reference solvents for CO_2 absorption [29]. The removal of CO_2 from diluted flue gas streams becomes efficient because of the chemical absorption [30].

Amine-based CO_2 absorption utilizes the principle of chemical reactions where the reaction between the amine (e.g., MEA) and the dissolved CO_2 leads to the formation of carbamate compounds (Equation (1)) [31]. The detailed mechanism, including the formation of zwitterions and side reactions with water, is, for simplicity, not shown but can be seen in the literature [32,33]. If heat is applied, the carbamate formation is reversed, and the CO_2 and amine are released.

$$CO_2 + RNH_2 \rightleftharpoons RNHCOOH \tag{1}$$

Although this technique is widely used, one major disadvantage is the high energy demand for solvent regeneration. Standard amine-based systems have an energy consumption rate from 2.5 to 3.5 GJ/ton of captured CO_2. The implementation of heat integration techniques has shown the potential to reduce this energy demand by a further 0.5 GJ per ton [18]. Moreover, using amine-based absorbents such as MEA leads to corrosion in the columns [34]. Due to a certain amount of oxygen in the flue gas, the solvent can also decompose into corrosive products, reducing the solvent volume and harming the environment and people [35].

Apart from the previous solvents for chemical absorption, several other liquid sorbents can be used, as seen in the literature. These include amine blends [36,37], aqueous alkali metal solutions [38], nanofluids (suspended nanoparticles) [39], and chilled aqueous ammonia [40].

Furthermore, the use of ionic liquids (ILs) is considered [41–43]. ILs are ionic organic compounds with melting points below 100 °C and advantages like an extremely low vapor pressure, good electrical conductivity, as well as thermal/chemical stability. In the case of carbon capture, they are favorable due to their high capacity and fast absorption. They can absorb the CO_2 physically, or, in the case of functionalized ILs, a chemisorption with CO_2-philic sides on the compounds is also possible and the topic of recent research [43].

Solvents for physical absorption work differently than solvents for chemical absorption. In such solvents, CO_2 dissolution is not controlled by chemical reactions. The efficacy of dissolving CO_2 depends primarily on the solubility of CO_2 in the solvent, which follows Henry's law and is influenced by partial pressures [44]. In industry, several solvents are used for physical absorption of acidic gases (CO_2 and H_2S), such as chilled methanol in the Rectisol process [45–48], dimethyl ether of polyethylene (DMPEG) in the Selexol process [49,50], propylene carbonate in the Fluor process [51], and *N*-methyl-2-pyrrolidone (NMP) in the Purisol process [52]. The Rectisol process is more efficient for carbon capture than the Selexol or Purisol process due to the higher solubility of CO_2 in methanol at high pressures and low temperatures. The Selexol process can be efficient. However, due to the usage of NMP, it is more expensive than methanol [47]. A challenge of these solvents is the selectivity for CO_2, which must be high to only absorb CO_2. If there are contaminants such as H_2S in the flue gas, they can be absorbed as well [46].

As a current application example, Shell has built the Shell CANSOLV CO_2 Capture System in 2014, integrated into the previously mentioned coal-fired power plant in Estevan, Canada. The capture system uses the DC-103 proprietary solvent [53,54].

3.1.2. Adsorption on Suitable Materials

In contrast to the absorption carbon capture, where a liquid dissolves CO_2, adsorption is the deposition of CO_2 on a solid surface. The adsorption can happen in two different ways: The first is chemical adsorption (chemisorption), where the CO_2 creates new covalent bonds with the surface. Second, the physical adsorption (physisorption) creates no new bonds between the solid and CO_2 on the surface. Physisorption is based on weak Van der Waals interactions, which influence the energy demand for the regeneration process. The regeneration energy demand for physisorption is lower than for chemisorption due to the interaction of the gas molecules and the solid surface [55,56].

While the processes of CO_2 adsorbing vary only with the used material, the desorption process in terms of material regeneration can be performed in different ways. It can be achieved with heat in a temperature swing adsorption (TSA) [57], with pressure differences in a pressure swing adsorption (PSA) or vacuum swing adsorption (VSA) [58,59], with a combination of temperature and pressure in temperature vacuum-pressure swing adsorption (TVPSA) [60] or with electricity in an electric swing adsorption (ESA) [61–63].

As for solid sorbent materials, many current investigations on different materials have been carried out in the literature. Physical adsorbents can be divided into carbonaceous materials and non-carbonaceous materials. Carbonaceous sorbents are graphene [64, 65], activated carbon (AC) [66,67], carbon molecular sieves (CMSs) [68,69], and carbon nanotubes (CNTs) [70,71]. Non-carbonaceous sorbents, on the other hand, are materials such as metal–organic frameworks (MOFs) [72–74], natural or synthetic zeolites [74,75], and silica-based materials [76,77]. Solid sorbents for chemisorption differ from those with physisorption because of the chemical reactions on the surface. This results in an energy-intense regeneration to break the newly formed bonds and causes sorbent degeneration. Possible and studied materials are amine-based materials, metal salts, or hydrotalcite [78–81]. Due to the usage of different materials, regeneration processes, and CO_2 concentrations in the flue gas, the energy demand is estimated to range from 1.5 to 4.5 MJ/kg CO_2 [7].

Current examples of carbon capture with solid adsorbents are the demonstration plant in Lashburn (Canada) operated by Svante Inc., which uses its VeloxoTherm™ process based on CO_2 adsorption [82].

3.1.3. Calcium Looping (CaL)

Another emerging chemisorption solid carbon capture process traps CO_2 via a carbonization step of calcium, called calcium looping or calcium cycling (Figure 3). Intended as a post-combustion carbon capture technology, this approach was first proposed in 1999 [83].

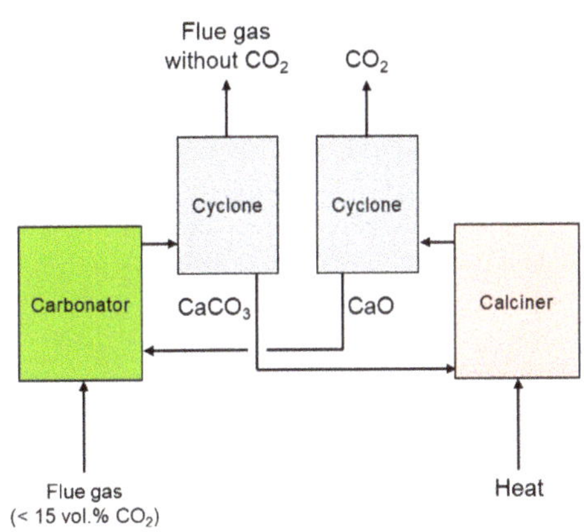

Figure 3. Simplified flowsheet of CO_2 capture via calcium looping (based on [84]).

The principle behind this carbon capture technique is the reversible reaction of CaO forming $CaCO_3$ with CO_2 in an exothermic reaction (Equation (2)). First, the flue gas is inserted in the carbonator reactor, which has temperatures ranging from 600 to 700 °C. The resulting carbonate is transferred in a cyclone to separate the solid from the cleaned flue gas. Furthermore, in the calciner, with temperatures from 900 to 950 °C, the $CaCO_3$ is then regenerated back to CaO, and the pure CO_2 is captured [85]. The reaction to form the carbonate is rapid, while the reversal reaction is slower due to the CO_2 diffusion through the $CaCO_3$ layer formed on the free CaO surfaces [86].

$$CaO + CO_2 \rightleftharpoons CaCO_3 \quad \Delta H_r = -178 \text{ kJ/mol} \tag{2}$$

An obvious challenge of this process is the energy required to reach the high temperatures for the endothermic calcination step. However, some of the heat generated from the carbonation step can be used to lower the energy requirement of the calcination. Another challenge is the loss of activity due to the fast sorbent decay after about 20 to 30 cycles, which leads to sintering processes in the calcination step [87]. The advantages of this carbon capture technique are the use of cheap material, e.g., limestone [88], and relatively low-efficiency penalty ranging from 7 to 8% [89,90]. In the case of retrofitted coal-fired power plants, it is also possible to achieve power efficiency penalties down to 2.75% [91]. Moreover, heat for the calciner can be generated by burning fuel together with pure oxygen. Due to the temperature range and the $CaO/CaCO_3$ equilibrium, the possibility of in situ carbon capture would be achievable while generating additional heat for power generation, thus increasing the plant's energy output [92].

For example, integration into high-temperature processes like cement or steel can be favorable. In 2015, a technical analysis with process modeling of a cement plant with integrated calcium looping carbon capture was performed. The plant has several benefits due to the handling of limestone and lime, raw products of the cement process. Lime (CaO) can be purged from the CaL process, replacing limestone in the kiln feed. The heat needed for the calcination in the CaL process can be obtained by burning solid fuels, like petcoke, a material which is also used in the cement plant, with pure oxygen. Furthermore, the exothermic reaction in the carbonator is used for steam generation to generate electricity via steam turbines. The latter can then be utilized for air separation to obtain pure oxygen for combustion and CO_2 capture, purification, and compression. Overall, the power yield of this integrated plant is calculated to be 426.66 kWh per ton produced clinker, taking the power consumption of all steps into account. Carbon capture efficiency can reach up to 94.7% [93].

Generally, it is estimated that the cost for calcium looping is around USD 29–50/ton of CO_2, which is 50% cheaper than carbon capture with amines [90].

Currently, the project CaLby2030 (started in 2022) aims to develop a commercial scale of calcium looping carbon capture in high-temperature industries such as the cement or steel industry, which account for more than 5 Gt of CO_2 emissions per year. The demonstration plants are planned for Germany, Sweden, and Spain. They claim to demonstrate >99% CO_2 capture rates [94]. For example, the La Pereda Hunosa power plant, a former 50 MW coal-fired power plant in the Asturias region in Spain, has been retrofitted to burn biomass and solid recovered fuel since 2023. Since 2012, an experimental CaL carbon capture facility called La Pereda Pilot has been operating with a small amount of flue gas (2000 N m^3/h) from the power plant. The calciner is powered by a 2 MW_{th} combustion of biomass pellets with enriched oxygen contents of around 37 vol.%. The facility achieves up to 99.5% carbon capture efficiency [95,96].

3.1.4. Membrane Separation

Using membranes to separate components from a mixture is a well-established and commercialized method. Current applications include natural gas or biogas purification [97,98], water purification [99], or reverse osmosis for water desalination [100]. Using membrane technology over other established technologies, such as distillation, offers some important advantages. These are the low energy consumption due to the absence of extensive heating or cooling processes (but for the regeneration of the sorbent), the possibility of easy linear scaling of the working system due to the modularity, and the incorporation of currently used technologies [101,102]. In the case of carbon capture from flue gases, much progress has been made over the last decades [20,28,103]. Three possible ways to operate a hollow fiber membrane, for example, in a continuous carbon capture setup, are shown in Figure 4.

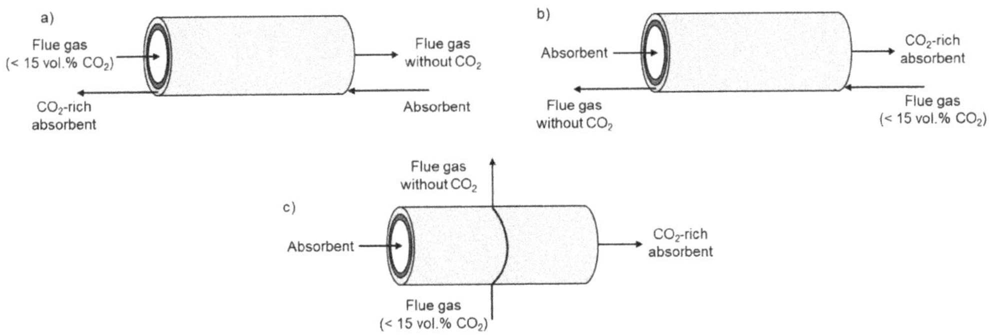

Figure 4. Simplified flowsheet of CO_2 absorption via membrane separation using a hollow fiber membrane: (**a**) flue gas stream inside the lumen side, (**b**) fluid absorbent inside the lumen side, and (**c**) transverse flow of flue gas stream (based on [102]).

The flue gas stream can flow parallel to the liquid on the shell side, parallel to the liquid inside the lumen side, and transverse to the liquid stream on the lumen side. In the case of the liquid absorbent stream flowing through the fiber lumen, absorption performance is significantly improved by 150 to 180%. This can be explained by the better phase contact between gas and liquid. Furthermore, a parallel flow operation is 20% more effective than a possible co-current operation [104].

The selection of membrane material is highly important, and studies of different materials can be found in the literature. Examples are mixed matrix membranes (MMMs) [105–107], polymers of intrinsic microporosity membranes (PIMs) [108], metal–organic-frameworks (MOFs) [109,110], ceramics [111,112], composites [113,114], or hollow fibers of different plastic materials [115]. Based on the material and process used, these membranes can be operated with or without a sorbent liquid on the other side of the membrane to absorb the CO_2 as an additional driving force. Well-established sorbents such as amines, aqueous alkali metal solutions, or ionic liquids are possible. A special type of membrane is the liquid membrane (LM), which can have a support (e.g., polyethylene) combined with a liquid (e.g., ionic liquids) or exist without any support. Bulk liquid membranes (without support) consist of a non-miscible liquid that acts as the membrane phase [116]. More detailed information can be found in the literature [117–120]. Disadvantages concerning mass transfer can occur depending on the material used for the membrane. The most prominent challenges in choosing membrane materials are the mass transfer resistance and membrane wetting due to the mostly aqueous liquid used for absorption [121]. Furthermore, for membrane separation of flue gases, other significant challenges occur. Due to the low concentration of CO_2 of around 10–15 vol.%, the driving force for separation is decreased because membrane separation depends on pressure differences. With low partial pressures of the CO_2 in the gas stream, the separation is challenging, resulting in inefficient separation for flue gas carbon capture [122]. Applications with higher partial pressures are more favorable, like pre-combustion carbon capture. Another challenge is the membrane selectivity for a specific component. In most cases, choosing the right material is a trade-off between permeability and selectivity [119]. In addition, impurities, like NO_x or SO_x, if not removed, can damage the membrane [122]. Despite these challenges, membrane carbon capture is a low-cost method. For a CO_2/N_2 containing diluted flue gas feed with 1% CO_2 in a membrane separation process, an achieved 99.8% purity of the CO_2 gas stream after separation was calculated. The results claim costs of USD 50.1/ton of CO_2 and an energy demand of 0.5 GJ per ton of CO_2 [123].

Several projects have utilized membrane carbon capture processes for post and pre-combustion [124]. A current example is the construction of a pilot plant for membrane separation at the Dry Fork Power Station in Wyoming (USA) by the company Membrane Technology and Research Inc. (MTR). They construct modular membrane modules using their Polaris™ polymeric membranes. The pilot plant aims to capture 55,000 tons of CO_2 per year. The construction was finished in October 2024. They claim to be the largest application of membrane carbon capture to date [125].

3.1.5. Cryogenic CO_2 Separation

The technology of cryogenic carbon capture is based on the simple concept of separating components based on their different physical properties, such as boiling point or, in the case of CO_2, desublimation properties. Therefore, carbon capture uses flue gas streams with highly diluted CO_2. Cryogenic separation can be difficult and energy-intensive due to the low concentration of CO_2 and the presence of other components like H_2O. As the latter can clog the system at these low temperatures [20,126], the technology performs better with highly concentrated CO_2. This can be achieved with different combustion setups (e.g., using the pre-combustion route) or additional carbon capture technologies like PSA to increase the CO_2 concentration in the gas stream [18].

Based on the used phase equilibria, this approach has two possible pathways (Figure 5). The conventional way uses the vapor/liquid equilibrium (VLE), and the unconventional way uses the vapor/solid equilibrium (VSE).

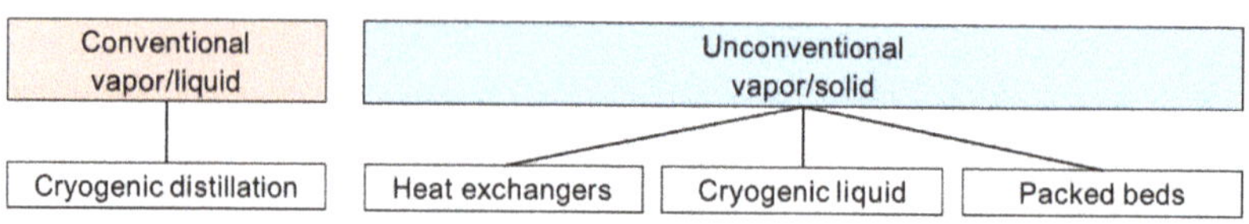

Figure 5. Techniques for cryogenic CO_2 capture in the conventional and unconventional pathways (based on [102]).

The conventional method, cryogenic distillation, is a complex and energy-intensive process due to the special properties of CO_2. The column must operate at a higher pressure because of the triple point of CO_2 at 5.11 bar to achieve a vapor/liquid separation. Due to the low CO_2 concentration in the flue gas, the efficiency is low, and the energy demand is high [127]. The energy required for a conventional cryonic distillation for a feed with 70% CO_2 by volume is estimated at around 1472 kJ/kg CO_2 [128]. For diluted CO_2 streams like flue gases, the energy demand is estimated to be from around 2160 to 2376 kJ/kg [35,129]. The resulting energy requirement for cryogenic distillation is therefore highly dependent on the concentration in the feed gas mixture.

In the unconventional way, there are three main technologies. The usage of heat exchangers, cryogenic liquids, or packed beds. The usage of heat exchangers was first developed in 2002 [130]. In this approach, the CO_2 desublimates on the fins of a heat exchanger, which is cooled by a blend of refrigerants (e.g., from methane to butane). The refrigerants evaporate to create the necessary cooling of around −120 °C. The system consists mostly of more than one heat exchanger in a cascade and can be operated with diluted flue gas streams but must be dried to avoid water ice forming on the fins. An advantage of this approach is the possibility of getting high purities of CO_2 after a sublimation step of 99.9%. The total electric energy consumption of this cooling method with an initial CO_2 concentration of around 15% is calculated to be from around 541 to 1119 kJ/kg CO_2, depending on the efficiency of the refrigerating system. It counts for all electrical systems such as fans, pumps of the cooling tower, and compressors of the refrigerating system [131,132]. It is, therefore, compatible with other carbon capture technologies and lower than the conventional way using cryogenic distillation.

Another interesting concept in the field of heat exchangers is the usage of Stirling coolers. The system consists of three coolers where the input stream is first precooled to remove any H_2O in the gas stream and a second cooler for the CO_2 freezing step. The solids formed during the second step are transported to a third cooler to create cryogenic storage for the collected CO_2. At a flow rate of 1.5 L/min and a CO_2 concentration of 13 vol.% in the flue gas, the energy consumption over the whole system was 1.5 MJ/kg CO_2. A capturing rate of 96% of CO_2 was achieved [133,134]. This concept receives more and more interest in the literature due to the high efficiency of Stirling coolers, high reliability, and compact size, among other reasons [135].

Another way of cryogenic carbon capture is the usage of cryogenic liquids like isopentane. The flue gas containing CO_2 is passed through the liquid in a bubble column and the dissolved CO_2 forms solids. The slurry liquid can then be filtered to obtain the pure CO_2 out of the liquid. The technique aims to create a closed loop to achieve a continuous carbon capture method [136]. An energy penalty of 0.74 MJ_{el}/kg CO_2 or, in other words, 15% of the plant's power output, is estimated through simulation for a coal-fired power plant with cryogenic carbon capture using cryogenic liquids [137].

The third approach uses a packed bed with a cold material to desublimate the CO_2 onto the surface. The packed beds are integrated into a column, and the flue gas moves through the column until every packed bed is covered in CO_2 solids. Then, the column must be regenerated to obtain the pure CO_2 [128,138]. An alternative to this process is using just one packed bed in a cycle of CO_2 desublimation onto the surface and subsequently transporting the packed bed out of the column into a regeneration unit to sublimate the CO_2. Afterward, the packed bed can be transported back into the column, and the capturing starts again [139]. The energy requirement for multiple beds at an initial CO_2 concentration of 10 vol.% is estimated to be 1.8 MJ_{el}/kg CO_2 for cooling down the packed beds to −70 °C. However, the process can recover >99% of the CO_2 from the flue gas [140]. For only one single bed operated in a cycle, energies were calculated for carbon capture systems of three examples: an oil-fired boiler, a combined cycle gas turbine (CCGT), and for biogas upgrading. For the oil-fired boiler, the energy consumption was calculated to be 1.20 MJ/kg (332 kWh/ton), for the CCGT, 2.99 MJ/kg (831 kWh/ton), and, for the biogas upgrading, 0.95 MJ/kg (263 kWh/ton) [139].

These techniques are still under development. In March 2024, Carbon America announced a facility with a cryogenic carbon capture technology (FrostCC) that compresses and expands the flue gas with recuperative heat integration. The facility has been operating for more than 1000 h and has a capture capacity from 500 tons up to 1000 tons of CO_2 per year, with a capture rate of 99% [141].

3.2. Pre-Combustion

In the last chapter, the post-combustion route was discussed, along with several techniques. However, instead of changing the approach for the carbon capture method from diluted CO_2 streams like flue gases, the combustion route can also be altered. The pre-combustion route is another important route in carbon capture technologies (Figure 6). Here, a commonly used fuel (e.g., coal) reacts with air (or pure oxygen) to produce syngas after a partial oxidation step (Equation (3)). A steam-reforming step (Equation (4)) can also be employed instead of partial oxidation, or in parallel; it is advantageous to use exothermic partial oxidation to heat the endothermic steam reforming. In parallel, it is called auto-thermal reforming [142].

$$C_nH_m + \frac{n}{2} O_2 \rightleftharpoons n\,CO + \frac{m}{2} H_2 \quad \Delta H_{CH4} = -36\ kJ/mol \tag{3}$$

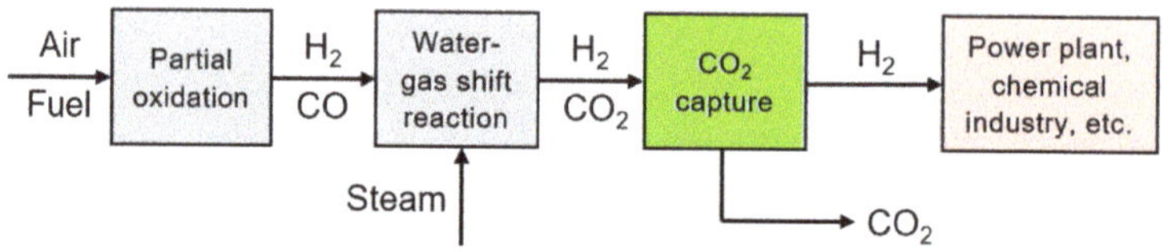

Figure 6. Simplified flowsheet of the pre-combustion route (based on [17]).

$$C_nH_m + n\,H_2O \rightleftharpoons n\,CO + \frac{2n+m}{2} H_2 \quad \Delta H_{CH4} = +206\ kJ/mol \tag{4}$$

$$CO + H_2O \rightleftharpoons CO_2 + H_2 \quad \Delta H_{CH4} = -41\ kJ/mol \tag{5}$$

When used alone, the reforming step typically produces about 43% H_2, 11% CO, 21% H_2O, and 6% CO_2 [143]. Following the step of syngas production, a water–gas shift reaction (Equation (5)) is necessary to transform any of the CO into CO_2. Afterward, the stream consists mainly of CO_2 and H_2. The syngas has to be cooled down to prevent damage or loss of efficiency in the following steps. Additionally, it has to be cleaned up from impurities such as hydrochloric acid (HCl), hydrogen sulfide (H_2S), carbonyl sulfide (COS), and trace metals like mercury (Hg) [144]. Due to the usage of only oxygen and not air, the partial pressure of CO_2 is much higher than in a diluted flue gas stream. Therefore, the CO_2 capture is easier and more cost-efficient. As the main reason for efficiency loss in the pre-combustion approach is the energy-intense water–gas shift reaction, this is of high importance [19]. However, despite this challenge, producing H_2 for power generation in, e.g., fuel cells, or the production of value-added chemicals, is an advantage [145].

To achieve a higher efficiency and counteract the efficiency loss of the water–gas shift reaction, an integrated gasification combined cycle can be added (Figure 7). The pure H_2 stream after the carbon capture is combusted with air in a gas turbine to generate heat and electricity. The waste heat is then recovered and converted into steam, which produces electricity in a steam turbine [146]. The principle behind a combined gas and steam turbine is called the Brayton–Rankine cycle [147]. Due to the possibility of additional carbon capture, the interest in integrated gasification combined cycle (IGCC) plants has increased in the 21st century [148]. There are also variations of the concept using other fuels and process design, but the integrated cycle stays for all approaches. For example, the biomass-based integrated gasification combined cycle (BIGCC) or the natural gas combined cycle (NGCC). For the NGCC, the natural gas is used directly in the gas turbine without the pre-combustion design.

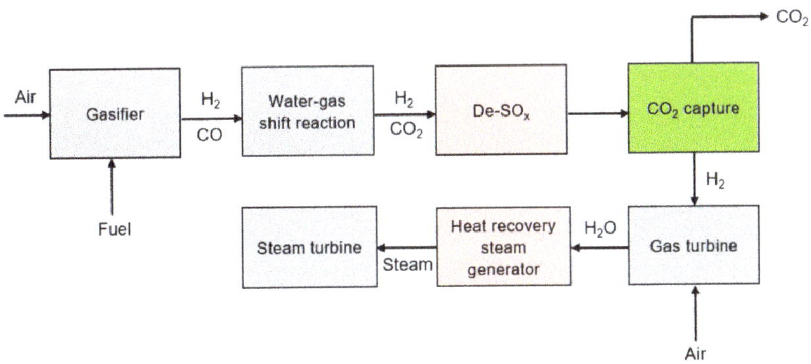

Figure 7. Simplified flowsheet of an integrated gasification combined cycle (based on [149]).

Currently, there are several IGCC power plants around the world operating with net electric power ranging from 127 to 2400 MW and efficiencies of up to 48% [148]. For example, the IGCC power plant unit of the Nakoso Thermal Power Station, run by Jōban Joint Power Co., Ltd. in Iwaki (Japan). It operates with an electric power of 250 MW and an efficiency of 48%. The application of carbon capture techniques is not completely clear from the source. A gas clean-up is described via MDEA and gypsum recovery (for desulfurization), but additional storage or utilization of the CO_2 is not further mentioned [150]. As a second example, the Jazan IGCC Complex is operated by Saudi Aramco in Jazan Economic City, Saudi Arabia. The electric power output is claimed to be 3800 MW. Additional CCSU technologies are not presented [151].

Due to the higher efficiency and the possibility of capturing CO_2 with a higher partial pressure, the cost of carbon capture for pre-combustion and additional IGCC technology is estimated to be around USD 77.0 (GBP 60.4) per ton of CO_2. Capturing CO_2 via pre-combustion is cheaper in comparison to the post-combustion and oxy-fuel technology, which costs around USD 162.7 (GBP 124.7) and USD 263.2 (GBP 206.6), respectively [10].

3.3. *Oxy-Fuel Combustion*

Oxy-fuel combustion is another combustion method apart from pre-combustion, where effective carbon capture technologies can be applied (Figure 8). Pure oxygen is used for combustion, which is obtained by mostly cryogenic air separation or can be obtained during water electrolysis. After burning the fuel and soot removal, the flue gas is split and recycled back into the combustion chamber. This is because temperature control can be achieved in the combustion chamber. Furthermore, flue gas is more concentrated with CO_2, which makes separation much easier [19,152].

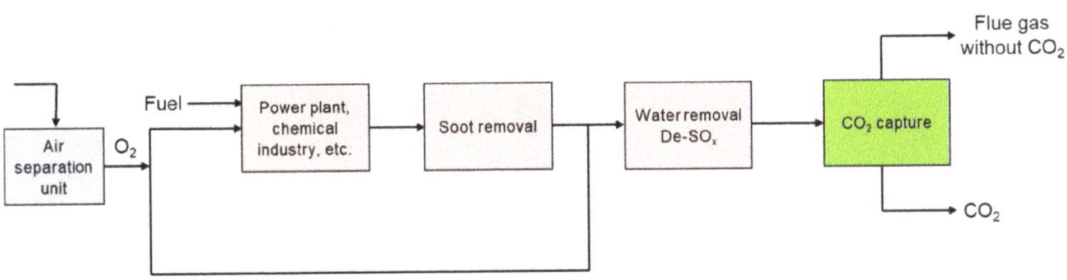

Figure 8. Simplified flowsheet of the oxy-fuel combustion route (based on [19]).

Oxy-fuel combustion was first postulated in the 1980s as a method for combustion, achieving high purities of CO_2 in the recycling flue gas of ~95% after water removal [153].

Two main applications of possible oxy-fuel process designs exist, i.e., coal-fired power plants, pulverized-coal-fired oxy-fuel combustion (oxy-PC) [154], and the usage of circulation fluidized beds (oxy-CFBs) [155]. As for the oxy-PC, the pulverized fuel is used in certain boilers, which have been extensively reviewed in the literature [154,156–158]. The oxy-CFB method, on the other hand, employs another approach utilizing a different boiler system with limestone as an additional component. An advantage of oxy-CFB with limestone is that in situ sulfur capture can be achieved [159,160]. Oxy-fuel combustion has the advantage of reduced NO_x emissions during the combustion compared to combustion with air due to the suppression of thermal NO_x formation as part of the temperature control. Moreover, due to the flue gas recycling, parts of the NO_x become decomposed after re-entering the combustion zone [158,161]. In the case of oxy-CFB, however, the presence of limestone can negate this advantage and increase NO_x emissions depending on the temperature [155]. The cost of oxy-fuel combustion can be high due to the necessary air separation. However, an energy and economic assessment in the literature shows that when used in coal-fired power plants with heat-integration management, the CO_2-avoiding cost can reach USD 43.24/ton of CO_2. A challenging drawback is the reduced net power efficiency of 10.41% for the power plant, due to energy-intense steps like the air separation [162].

Using oxy-fuel combustion for ships is evaluated with a techno–economic analysis to reduce CO_2 emissions by onboard carbon capture. However, onboard O_2 production is one of the biggest challenges due to higher energy demand and reduced engine effectiveness. It is also evaluated that ships can obtain liquified O_2 from harbors, avoiding the on-board O_2 production, which increases efficiency but requires additional storage capacity on ships [163].

Currently, there are only a few operating oxy-fuel power plants worldwide. Many of them are no longer operating [124]. One oxy-fuel plant is the NET Power demonstration plant in La Porte, Texas (USA). The project started in 2018, operating with an output power of 50 MW. Currently, there is a planned scale-up to 300 MW, with a carbon capture capacity of 850 Mt of CO_2 per year, starting in 2027/28 [164,165]. As for 2024, China plans to construct an oxy-fuel combustion project based in the cement sector, which they claim to be the world's largest project of its kind. The plant aims for an annual carbon capture of 0.2 Mt CO_2 [166].

The Fraunhofer UMSICHT Institute takes a different approach. They combine a biogas plant, which produces mainly CH_4 and CO_2, with an oxy-fuel combustion plant. Due to the usage of biomass, carbon capture, and oxy-fuel, the project name is BO_2CCS. Furthermore, they want to utilize the O_2 produced during electrolysis for H_2 production and create a market for this unused by-product [167].

3.4. Chemical Looping Combustion (CLC)

An emerging technology for carbon capture is chemical looping combustion (CLC) (Figure 9). It was first patented in 1954 and intended as a production method for pure CO_2 [168]. Here, the air and fuel are separated from each other in the combustion process. The metals/metal oxides, which are transferred between the fuel reactor (reducer), and the air reactor (oxidizer) act as oxygen carriers. In the air reactor, the metals/metal oxides are loaded with new oxygen atoms due to the reaction with air. Then, the metal oxides react in the fuel reactor with the solid or fluid fuel and release some of the oxygen for combustion to form CO_2 and water, which can then be separated after condensation of water. A pure CO_2 stream for additional storage or utilization is achieved. The metal oxides used can be based on different metals like Fe, Ni, or Cu [169]. Due to of the usage of air instead of pure oxygen, the need for energy-intense air separation is obsolete. Furthermore, a dilution of the CO_2 stream with air is not possible.

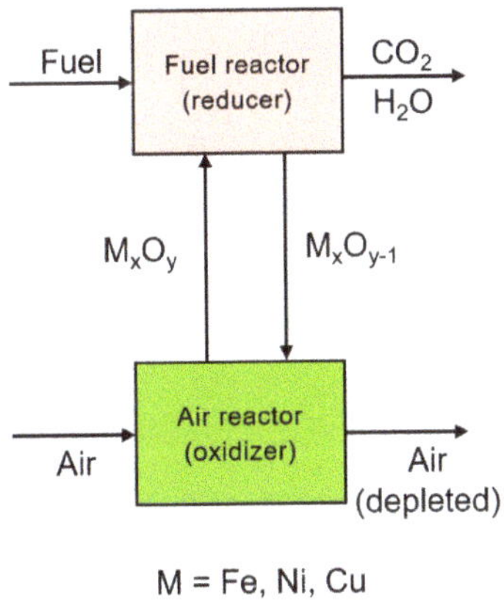

Figure 9. Simplified flowsheet of chemical looping combustion route (based on [170,171]).

Equation (6) shows the simplified oxidation reaction in the air reactor between the metal oxides and oxygen (air). Equation (7) depicts the simplified reduction reaction in the fuel reactor with gaseous, liquid, or solid fuel. In the case of solid fuel, additional steam or CO_2 is necessary to fluidize the fuel [169]. In reality, the oxidation or reduction of the metal oxides is not ideal due to effects such as phase transformations or surface and bulk oxygen defects [172].

$$M_xO_{y-1} + ½\, O_2 \rightarrow M_xO_y \tag{6}$$

$$C_nH_{2m} + (2n + m)\, M_xO_y \rightarrow (2n + m)\, M_xO_{y-1} + n\, CO_2 + m\, H_2O \tag{7}$$

Specific and important properties are necessary for the selection of metal oxides as oxygen carriers due to the cycling process and high temperatures needed for CLC. These properties are, for example, a high melting point, high resistance to attrition and agglomeration, high reactivity with oxygen and the fuel, as well as high oxygen transport capacities [170]. The selection is of great importance, and many investigations of different metal oxides can be seen in the literature [173–177]. The carbon capture cost highly depends on the materials used and the process design, but it is estimated to be around USD 39 (EUR 37) per ton of CO_2 for a power plant with a net electric power of 630 MW and a capture rate of over 90% [178] Currently, possible metal oxides are, for example, Fe_2O_3, Mn_3O_4, CuO, NiO, CoO, or $FeTiO_3$ (ilmenite) [178]. The choice of the oxygen carrier used in a CLC process also depends on environmental aspects such as the processing of the material, recycling of used oxygen carriers, handling of plant waste as well as material loss during processing and operation. Considering these challenges, CLC carbon capture processes can optimized regarding sustainability [179].

Currently, no commercial plants use CLC for carbon capture, but there are some projects that explore further up-scaling. For example, the TU Vienna investigated and successfully operated a 100 kW pilot plant with CLC over three years. Now, they are up-scaling the process, aiming for a demonstration plant with an electric power of 10 MW [178]. Furthermore, a consortium between China and Europe (CHEERS) is constructing and testing a chemical looping combustion demonstration plant near Chengdu, China. It is currently in the testing phase after start-up. For the oxygen carrier, the plant uses ilmenite ($FeTiO_3$). The plant will have an electric power output of 3 MW with a carbon capture rate of 96% [180]. As for 3 MW power plants using CLC, there are also two demonstration plants under construction in Ulsan (South Korea) [181] and at Ningxia University in China [182].

4. Direct Air Capture (DAC) Approach to Capture Airborne CO_2

In addition to capturing CO_2 from point sources like power plants or chemical industries, another approach is to lower the CO_2 content in the atmosphere is by directly capturing airborne CO_2 from the atmosphere (Figure 10).

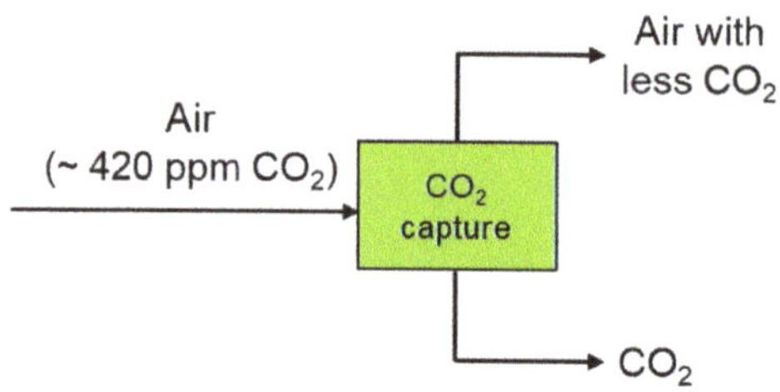

Figure 10. Simplified flowsheet of direct air capture.

The approach to capture CO_2 regardless of the CO_2 emission source has gained more and more interest in recent years, despite the low concentration of CO_2 in the air of currently around 420 ppm [183]. In a projection by the International Energy Agency (IEA) for reaching the net-zero emission goal by 2025, the DAC removal target is ~980 Mt CO_2 per year [184]. The benefit of directly removing CO_2 from the atmosphere with additional carbon storage technologies is an important alternative to the general approaches of capturing CO_2 at fixed locations [185]. Due to the very low concentration of CO_2, a large air volume must be moved through an additional carbon capture technology, which makes this technology expensive, with up to USD 1065/ton of CO_2. Recent developments in DAC systems could reduce that cost to less than USD 100/ton of CO_2 [186]. The cost is highly determined by the capturing system and the energy cost. Depending on the method, the energy usage can range from 3.4 to 4.8 GJ per ton of CO_2 (thermal) and from 0.55 to 1.12 GJ per ton of CO_2 (electrical) [186].

Furthermore, the compatibility and profitability of DAC systems depend on government policy and funding [187]. Policies like carbon credits or a carbon tax could motivate industries to invest more in DAC systems for decentralized carbon capture. This is even more relevant for sectors that cannot capture their CO_2 emissions at point sources. While carbon credit prices are not well-regulated and standardized, a carbon tax is easier to implement [188]. For example, a tax higher than the amount of the carbon capturing cost could create a market for such technologies. In that case, industries can avoid paying the tax by investing in cheaper solutions like carbon capture. The DAC company Skytree suggested a carbon tax of more than USD 200/ton. Together with added value to the captured CO_2, selling it to other industries can turn DAC systems profitable because the revenue is greater than the capturing cost. [188]. As a current example, the carbon tax in Sweden for 2024 is USD 131.73 (SEK 1450) [189]. While such policies like a carbon tax can reduce fossil energy consumption and lower carbon emissions, they can also have negative aspects that should be taken into account as well. For example, a tax that is too high can lead to social–economic consequences like industries relocating to countries with lower carbon tax [190,191].

While policy measures can make DAC systems profitable, technological solutions are required to reduce the high energy penalties even further thus improving the efficiency of DAC systems. Of the several available carbon capture techniques, absorption and adsorption are the most promising because they can capture CO_2 even from very dilute streams [192]. When adsorption processes are used, temperature swing adsorption, compared to pressure swing adsorption, is thermodynamically more effective for ultra-dilute systems, whereas pressure swing adsorption is better for bulk gas separations [193].

Previously discussed materials, such as MOFs or zeolites, can be used as solid sorbent material. An example is the zeolite Faujasite, which concentrates CO_2 from 400 ppm (0.04%) to 95% in a temperature vacuum swing adsorption (TVSA) [194]; the only downside is the high regeneration temperatures needed to desorb the CO_2 from the surface. Commercial utilization of the DAC system with adsorption is tested by Climeworks with their Orca facility in Iceland. The system can capture up to 4000 tons of CO_2 per year and uses geothermal energy produced by the Hellisheidi geothermal power plant to regenerate its sorbents. They also cooperate with Carbfix to store their captured CO_2 underground [195]. Since 2022, Climeworks has constructed a new DAC plant in Iceland, which will be about ten times bigger than the Orca plant and has a capturing capacity of up to 36,000 tons of CO_2 per year. The plant is currently in operational ramp-up [196].

As for absorption with liquid sorbents, amine-based sorbents or aqueous alkali metal solutions can be used due to their efficiency in absorbing CO_2 from dilute streams mboxciteB185-processes-3405354,B197-processes-3405354. However, absorption has some drawbacks, such as the liquid amine-based sorbents degradation and a lower rate of CO_2 absorption compared to alkaline solutions because of present oxygen, high energy demand, and environmental hazards [198]. Hydroxide-based aqueous solutions have an even higher energy demand for regeneration with temperatures of up to 900 °C [21]. However, another step can be implemented to solve the problem of the high regeneration energy for a hydroxide-based aqueous solution. The dissolved CO_2 in a NaOH solution forms a carbonate, which can react with additional $Ca(OH)_2$ to form $CaCO_3$. The solid $CaCO_3$ precipitates. After filtration and calcination, pure CO_2 and CaO are obtained. The CaO can be transferred back into the solution to form again $Ca(OH)_2$. Using this cycle, a total energy penalty of 2.6 GJ/ton of CO_2 can be achieved [199]. The company 1PointFive, in cooperation with Carbon Engineering, uses this technology for its DAC systems and is currently constructing a DAC plant in Ector County, Texas (USA). The plant is expected to be commercially operational in mid-2025 and is designed to capture up to 0.5 Mt CO_2 per year. The captured CO_2 will then be stored in underground saline formations [199].

5. Storing the Captured CO_2 (CCS)

A crucial part of carbon capture and storage (CCS) technologies is storing CO_2 underground after capturing it from a point source or the atmosphere to achieve net-zero or net-negative emissions of CO_2 to prevent further climate change [1].

After the CO_2 has been captured, it is compressed and liquefied for better transport via pipelines, ships, or trucks. At temperatures and pressures greater than 31.1 °C and 7.38 MPa, the CO_2 is in its supercritical state, which is favorable for transportation or injection. In its supercritical state, CO_2 still behaves like a gas but has a much higher density than in its gaseous state. Depending on temperature and pressure, the density can range from 200 to 900 kg/m^3 [200]. There are different possible approaches for storing the CO_2 (Figure 11). It can be either stored physically in geological formations, e.g., deep saline aquifers [201], or in depleted oil and gas reservoirs [202], used for enhanced oil recovery (EOR) [203], injected into mineral formations to mineralize the CO_2 to carbonates, trapping it chemically (CCMC) [204], or in unconventional storage location like coal beds or gas hydrates.

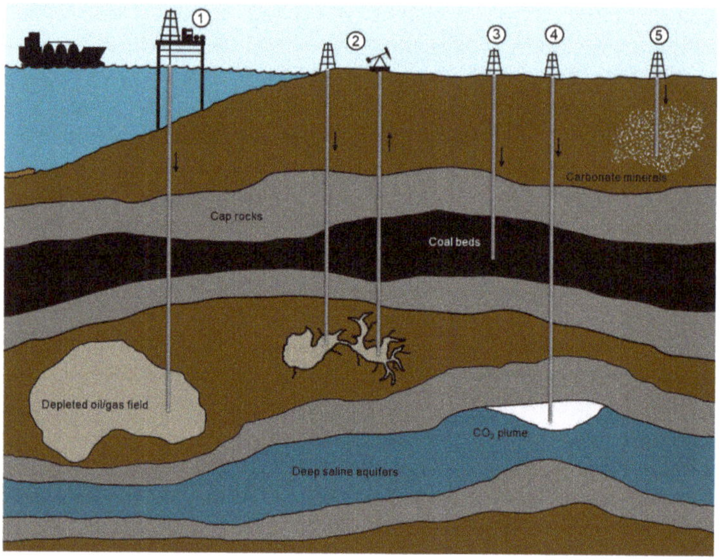

Figure 11. Simplified possibilities for carbon storage solutions: (1) depleted oil/gas fields, (2) enhanced oil recovery, (3) coal beds, (4) deep saline aquifers, and (5) carbonate materials (based on [205]).

Moreover, when bio-based fuels (biomass) are used for combustion processes and the resulting CO_2 is then captured and combined with storage, it is called BioCCS or BECCS [19]. A big advantage of using biomass is that CO_2 from the atmosphere is fixated in vegetation and is, therefore, removed from the atmosphere. The released CO_2 after combustion can then be captured and stored, which is necessary to achieve real net-negative CO_2 emissions (NETs) [206]. In the technical summary of the IPCC Special Report on Climate Change and Land (SRCCL), BECCS is stated as one of the most effective migration activities in the special topic of land use, which has the potential to lower CO_2 emissions by 0.4 to 11.3 Gt CO_2 per year between 2020 and 2050 [207].

5.1. Deep Saline Aquifers

Deep saline aquifers are underground fractured spaces filled with water that flows at a very low velocity, in the order of a couple of centimeters per year. They are highly saline, making them unsuitable for extraction and use for industry or human consumption. They are widely available and have a high potential for global carbon storage [200]. It is estimated that at least 1000 Gt of CO_2 or more can be stored in those [208]. However, it should be considered that not all deep aquifers are suitable for long-term carbon storage, as most can have some routes to the surface, enabling leakage over long-term periods. Favorable deep aquifers are located at depths of 800 to 3000 m below the surface and are covered with a low permeable rock (cap rock), preventing diffusion to the surface [209]. Generally, for trapping CO_2 inside geological formations, several mechanisms are identified in the literature, including structural and stratigraphic trapping, dissolution trapping, residual trapping, and mineral trapping [210].

Structural and stratigraphic trapping relies on the previously mentioned impermeable or low permeable cap rocks, preventing CO_2 in an upward migration leakage. The CO_2 is sequenced in a supercritical state into the aquifer, forming a CO_2 plume. This plume can be held in place if the aquifer has a concave-down geometry at the injection site (Figure 12). The CO_2 can also flow slowly to such geometry near the injection site [210,211].

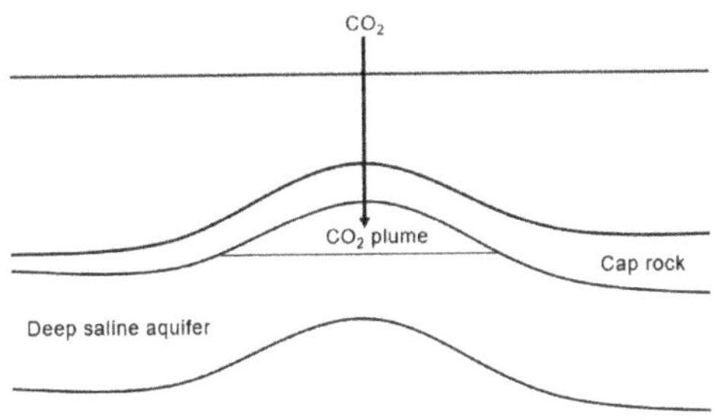

Figure 12. Concave-down geometry of a deep saline aquifer with cap rock sealing (based on [211]).

Possible leakage can occur due to the mobile CO_2 phase. If the cap rock has faults or fractures, the CO_2 can easily escape into shallow aquifers or abandoned wells, for instance, and, over a period of time, back to the surface [211]. A deeper understanding of the subsurface conditions, such as cap rock integrity and potential pathways for CO_2 migration back to the surface, is important to ensure the safety of storage sites. This applies also to other geological formations [212,213].

Furthermore, the CO_2 can be stored via residual or capillary trapping. The literature considers it a fast and safe method for long-term carbon storage [214–216]. After the CO_2 is injected and forms a plume, capillary forces of the porous rocks can snap off parts of the plume, creating isolated bubbles or ganglia and trapping it inside the pores, making the CO_2 immobilized and unable to escape for hundreds of years. After immobilization, the CO_2 will slowly be dissolved or mineralized [217,218]. The advantage of immobilized CO_2 due to capillary trapping is substantial for storage security [219].

Solubility trapping works by the slow process of CO_2 dissolution from the injected CO_2 plume into the saline water and is considered the most effective storage mechanism, because the forming CO_2–brine mixture has a higher density and, therefore, sinks to the ground, removing the CO_2 from the top [218]. The CO_2 is first only dissolved as a gas into the brine solution. This applies to 99% of the total dissolved CO_2. Subsequently, carbonic acid will be formed (Equation (8)), which will then dissociate into hydrogen carbonate (Equation (9)) and further into carbonate (Equation (10)) [220].

$$CO_2 + H_2O \rightleftharpoons H_2CO_3 \quad (8)$$

$$H_2CO_2 \rightleftharpoons H^+ + HCO_3^- \quad (9)$$

$$HCO_3^- \rightleftharpoons H^+ + CO_3^{2-} \quad (10)$$

The density difference results in a natural convection and is only controlled by molecular diffusion [220]. It has been simulated that, for a 30-year injection of 1 Mt of CO_2 per year in a saline aquifer, roughly 8% of the injected CO_2 will be dissolved after 200 years if natural convection is the only mechanism. To optimize and accelerate this process, an additional brine injection after the CO_2 injection period is presented, resulting in a 50% dissolution of the CO_2 after the 200-year period [220].

Additionally, mineral trapping can be a consequence of solubility trapping, because the formed carbonic acid may react with minerals inside the brine solution or mineral formations surrounding the saline aquifer. A major advantage of mineral capture is that the CO_2 is bound chemically and precipitates from the solution, allowing for long-term storage. However, the reaction of the CO_2 with the minerals in the water can be fast compared to other traps or very slow, up to thousands of years, depending on the minerals present [218]. It also depends on the pH value of the brine solution, which can be modified to enhance the mineralization processes [221].

Storing CO_2 or, more specifically, acid gases, in deep saline aquifers was first operated in the 1990s by several gas and oil producers in the Alberta basin, Canada. Due to regulations and the uneconomic separation of acid gases (containing H_2S and CO_2), mixtures with gas contents ranging from 85% H_2S and 15% CO_2 to 2% H_2S and 98% CO_2 were injected into deep saline aquifers. By the end of 2002, nearly 1.5 Mt CO_2 and 1 Mt H_2S were injected successfully in this region [222]. The first commercial project dedicated to only injecting CO_2 on a bigger scale into deep saline aquifers started in 1996. The CO_2 was captured at the Sleipner gas field in the North Sea. By 2004, the project had successfully stored nearly 8 Mt of CO_2 [222]. Currently, they claim to have stored around 20 Mt CO_2 [223] with a carbon removal of around 0.8 Mt CO_2 per year (as of 2016) [224].

5.2. Depleted Oil and Gas Reservoirs (DOGR)

Another way to store CO_2 in geological formations is in depleted oil and gas reservoirs. It is considered a safer option than deep saline aquifers because of the proven ability to store hydrocarbons, such as natural gas, over many years [202]. The IPCC Special Report on Carbon Dioxide Capture and Storage estimates the storage capacity of depleted oil and gas reservoirs in a range from 675 to 900 Gt CO_2 [208]. The reservoirs are mostly sealed with impermeable rocks, which provide a sealed storage location [202]. Due to the presence of (brine) water or residual oil in these reservoirs, previously discussed trapping mechanisms can also be applied, including structural trapping, capillary trapping, solubility trapping, and mineralization trapping [225]. Despite the possible leakage pathways, such as faults or fractures in the cap rock, well-integrity is a high-risk leakage pathway for this storage location [226]. Wellbores consist of steel tubes embedded in cement, while old abandoned wells are sealed with cement plugs. Corrosion can cause fractures in the cement or even the steel due to the acidic nature of CO_2 allowing CO_2 to escape into shallow formations or even into the atmosphere. In addition, old wells may have outdated wellbore standards and little information about the reservoir geology, posing a potential leakage threat. Furthermore, most of the old wells were not designed to handle CO_2 [226,227]. If enhanced oil/gas recovery has been used in these reservoirs, the infrastructure for handling supercritical CO_2 and data on potential leakage risks are available, facilitating CO_2 storage activities [228].

The company Wintershall Dea, a leading member of the Greensand consortium, plans to store from 0.5 to 1 Mt CO_2 per year by 2025 and from 4 to 8 Mt CO_2 per year by 2030 in the old offshore oil field Nini West (Danish North Sea) [229]. Another example is Beach Energy's Moomba CCS project, which stores CO_2 in depleted gas reservoirs near the Moomba Gas Plant, Australia. They achieved their first injection in 2024 and can store up to 1.7 Mt CO_2 per year [230].

5.3. Enhanced Oil Recovery (EOR)

Enhanced oil recovery is closely related to the prior discussed method to store CO_2 in depleted oil/gas reservoirs. The difference is that EOR is used at active wells, maximizing the yield of oil or gas from the site. The decline in new oil and gas field discoveries over the last few decades has highlighted EOR as an important strategy for the future [231]. Specifically, it is a tertiary oil recovery method used to extract residual oil from oil fields that cannot be recovered by conventional primary or secondary methods. In numbers, the leftover oil is around two-thirds of all original oil in place (OOIP) [232].

There are three possible EOR strategies, whereas only one can be used to store CO_2 underground. The first is thermal EOR, in which heated steam or hot water is injected into the well to lower the viscosity of the oil and enhance the oil extraction. In situ heat generation from the combustion of tiny amounts of oil within the oil reservoir can also be applied [233]. However, this technique of in situ heat generation is not easy to implement due to the complex nature of combustion [231]. Another method is the chemical EOR, where chemicals such as polymers or surfactants are injected to enhance oil properties like viscosity or lower interfacial tension. For polymers, naturally derived polymers like xanthan gum or synthetically based polymers like polyacrylamides can be used. For surfactants, cationic, anionic, nonionic, or zwitterionic surfactants, like the anionic alkyl benzene sulfonate, can be used [234–236].

The last EOR method and practical for CO_2 storage is the CO_2-EOR. For this, CO_2 with a purity from around 90 to 98% [237] is pumped in its supercritical state into the oil fields to extract the remaining oil, whereas most of the CO_2 stays deep underground [203]. By injecting the CO_2, it becomes miscible with the oil, lowering its viscosity and improving the volumetric sweep and displacement efficacies. Consequently, oil production is enhanced [238,239].

In applications where miscible CO_2 flooding is not practical, e.g., due to technical requirements, CO_2 can also be injected at lower pressures to act as an immiscible fluid, displacing the oil in the reservoir throughout the porous rock structure. The oil is then transported to the well for extraction. Knowledge of the rock wettability and interfacial tensions between CO_2 and oil is crucial for this progress [240].

In 2010, the project at the Petrobras Santos Basin Pre-Salt Oil Field (Brazil) started, which is currently the largest CCS facility on the planet with a yearly carbon capture of over 10 Mt CO_2 and enhanced oil recovery for storage. As for 2022, they claim to have 40.8 Mt of CO_2 reinjected since the beginning of the project [241].

5.4. Mineralizing the CO_2 Underground

Besides large geological formations, such as saline aquifers and depleted oil/gas reservoirs, CO_2 can be injected directly into mineral formations to carbonize the CO_2 to carbonates and immobilize it for long-term storage. If the operating humidity is less than 0.2 relative humidity, the process is known as the dry route. In contrast, the water route is the mineralization of CO_2 under wet conditions [84]. To form carbonates with acid gases like CO_2, the rocks should be rich in calcium and magnesium or iron silicate [242], like in mafic or ultramafic lithologies [208]. The term mafic describes ferromagnesium minerals of all kinds, for example, basalt [243]. Basalt formations are widely available. Most of the ocean floor and more than 5% of the continents are basaltic [208]. The mineralization also highly depends on the pressure of the gas stream and time, resulting in slow kinetics and low conversions in most cases [84]. They can form carbonates in combination with CO_2 according to this general Equation (Equation (11)). In the water route, the CO_2 forms carbonic acid (Equations (8)–(10)), which improves the dissolution of the minerals and the carbonation process [84].

$$M_xSi_yO_{x+2y+z}H_{2z} + x\,CO_2 \rightarrow x\,MCO_3 + y\,SiO_2 + z\,H_2O \quad (11)$$

In contrast to the direct carbonization route of minerals, the indirect route includes two steps. In the first step, the minerals are dissolved with an acid (Equation (12)) and then converted to carbonates with CO_2 in alkaline conditions (Equations (13)–(14)). It has higher carbonation efficiencies and purity of the by-products compared to the direct route [84].

$$M_xSi_yO_{x+2y+z}H_{2z} + 2x\,H^+ \rightarrow x\,M^{2+} + y\,SiO_2 + (x+z)\,H_2O \quad (12)$$

$$CO_2 + 2\,OH^- \rightarrow CO_3^{\,2-} + H_2O \quad (13)$$

$$M^{2+} + CO_3^{\,2-} \rightarrow MCO_3 \quad (14)$$

Other rock types such as sandstone, carbonate, and shale can also be used for CO_2 mineralization methods. Mineralization processes generally depend highly on temperature, permeability, rock type, porosity, and mineral precipitation. A deeper understanding of these factors for a specific storage location strengthens the effectiveness of carbon storage [244].

Commercial carbon storage via mineralization is being carried out by the company Climeworks and its DAC facility Orca. By injecting the captured CO_2 into the basaltic rock formation in Iceland, the CO_2 is transformed chemically into carbonates [245].

5.5. Coal Seams

An alternative approach to carbon storage could be the adsorption of CO_2 in unmineable coal seams. Estimations of global capacity are uncertain and range from 3 to 200 Gt of CO_2 [246]. Due to the coalification process, these coal seams contain methane, mostly adsorbed on the surface of the coal. By injecting CO_2 into these coal seams, the coal bed methane (CBM) can be replaced and extracted, which is called enhanced coalbed methane recovery (CO_2-ECBM) [247]. This process is analogous to the CO_2-EOR process. Owing to the higher affinity of CO_2 for adsorption on the coal in comparison to methane, with ratios ranging from 2:1 up to 10:1 for low-rank coal, the CO_2-ECBMR method has the potential to serve as a viable carbon storage method [248]. A field test in Japan concluded that permeability is a key factor for technical feasibility, allowing economical and large-scale operations [249]. The effects of different parameters on the permeability, such as temperature, pressure, stress changes in the coal bed, and different coal ranks, are currently investigated in the literature [250–253]. Ongoing applications are rare, but successful projects still in operation include the San Juan Basin (NM, USA), Fenn Big Valley (AB, Canada), or Uinta Basin (UT, USA). The project in San Juan Basins started in 1996, and the project in the Fenn Big Valley started in 1998, having maximum injection rates of 0.1 and 0.02 Mt CO_2 per year, respectively. The Uinta Basin project, which started in 2005, operates at a significantly higher maximum rate, injecting 0.9 Mt of CO_2 per year [35].

5.6. Global Status of CCS Facilities

Currently, there are about 50 operational CCS projects with a CO_2 capturing capacity of 51 Mt CO_2 per year. A total of 44 are in construction, adding an additional 51 Mt CO_2 capture capacity. Furthermore, 534 more facilities are in development. In total, all facilities will have a carbon capture capacity of around 416 Mt CO_2 per year, which is an average growth rate of 32% since 2017 [254].

The issue of global capacity for CO_2 storage is of significant importance. However, due to the absence of sufficient data for individual storage locations and the complexity of calculating the size of different storage possibilities, it is currently challenging to determine an exact global capacity for CO_2 storage. Several studies on this topic are present in the literature, and it is possible to derive estimations from them. However, their validity depends on the scale in question. Regional estimations are mostly more robust and accurate than global ones. By 2007, global CO_2 storage was ascertained to be up to 200,000 Gt. For Europe and the USA alone, it was up to 2499 Gt and 3747 Gt, respectively [255,256]. In the Global Status of CCS 2021 report, they estimated a higher storage capacity for the USA of 8061.8 Gt [254]. It is highly dependent on the given data and calculation methods, which can differ year by year. Nevertheless, it does show the potential of CO_2 storage possibilities, especially because of our annual emission rate of 37.5 Gt (2023) [3]. Yet storage capacity is not infinite and is only a medium-term solution. The importance of techniques to avoid CO_2 emissions on a larger scale should be the main goal. During this transition phase, CO_2 storage seems like an acceptable solution.

6. Utilizing the Captured CO_2 (CCU)

Another possible approach after CO_2 capture besides storing is the utilization of CO_2 as a valuable product. A report by the International Energy Agency (IEA) states that, in 2015, the market for CO_2 consumption was around 230 Mt CO_2 per year. Five years later, in 2020, it was expected to be 250 Mt CO_2 per year, with a rising trend. The three largest CO_2 consumers are the USA (33%), China (22%), and Europe (16%). By sectors, the fertilizer industry, which produces urea, accounts for 57% of total CO_2, the oil industry 34%, including EOR, the food and beverage sector with 3% each, metal fabrication with 2%, and others with 4% [257]. There are many CO_2 utilization possibilities that are presented in the literature [258–260]. For example, CO_2 utilization can be used to add value to products like chemicals or fuels. CO_2 can also act as a C_1 building block for the synthesis of chemicals such as formic acid [261], methanol [262,263], methane [264], polycarbonates [265,266], syngas for industry [267], and urea [268]. It is important to note that every utilization method only captures CO_2 in the short term, leading to a net-zero but not a net-negative emissions scenario. After a period of time, the utilized CO_2 will be released back into the atmosphere and thus must be captured again. Despite this fact, the recycling of already available CO_2 instead of creating new CO_2 by burning fossil fuels is an important factor for a circular economy and a sustainable future.

6.1. Direct Utilization of CO_2

The direct utilization of CO_2 as gas or in a supercritical state plays an important role in industries such as food, beverages, oil, and welding applications as shielding gas [269]. For example, in food processing, supercritical CO_2 can be used as a blowing agent but also for essential oil or caffeine extraction, and for preservation or freeze drying [270]. Bottled water, as an example of beverages, reached 2023 a market size of USD 267.15 billion [271], with sparkling (carbonated) water accounting for USD 29.71 billion in 2020 [272]. With other carbonated beverages like beer, soft drinks, and sparkling wine not even included, the demand for CO_2 in the beverages industry is high and keeps growing.

6.2. Bulk Chemicals

Various bulk chemicals can be produced using CO_2 from capturing processes, enhancing the economic value of CO_2 as industrial feedstock [269]. Many projects with this purpose exist, like the EU Horizon 2020 funded project RECO$_2$DE, which aims to produce value-added additives for the cement industry, like $CaCO_3$, different acids, or glycine from cement industry flue gas CO_2 [273].

6.2.1. Formic Acid

Formic acid is an important and valuable bulk chemical for industry, which has many fields of application such as pharmaceutics, textiles, or agriculture [274]. Current research considers formic acid additionally as a liquid organic hydrogen carrier (LOHC) due to its high H_2 capacity of 53 g H_2/L [275]. Global production of formic acid via hydrolysis of methyl formate was around 770,000 tons in 2014 [274]. Formic acid is conventionally produced by reacting methanol with CO (Equation (15)) in an exothermic reaction and following endothermic hydrolysis (Equation (16)) [276,277].

$$CH_3OH + CO \rightarrow HCOOCH_3 \quad \Delta H_r = -29\ kJ/mol \tag{15}$$

$$HCOOCH_3 + H_2O \rightleftharpoons CH_3OH + HCOOH \quad \Delta H_r = +16.3\ kJ/mol \tag{16}$$

Utilizing captured CO_2 represents a different approach to the production of formic acid by directly reacting CO_2 with H_2, preferably sustainable H_2 (Equation (17)) [278].

$$CO_2 + H_2 \rightarrow HCOOH \quad \Delta H_r = -31.2\ kJ/mol \tag{17}$$

The CO_2 hydrogenation is thermodynamically not favored in contrast to the hydrolysis of methyl formate. Hence, a catalyst for direct hydrogenation should be utilized. In the literature, homogenous catalysts are the most studied systems for direct CO_2 hydrogenation [279], first described in 1976 by a combination of VIII transition metals and a base [280]. Currently, more advanced catalysts like noble metals with pincer ligands [281] or catalysts with non-platinum metals like Fe or Co [282,283] are showing high activity in basic solutions. However, the use of basic agents to remove the acidic product produces significant amounts of salt waste [279]. Research on catalysts in acidic conditions has already been carried out to counteract this [284]. A challenge for this pathway is the lower activity for catalysts used under acidic conditions like Ru complexes with bipyridine ligands [285].

6.2.2. Urea

The production of sustainable urea is highly important as it is the most important nitrogenous fertilizer [286]. The industrial process for the production of urea via the Bosch-Meiser process [253] proceeds by the reaction of NH_3 with CO_2 to form ammonium carbamate (Equation (18)) and further urea (Equation (19)). The main issue is currently the fossil fuel-based production of NH_3 by steam reforming CH_4 and additional water–gas shift reactions to produce CO_2 and H_2 [287]. A key role in turning urea production sustainable is the sustainable production of H_2 via electrolysis for NH_3 production and captured CO_2 utilization as feedstock [288].

$$2\ NH_3 + CO_2 \rightleftharpoons NH_2COONH_4 \quad \Delta H_r = -117\ kJ/mol \tag{18}$$

$$NH_2COONH_4 \rightleftharpoons NH_2CONH_2 + H_2O \quad \Delta H_r = +16\ kJ/mol \tag{19}$$

A techno–economic and environmental analysis of a urea production plant was performed in 2024. The plant consists of an air separation unit for N_2, a water electrolysis unit for H_2, a carbon capture unit for CO_2, an ammonia synthesis unit for NH_3, and a urea synthesis unit for final urea production. The carbon capture unit uses MEA absorption to capture 90% CO_2 from cement flue gas. The study reports that, for a plant with an annual urea production of 220,000 tons of urea, an energy consumption of around 8.18 million GJ per year (2272 GWh), where the electrolysis consumes roughly half of the total energy demand. The energy demand is in the same range as the conventional process. The global warming potential is significantly lower at 326 kg CO_2-eq per ton of urea. This is due to the purging of some NH_3 during the process, not capturing CO_2 in the flue gas stream, and the released, unreacted CO_2 after urea production. However, the produced green urea would cost USD 571/ton and would therefore be 62.2% more expensive than fossil fuel-based urea (USD 351.25/ton). With carbon credit systems or lower electricity prices, the cost of urea could be lowered to USD 417.72/ton or even USD 200/ton [268].

Research is also ongoing to substitute NH_3 with only N_2. By using frustrated Lewis pairs (FLPs) on rice-like InOOH nanocrystals, a reaction of N_2 and CO_2 can produce urea electrochemically at a yield rate of 6.85 mmol/h per gram [286]. To provide more context, using 1 kg of a catalyst, daily urea production would be 9.87 kg. Despite current low production rates, replacing NH_3 with N_2 may be a possible pathway to more sustainable urea. However, potential environmental issues by handling indium, e.g., production and waste management, in large quantities, should be taken into consideration.

6.3. Fuels

Besides bulk chemicals, the production of fuels via captured CO_2 is an important topic for a net-zero emission future. Next to single components such as methanol and methane used as fuels, the possibility for syngas production via reversed water–gas shift reaction (Equation (5)) can be achieved. The only drawback is the need for hydrogen as feedstock because it is also a part of syngas. The hydrogen should be derived from renewable sources to transform such syngas production sustainably. If achieved in the industry, syngas can be used to serve as a renewable feedstock for fuel production or ammonia production, among others.

As an example, the European Horizon 2020 project SUN-to-LIQUID produces Fisher-Tropsch fuels via syngas from the thermochemical conversion of captured CO_2 and water [289].

6.3.1. Methanol

Methanol, with a current production capacity of 181 million tons in 2024 [290], is an important chemical for fuel applications due to its high energy content of 726.3 kJ/mol [291]. It can be mixed with conventional fuels without technical modifications up to 85% [292]. Besides formic acid, methanol is also considered a good liquid organic hydrogen carrier for H_2 transportation and serves as an important C_1 building block for industry [293].

Conventionally, methanol is produced by a reaction of CO and H_2 (Equation (20)) via Cu/ZnO catalyst, both derived from fossil-based syngas. For CO_2 utilization, methanol can be directly synthesized via hydrogenation of CO_2 (Equation (21)) [294,295].

$$CO + 2\,H_2 \rightleftharpoons CH_3OH \quad \Delta H_r = -90.7\ kJ/mol \tag{20}$$

$$CO_2 + 3\,H_2 \rightarrow CH_3OH + H_2O \quad \Delta H_r = -49.5\ kJ/mol \tag{21}$$

The direct hydrogenation of CO_2 with H_2 to form methanol is thermodynamically favorable, but kinetics are a major drawback of this reaction. Industrial applications are therefore currently not economically feasible without high active catalysts. The conventional route is additionally more attractive because of the methanol yield and high equipment/hydrogen production cost. If large scale production of renewable hydrogen can be achieved, processes for direct hydrogenation, such as methanol production, can become more favorable [296,297].

Methanol is also an important starting material for several other products, such as formaldehyde, fuel additives (MTBE), acetic acid, methacrylate, and several other chemicals [292]. It can also be used to produce dimethyl ether (DME), another important fuel alternative, via a dehydration reaction (Equation (22)) [269,295].

$$2\,CH_3OH \rightarrow CH_3OCH_3 + H_2O \quad \Delta H_r = -23.4\ kJ/mol \tag{22}$$

Due to all the advantages of producing methanol from captured CO_2 and renewable H_2 sources, it is considered as an important key factor for transitioning from fossil-based to renewable energy usage [294].

Current example is the company Carbon Recycling International (CRI), which has been operating the George Olah Plant in Iceland since 2012. They claim to be the world's first e-methanol plant, with a production of 214 Mt methanol per year from captured CO_2 [298].

6.3.2. Methane

Methane (CH_4) is in high demand as a fuel, for heat generation in buildings, and as a feedstock for the chemical industry. Currently, CH_4 is used in steam reforming for syngas production (Equation (23)), resulting in H_2 and CO for further processes. This is also the most common method for H_2 production [295].

$$CH_4 + H_2O \rightleftharpoons CO + 3\,H_2 \quad \Delta H_r = +205.8\ kJ/mol \tag{23}$$

To produce CH_4 from CO_2, direct hydrogenation can be applied (Equation (24)). The so-called Sabatier reaction is named after the French chemist Paul Sabatier [269,295].

$$CO_2 + 4\,H_2 \rightleftharpoons CH_4 + 2\,H_2O \quad \Delta H_r = -164.63\ kJ/mol \tag{24}$$

The reaction can be carried out using precious metal catalysts, like Ru or Pd, which can be costly and are limited in their reserves. A different approach is the usage of non-noble metal catalysts like Co. These catalysts have lower activity, but experiments have shown that by combining them with, e.g., porous molecular sieves like KIT-6 as a support, the activity and selectivity can be increased [299]. Moreover, Ni catalysts can be used, which show high activity and selectivity and are also much cheaper than precious metal catalysts [299].

It is also possible to produce methane biologically via hydrogenotrophic methanogens by feeding the microorganisms with captured CO_2 and H_2 [300].

A recent study investigated a compressed natural gas (CNG) pilot plant in Poland that continuously produces CH_4 via hydrogenation of CO_2 from flue gas, which is captured with amines. The H_2 needed for the process was produced by water electrolysis from renewable energy. The plant uses a two-stage catalytic reactor filled with a Ni/Al_2O_3 catalyst. The plant carried out a 99.4% conversion of the CO_2 at a gas flow of 8.8 kg/h with temperatures in the reactors of about 300 °C and 9.3 bar. The produced CNG has a CH_4 content of ~94.6% and can thus be injected into the local gas grid. However, the study also describes that catalyst wetting was a problem during long-term hydrogenation, and therefore, a decrease in CH_4 production was observed [301].

6.4. Algae Biomass

Another big part of CO_2 utilization is the use of algae fed with CO_2 from flue gases for biomass cultivation. The biomass can then be transformed into other value-added products, such as fuels [302–304]. For example, algae like *Chlorella* sp. can be cultivated in an outdoor photobioreactor (55 m^2) to produce 1 kg dry biomass from 4.4 kg CO_2 with a daily production rate ranging from 14.6 to 28.1 g/m^2. The results in this experiment were generated in summer, which can differ from experiments in winter due to seasonal or weather-related changes in sunlight. The impurities in the used flue gas (produced by natural gas combustion), such as NO_x and CO with contents up to 45 mg/m^3 and 3 mg/m^3, respectively, had no negative influences on the growth [305].

A commercial example is the company Algaecytes, which claims to have built the world's largest commercial-scale microalgae production plant located in Dessau (Germany) to harvest essential omega-3 fatty acids. The plant with an operation start planned for 2024 consists of an enclosed facility with LED lights inside to achieve 24/7 production [306].

7. Conclusions

This work has reviewed a selection of both established and emerging CO_2 capture technologies, as well as storage and utilization options. Processes like absorption using MEA are already well-established in industrial practices, and emerging technologies like calcium looping and membrane carbon capture are being tested on a bigger scale. Techniques like cryogenic carbon capture, initially considered as an uneconomic method of capturing CO_2 from flue gases, are currently competitive with other carbon capture technologies. This shows that every technology has potential if it is sufficiently researched and tested in practice. Furthermore, direct air capture is no longer a concept; it has crossed over into practical experimentation and commercialization.

Several use cases for carbon utilization are reviewed for important products like methanol or formic acid, which are C_1 building blocks for the chemical industry. Implementing such production chains for a sustainable future regarding a cyclic economy and CO_2 as a new alternative carbon source instead of oil or natural gas is discussed. The potential of applied carbon storage is presented, and storage capacities for mid-term storage to achieve net-zero or net-negative emissions are shown.

In total, with all facilities operating, under construction, or in development, a global carbon removal capacity of 416 Mt CO_2 per year was reached in 2024. This is, compared to our annual CO_2 emissions of 37.5 GT CO_2 per year, only 1.11%. However, given the rapid growth over the past few years and the ongoing research in novel technologies, carbon capture technologies will be more implemented in current applications.

It is obvious that all the discussed technologies need continued research and development to improve their performance, reduce their cost, and enable them to be more universally applicable. Real-life applications of some of the technologies discussed in this article, like storage solutions and enhanced capture strategies, are currently being tested and up-scaled for experimental and pilot scenarios or even demonstration plants.

Author Contributions: T.M.T.: conceptualization, investigation, writing—original draft. M.W.: supervision, proofreading. All authors reviewed the manuscript. All authors have read and agreed to the published version of the manuscript.

Funding: This research received no external funding.

Data Availability Statement: This manuscript does not report data generation or analysis.

Acknowledgments: We would like to thank Melanie Walther for additional proofreading of the manuscript.

Conflicts of Interest: The authors declare that they have no known competing financial interests or personal relationships that could have appeared to influence the work reported in this paper.

References

1. IPCC. *2023: Climate Change 2023: Synthesis Report. Contribution of Working Groups I, II and III to the Sixth Assessment Report of the Intergovernmental Panel on Climate Change*; Core Writing Team, Lee, H., Romero, J., Eds.; IPCC: Geneva, Switzerland, 2023. [CrossRef]
2. Cuartas, J.; Mccoy, D.C.; Torres, I.; Burghardt, L.; Shonkoff, J.P.; Yoshikawa, H. The Developmental Consequences of Early Exposure to Climate Change-Related Risks. *Child Dev. Perspect.* **2024**, *18*, 145–154. [CrossRef]
3. Global Carbon Project 2024. Available online: https://globalcarbonbudget.org/gcb-2024/ (accessed on 5 October 2024).
4. European Commission. Communication from the Commission to the European Parliament: The European Green Deal. Available online: https://eur-lex.europa.eu/legal-content/EN/TXT/?uri=CELEX:52019DC0640 (accessed on 15 October 2024).
5. Chandran Govindaraju, V.G.R.; Tang, C.F. The Dynamic Links between CO_2 Emissions, Economic Growth and Coal Consumption in China and India. *Appl. Energy* **2013**, *104*, 310–318. [CrossRef]
6. Olabi, A.G.; Wilberforce, T.; Elsaid, K.; Sayed, E.T.; Maghrabie, H.M.; Abdelkareem, M.A. Large Scale Application of Carbon Capture to Process Industries—A Review. *J. Clean. Prod.* **2022**, *362*, 132300. [CrossRef]
7. Raganati, F.; Ammendola, P. CO_2 Post-Combustion Capture: A Critical Review of Current Technologies and Future Directions. *Energy Fuels* **2024**, *38*, 13858–13905. [CrossRef]
8. Vasudevan, S.; Farooq, S.; Karimi, I.A.; Saeys, M.; Quah, M.C.G.; Agrawal, R. Energy Penalty Estimates for CO_2 Capture: Comparison between Fuel Types and Capture-Combustion Modes. *Energy* **2016**, *103*, 709–714. [CrossRef]

9. Sun, N.; Tang, Z.; Wei, W.; Snape, C.E.; Sun, Y. Solid Adsorbents for Low-Temperature CO_2 Capture with Low-Energy Penalties Leading to More Effective Integrated Solutions for Power Generation and Industrial Processes. *Front. Energy Res.* **2015**, *3*, 9. [CrossRef]
10. Kheirinik, M.; Ahmed, S.; Rahmanian, N. Comparative Techno-Economic Analysis of Carbon Capture Processes: Pre-Combustion, Post-Combustion, and Oxy-Fuel Combustion Operations. *Sustainability* **2021**, *13*, 13567. [CrossRef]
11. Garcia, J.A.; Villen-Guzman, M.; Rodriguez-Maroto, J.M.; Paz-Garcia, J.M. Technical Analysis of CO_2 Capture Pathways and Technologies. *J. Environ. Chem. Eng.* **2022**, *10*, 108470. [CrossRef]
12. Karayil, A.; Elseragy, A.; Aliyu, A.M. An Assessment of CO_2 Capture Technologies towards Global Carbon Net Neutrality. *Energies* **2024**, *17*, 1460. [CrossRef]
13. Zhang, Z.; Vo, D.-N.; Kum, J.; Hong, S.-H.; Lee, C.-H. Enhancing Energy Efficiency of Chemical Absorption-Based CO_2 Capture Process with Advanced Waste-Heat Recovery Modules at a High Capture Rate. *Chem. Eng. J.* **2023**, *472*, 144918. [CrossRef]
14. Rubin, E.S.; Chen, C.; Rao, A.B. Cost and Performance of Fossil Fuel Power Plants with CO_2 Capture and Storage. *Energy Policy* **2007**, *35*, 4444–4454. [CrossRef]
15. Obi, D.; Onyekuru, S.; Orga, A. Review of Recent Process Developments in the Field of Carbon Dioxide (CO_2) Capture from Power Plants Flue Gases and the Future Perspectives. *Int. J. Sustain. Energy* **2024**, *43*, 2317137. [CrossRef]
16. Bravo, J.; Drapanauskaite, D.; Sarunac, N.; Romero, C.; Jesikiewicz, T.; Baltrusaitis, J. Optimization of Energy Requirements for CO_2 Post-Combustion Capture Process through Advanced Thermal Integration. *Fuel* **2021**, *283*, 118940. [CrossRef]
17. Madejski, P.; Chmiel, K.; Subramanian, N.; Kuś, T. Methods and Techniques for CO_2 Capture: Review of Potential Solutions and Applications in Modern Energy Technologies. *Energies* **2022**, *15*, 887. [CrossRef]
18. Rezaei, S.; Liu, A.; Hovington, P. Emerging Technologies in Post-Combustion Carbon Dioxide Capture & Removal. *Catal. Today* **2023**, *423*, 114286. [CrossRef]
19. Osman, A.I.; Hefny, M.; Abdel Maksoud, M.I.A.; Elgarahy, A.M.; Rooney, D.W. Recent Advances in Carbon Capture Storage and Utilisation Technologies: A Review. *Environ. Chem. Lett.* **2021**, *19*, 797–849. [CrossRef]
20. Mondal, M.K.; Balsora, H.K.; Varshney, P. Progress and Trends in CO_2 Capture/Separation Technologies: A Review. *Energy* **2012**, *46*, 431–441. [CrossRef]
21. Yagmur Goren, A.; Erdemir, D.; Dincer, I. Comprehensive Review and Assessment of Carbon Capturing Methods and Technologies: An Environmental Research. *Environ. Res.* **2024**, *240*, 117503. [CrossRef]
22. Car, A.; Stropnik, C.; Yave, W.; Peinemann, K.V. Pebax®/Polyethylene Glycol Blend Thin Film Composite Membranes for CO_2 Separation: Performance with Mixed Gases. *Sep. Purif. Technol.* **2008**, *62*, 110–117. [CrossRef]
23. Figueroa, J.D.; Fout, T.; Plasynski, S.; McIlvried, H.; Srivastava, R.D. Advances in CO_2 Capture Technology—The U.S. Department of Energy's Carbon Sequestration Program. *Int. J. Greenh. Gas Control* **2008**, *2*, 9–20. [CrossRef]
24. Giannaris, S.; Janowczyk, D.; Ruffini, J.; Hill, K.; Jacobs, B.; Bruce, C.; Feng, Y.; Srisang, W. SaskPower's Boundary Dam Unit 3 Carbon Capture Facility—TheJourney to Achieving Reliability. In Proceedings of the 15th International Conference on Greenhouse Gas Control Technologies, GHGT-15, Abu Dhabi, UAE, 15 March 2021.
25. SaskPower—Boundary Dam Power Station. Available online: https://www.saskpower.com/our-power-future/our-electricity/electrical-system/system-map/boundary-dam-power-station (accessed on 31 December 2024).
26. Lau, H.C. The Contribution of Carbon Capture and Storage to the Decarbonization of Coal-Fired Power Plants in Selected Asian Countries. *Energy Fuels* **2023**, *37*, 15919–15934. [CrossRef]
27. Spigarelli, B.P.; Kawatra, S.K. Opportunities and Challenges in Carbon Dioxide Capture. *J. CO2 Util.* **2013**, *1*, 69–87. [CrossRef]
28. Yang, H.; Xu, Z.; Fan, M.; Gupta, R.; Slimane, R.B.; Bland, A.E.; Wright, I. Progress in Carbon Dioxide Separation and Capture: A Review. *J. Environ. Sci.* **2008**, *20*, 14–27. [CrossRef] [PubMed]
29. El Hadri, N.; Quang, D.V.; Goetheer, E.L.V.; Abu Zahra, M.R.M. Aqueous Amine Solution Characterization for Post-Combustion CO_2 Capture Process. *Appl. Energy* **2017**, *185*, 1433–1449. [CrossRef]
30. Rao, A.B.; Rubin, E.S. A Technical, Economic, and Environmental Assessment of Amine-Based CO_2 Capture Technology for Power Plant Greenhouse Gas Control. *Environ. Sci. Technol.* **2002**, *36*, 4467–4475. [CrossRef] [PubMed]
31. Lv, B.; Guo, B.; Zhou, Z.; Jing, G. Mechanisms of CO_2 Capture into Monoethanolamine Solution with Different CO_2 Loading during the Absorption/Desorption Processes. *Environ. Sci. Technol.* **2015**, *49*, 10728–10735. [CrossRef] [PubMed]
32. Caplow, M. Kinetics of Carbamate Formation and Breakdown. *J. Am. Chem. Soc.* **1968**, *90*, 6795–6803. [CrossRef]
33. Danckwerts, P.V. The Reaction of CO_2 with Ethanolamines. *Chem. Eng. Sci.* **1979**, *34*, 443–446. [CrossRef]
34. Kittel, J.; Idem, R.; Gelowitz, D.; Tontiwachwuthikul, P.; Parrain, G.; Bonneau, A. Corrosion in MEA Units for CO_2 Capture: Pilot Plant Studies. *Energy Procedia* **2009**, *1*, 791–797. [CrossRef]
35. Leung, D.Y.C.; Caramanna, G.; Maroto-Valer, M.M. An Overview of Current Status of Carbon Dioxide Capture and Storage Technologies. *Renew. Sustain. Energy Rev.* **2014**, *39*, 426–443. [CrossRef]
36. He, X.; He, H.; Barzagli, F.; Amer, M.W.; Li, C.; Zhang, R. Analysis of the Energy Consumption in Solvent Regeneration Processes Using Binary Amine Blends for CO_2 Capture. *Energy* **2023**, *270*, 126903. [CrossRef]

37. Kumari, M.; Vega, F.; Gallego Fernández, L.M.; Prasad Shadangi, K.; Kumar, N. Liquid Amine Functional, Aqueous Blends and the CO_2 Absorption Capacity: Molecular Structure, Size, Interaction Parameter and Mechanistic Aspects. *J. Mol. Liq.* **2023**, *384*, 122288. [CrossRef]
38. Valluri, S.; Kawatra, S.K. Use of Frothers to Improve the Absorption Efficiency of Dilute Sodium Carbonate Slurry for Post Combustion CO_2 Capture. *Fuel Process. Technol.* **2021**, *212*, 106620. [CrossRef]
39. Yu, W.; Wang, T.; Park, A.-H.A.; Fang, M. Review of Liquid Nano-Absorbents for Enhanced CO_2 Capture. *Nanoscale* **2019**, *11*, 17137–17156. [CrossRef] [PubMed]
40. Sibhat, M.; Zhu, Q.; Tsegay, G.; Yao, G.; Yin, G.; Zhou, Y.; Zhao, J. Enhancement Technologies of Ammonia-Based Carbon Capture: A Review of Developments and Challenges. *Int. J. Greenh. Gas Control* **2024**, *136*, 104196. [CrossRef]
41. Qu, Y.; Zhao, Y.; Li, D.; Sun, J. Task-Specific Ionic Liquids for Carbon Dioxide Absorption and Conversion into Value-Added Products. *Curr. Opin. Green Sustain. Chem.* **2022**, *34*, 100599. [CrossRef]
42. Liu, F.; Shen, Y.; Shen, L.; Sun, C.; Chen, L.; Wang, Q.; Li, S.; Li, W. Novel Amino-Functionalized Ionic Liquid/Organic Solvent with Low Viscosity for CO_2 Capture. *Environ. Sci. Technol.* **2020**, *54*, 3520–3529. [CrossRef]
43. Zhang, R.; Ke, Q.; Zhang, Z.; Zhou, B.; Cui, G.; Lu, H. Tuning Functionalized Ionic Liquids for CO_2 Capture. *Int. J. Mol. Sci.* **2022**, *23*, 11401. [CrossRef] [PubMed]
44. Soo, X.Y.D.; Lee, J.J.C.; Wu, W.-Y.; Tao, L.; Wang, C.; Zhu, Q.; Bu, J. Advancements in CO_2 Capture by Absorption and Adsorption: A Comprehensive Review. *J. CO2 Util.* **2024**, *81*, 102727. [CrossRef]
45. Cui, X.; Ye, H.; Dong, H. *A Novel Rectisol Process for H2s Enrichment and CO_2 Capture: Process Optimization and Thermodynamic Analysis*; SSRN: Rochester, NY, USA, 2024.
46. Ban, Z.H.; Keong, L.K.; Mohd Shariff, A. Physical Absorption of CO_2 Capture: A Review. *Adv. Mater. Res.* **2014**, *917*, 134–143. [CrossRef]
47. Yang, S.; Qian, Y.; Yang, S. Development of a Full CO_2 Capture Process Based on the Rectisol Wash Technology. *Ind. Eng. Chem. Res.* **2016**, *55*, 6186–6193. [CrossRef]
48. Sharma, I.; Hoadley, A.F.A.; Mahajani, S.M.; Ganesh, A. Multi-Objective Optimisation of a Rectisol^{TM} Process for Carbon Capture. *J. Clean. Prod.* **2016**, *119*, 196–206. [CrossRef]
49. Kapetaki, Z.; Brandani, P.; Brandani, S.; Ahn, H. Process Simulation of a Dual-Stage Selexol Process for 95% Carbon Capture Efficiency at an Integrated Gasification Combined Cycle Power Plant. *Int. J. Greenh. Gas Control* **2015**, *39*, 17–26. [CrossRef]
50. Chen, W.H.; Chen, S.M.; Hung, C.I. Carbon Dioxide Capture by Single Droplet Using Selexol, Rectisol and Water as Absorbents: A Theoretical Approach. *Appl. Energy* **2013**, *111*, 731–741. [CrossRef]
51. Acid Gas Treating. In *Handbook of Natural Gas Transmission and Processing*; Elsevier: Amsterdam, The Netherlands, 2006; pp. 261–294.
52. Burr, B.; Lyddon, L. A Comparison of Physical Solvents for Acid Gas Removal. Gas Processors' Association Convention, Grapevine, TX. 2008. Available online: https://www.bre.com/PDF/A-Comparison-of-Physical-Solvents-for-Acid-Gas-Removal-REVISED.pdf (accessed on 16 October 2024).
53. Hakka, L.E.; Ouimet, M.A. Method for Recovery of CO2 from Gas Streams. U.S. Patent US7056482B2, 6 June 2006.
54. Ouimet, M.A. Process for the Recovery of Carbon Dioxide from a Gas Stream. U.S. Patent US7601315B2, 13 October 2009.
55. Ben-Mansour, R.; Habib, M.A.; Bamidele, O.E.; Basha, M.; Qasem, N.A.A.; Peedikakkal, A.; Laoui, T.; Ali, M. Carbon Capture by Physical Adsorption: Materials, Experimental Investigations and Numerical Modeling and Simulations—A Review. *Appl. Energy* **2016**, *161*, 225–255. [CrossRef]
56. Abd, A.A.; Naji, S.Z.; Hashim, A.S.; Othman, M.R. Carbon Dioxide Removal through Physical Adsorption Using Carbonaceous and Non-Carbonaceous Adsorbents: A Review. *J. Environ. Chem. Eng.* **2020**, *8*, 104142. [CrossRef]
57. Clausse, M.; Bonjour, J.; Meunier, F. Adsorption of Gas Mixtures in TSA Adsorbers under Various Heat Removal Conditions. *Chem. Eng. Sci.* **2004**, *59*, 3657–3670. [CrossRef]
58. Krishna, R. Adsorptive Separation of $CO_2/CH_4/CO$ Gas Mixtures at High Pressures. *Microporous Mesoporous Mater.* **2012**, *156*, 217–223. [CrossRef]
59. Chue, K.T.; Kim, J.N.; Yoo, Y.J.; Cho, S.H.; Yang, R.T. Comparison of Activated Carbon and Zeolite 13X for CO_2 Recovery from Flue Gas by Pressure Swing Adsorption. *Ind. Eng. Chem. Res.* **1995**, *34*, 591–598. [CrossRef]
60. Zhao, R.; Deng, S.; Wang, S.; Zhao, L.; Zhang, Y.; Liu, B.; Li, H.; Yu, Z. Thermodynamic Research of Adsorbent Materials on Energy Efficiency of Vacuum-Pressure Swing Adsorption Cycle for CO_2 Capture. *Appl. Therm. Eng.* **2018**, *128*, 818–829. [CrossRef]
61. Grande, C.A.; Ribeiro, R.P.L.; Oliveira, E.L.G.; Rodrigues, A.E. Electric Swing Adsorption as Emerging CO_2 Capture Technique. *Energy Procedia* **2009**, *1*, 1219–1225. [CrossRef]
62. Verougstraete, B.; Gholami, M.; Gomez-Rueda, Y.; Pérez-Botella, E.; Schoukens, M.; Van Assche, T.R.C.; Denayer, J.F.M. Advancements and Challenges in Electric Heating for Enhanced Temperature Swing Adsorption Processes. *Sep. Purif. Technol.* **2025**, *353*, 128522. [CrossRef]
63. Grande, C.; Rodrigues, A. Electric Swing Adsorption for CO_2 Removal from Flue Gases. *Int. J. Greenh. Gas Control* **2007**, *2*, 194–202. [CrossRef]

64. Chowdhury, S.; Balasubramanian, R. Highly Efficient, Rapid and Selective CO_2 Capture by Thermally Treated Graphene Nanosheets. *J. CO2 Util.* **2016**, *13*, 50–60. [CrossRef]
65. Wang, Y.; Guan, C.; Wang, K.; Guo, C.X.; Li, C.M. Nitrogen, Hydrogen, Carbon Dioxide, and Water Vapor Sorption Properties of Three-Dimensional Graphene. *J. Chem. Eng. Data* **2011**, *56*, 642–645. [CrossRef]
66. Malini, K.; Selvakumar, D.; Kumar, N.S. Activated Carbon from Biomass: Preparation, Factors Improving Basicity and Surface Properties for Enhanced CO_2 Capture Capacity—A Review. *J. CO2 Util.* **2023**, *67*, 102318. [CrossRef]
67. Serafin, J.; Dziejarski, B. Activated Carbons—Preparation, Characterization and Their Application in CO_2 Capture: A Review. *Environ. Sci. Pollut. Res.* **2024**, *31*, 40008–40062. [CrossRef]
68. Ghaemi, A.; Karimi Dehnavi, M.; Khoshraftar, Z. Exploring Artificial Neural Network Approach and RSM Modeling in the Prediction of CO_2 Capture Using Carbon Molecular Sieves. *Case Stud. Chem. Environ. Eng.* **2023**, *7*, 100310. [CrossRef]
69. Wei, Y.; Zhao, T.; Wang, J.; Chen, Y.; Wang, Q.; Liu, X.; Zhao, Y. Ultramicroporous Carbon Molecular Sieve for Air Purification by Selective Adsorption Low-Concentration CO_2 and VOC Molecules. *Ind. Eng. Chem. Res.* **2023**, *62*, 7635–7641. [CrossRef]
70. Jena, K.K.; Panda, A.P.; Verma, S.; Mani, G.K.; Swain, S.K.; Alhassan, S.M. MWCNTs-ZnO-SiO_2 Mesoporous Nano-Hybrid Materials for CO_2 Capture. *J. Alloys Compd.* **2019**, *800*, 279–285. [CrossRef]
71. Hsu, C.-Y.; Rheima, A.M.; Mohammed, M.S.; Kadhim, M.M.; Mohammed, S.H.; Abbas, F.H.; Abed, Z.T.; Mahdi, Z.M.; Abbas, Z.S.; Hachim, S.K.; et al. Application of Carbon Nanotubes and Graphene-Based Nanoadsorbents in Water Treatment. *Bionanoscience* **2023**, *13*, 1418–1436. [CrossRef]
72. Zhao, M.; Yang, Y.; Gu, X.-S. MOF Based CO_2 Capture: Adsorption and Membrane Separation. *Inorg. Chem. Commun.* **2023**, *152*, 110722. [CrossRef]
73. Gebremariam, S.K.; Dumée, L.F.; Llewellyn, P.L.; AlWahedi, Y.F.; Karanikolos, G.N. Metal-Organic Framework Hybrid Adsorbents for Carbon Capture—A Review. *J. Environ. Chem. Eng.* **2023**, *11*, 109291. [CrossRef]
74. Tao, Z.; Tian, Y.; Wu, W.; Liu, Z.; Fu, W.; Kung, C.-W.; Shang, J. Development of Zeolite Adsorbents for CO_2 Separation in Achieving Carbon Neutrality. *NPJ Mater. Sustain.* **2024**, *2*, 20. [CrossRef]
75. Deng, Z.; Gopalan, A.; Sarkisov, L. In Silico Engineering of Ion-Exchanged Zeolites for High-Performance Carbon Capture in PSA Processes. *Chem. Eng. J.* **2023**, *459*. [CrossRef]
76. Reddy, K.S.K.; Varghese, A.M.; Ogungbenro, A.E.; Karanikolos, G.N. Aminosilane-Modified Ordered Hierarchical Nanostructured Silica for Highly-Selective Carbon Dioxide Capture at Low Pressure. *ACS Appl. Eng. Mater.* **2023**, *1*, 720–733. [CrossRef]
77. Alves, A.A.; De Oliveira Notório Ribeiro, J.; Vasconcelos, W.L. Development of Silica-Based Monoliths for the Capture of CO_2. *Mater. Res.* **2019**, *22*, e20190285. [CrossRef]
78. Wang, S.; Yan, S.; Ma, X.; Gong, J. Recent Advances in Capture of Carbon Dioxide Using Alkali-Metal-Based Oxides. *Energy Environ. Sci.* **2011**, *4*, 3805. [CrossRef]
79. González-Varela, D.; Hernández-Fontes, C.; Wang, N.; Pfeiffer, H. State of the Art and Perspectives of the CO_2 Chemisorption in Ceramics with Its Simultaneous or Subsequent Chemical Transformation. *Carbon Capture Sci. Technol.* **2023**, *7*, 100101. [CrossRef]
80. Duan, X.; Song, G.; Lu, G.; Wang, Y.; Sun, J.; Chen, A.; Xie, X. Chemisorption and Regeneration of Amine-Based CO_2 Sorbents in Direct Air Capture. *Mater. Today Sustain.* **2023**, *23*, 100453. [CrossRef]
81. Huhe, F.; King, J.; Chuang, S.S.C. Amine-Based Sorbents for CO_2 Capture from Air and Flue Gas—A Short Review and Perspective. *Res. Chem. Intermed.* **2023**, *49*, 791–817. [CrossRef]
82. Emissions Reduction Alberta. Available online: https://www.eralberta.ca/projects/details/veloxotherm-carbon-capture/ (accessed on 23 November 2024).
83. Shimizu, T.; Hirama, T.; Hosoda, H.; Kitano, K.; Inagaki, M.; Tejima, K. A Twin Fluid-Bed Reactor for Removal of CO_2 from Combustion Processes. *Chem. Eng. Res. Des.* **1999**, *77*, 62–68. [CrossRef]
84. Liu, H.; Lu, H.; Hu, H. CO_2 Capture and Mineral Storage: State of the Art and Future Challenges. *Renew. Sustain. Energy Rev.* **2024**, *189*, 113908. [CrossRef]
85. Erans, M.; Manovic, V.; Anthony, E.J. Calcium Looping Sorbents for CO_2 Capture. *Appl. Energy* **2016**, *180*, 722–742. [CrossRef]
86. Alonso, M.; Rodríguez, N.; Grasa, G.; Abanades, J.C. Modelling of a Fluidized Bed Carbonator Reactor to Capture CO_2 from a Combustion Flue Gas. *Chem. Eng. Sci.* **2009**, *64*, 883–891. [CrossRef]
87. Blamey, J.; Anthony, E.J.; Wang, J.; Fennell, P.S. The Calcium Looping Cycle for Large-Scale CO_2 Capture. *Prog. Energy Combust. Sci.* **2010**, *36*, 260–279. [CrossRef]
88. Mantripragada, H.C.; Rubin, E.S. Calcium Looping Cycle for CO_2 Capture: Performance, Cost And Feasibility Analysis. *Energy Procedia* **2014**, *63*, 2199–2206. [CrossRef]
89. Romano, M.C. Modeling the Carbonator of a Ca-Looping Process for CO_2 Capture from Power Plant Flue Gas. *Chem. Eng. Sci.* **2012**, *69*, 257–269. [CrossRef]
90. Hanak, D.P.; Anthony, E.J.; Manovic, V. A Review of Developments in Pilot-Plant Testing and Modelling of Calcium Looping Process for CO_2 Capture from Power Generation Systems. *Energy Environ. Sci.* **2015**, *8*, 2199–2249. [CrossRef]

91. Martínez, I.; Grasa, G.; Parkkinen, J.; Tynjälä, T.; Hyppänen, T.; Murillo, R.; Romano, M.C. Review and Research Needs of Ca-Looping Systems Modelling for Post-Combustion CO_2 Capture Applications. *Int. J. Greenh. Gas Control* **2016**, *50*, 271–304. [CrossRef]
92. Abanades, J.C.; Anthony, E.J.; Wang, J.; Oakey, J.E. Fluidized Bed Combustion Systems Integrating CO_2 Capture with CaO. *Environ. Sci. Technol.* **2005**, *39*, 2861–2866. [CrossRef] [PubMed]
93. Atsonios, K.; Grammelis, P.; Antiohos, S.K.; Nikolopoulos, N.; Kakaras, E. Integration of Calcium Looping Technology in Existing Cement Plant for CO_2 Capture: Process Modeling and Technical Considerations. *Fuel* **2015**, *153*, 210–223. [CrossRef]
94. Project—CaLby2030—Europe. Available online: https://www.calby2030.eu/overview-of-the-project.html (accessed on 11 November 2024).
95. Calby2030 Project. Available online: https://www.calby2030.eu/project/demonstration-sites.html (accessed on 31 December 2024).
96. Arias, B.; Alvarez Criado, Y.; Méndez, A.; Marqués, P.; Finca, I.; Abanades, J.C. Pilot Testing of Calcium Looping at TRL7 with CO_2 Capture Efficiencies toward 99%. *Energy Fuels* **2024**, *38*, 14757–14764. [CrossRef] [PubMed]
97. Kárászová, M.; Sedláková, Z.; Izák, P. Gas Permeation Processes in Biogas Upgrading: A Short Review. *Chem. Pap.* **2015**, *69*, 1277–1283. [CrossRef]
98. Baker, R.W.; Freeman, B.; Kniep, J.; Wei, X.; Merkel, T. CO_2 Capture from Natural Gas Power Plants Using Selective Exhaust Gas Recycle Membrane Designs. *Int. J. Greenh. Gas Control* **2017**, *66*, 35–47. [CrossRef]
99. Pospiech, B.; Kujawski, W. Ionic Liquids as Selective Extractants and Ion Carriers of Heavy Metal Ions from Aqueous Solutions Utilized in Extraction and Membrane Separation. *Rev. Chem. Eng.* **2015**, *31*, 179–191. [CrossRef]
100. Van der Bruggen, B. Desalination by Distillation and by Reverse Osmosis—Trends towards the Future. *Membr. Technol.* **2003**, *2003*, 6–9. [CrossRef]
101. Vakharia, V.; Salim, W.; Wu, D.; Han, Y.; Chen, Y.; Zhao, L.; Ho, W.S.W. Scale-up of Amine-Containing Thin-Film Composite Membranes for CO_2 Capture from Flue Gas. *J. Memb. Sci.* **2018**, *555*, 379–387. [CrossRef]
102. Luis, P.; Van der Bruggen, B.; Van Gerven, T. Non-Dispersive Absorption for CO_2 Capture: From the Laboratory to Industry. *J. Chem. Technol. Biotechnol.* **2011**, *86*, 769–775. [CrossRef]
103. Sasikumar, B.; Arthanareeswaran, G.; Ismail, A.F. Recent Progress in Ionic Liquid Membranes for Gas Separation. *J. Mol. Liq.* **2018**, *266*, 330–341. [CrossRef]
104. de Montigny, D.; Tontiwachwuthikul, P.; Chakma, A. Using Polypropylene and Polytetrafluoroethylene Membranes in a Membrane Contactor for CO_2 Absorption. *J. Memb. Sci.* **2006**, *277*, 99–107. [CrossRef]
105. Guerrero, G.; Venturi, D.; Peters, T.; Rival, N.; Denonville, C.; Simon, C.; Henriksen, P.P.; Hägg, M.-B. Influence of Functionalized Nanoparticles on the CO_2/N_2 Separation Properties of PVA-Based Gas Separation Membranes. *Energy Procedia* **2017**, *114*, 627–635. [CrossRef]
106. Solimando, X.; Lherbier, C.; Babin, J.; Arnal-Herault, C.; Romero, E.; Acherar, S.; Jamart-Gregoire, B.; Barth, D.; Roizard, D.; Jonquieres, A. Pseudopeptide Bioconjugate Additives for CO_2 Separation Membranes. *Polym. Int.* **2016**, *65*, 1464–1473. [CrossRef]
107. Nikolaeva, D.; Azcune, I.; Tanczyk, M.; Warmuzinski, K.; Jaschik, M.; Sandru, M.; Dahl, P.I.; Genua, A.; Loïs, S.; Sheridan, E.; et al. The Performance of Affordable and Stable Cellulose-Based Poly-Ionic Membranes in CO_2/N_2 and CO_2/CH4 Gas Separation. *J. Memb. Sci.* **2018**, *564*, 552–561. [CrossRef]
108. Lasseuguette, E.; Ferrari, M.-C.; Brandani, S. Humidity Impact on the Gas Permeability of PIM-1 Membrane for Post-Combustion Application. *Energy Procedia* **2014**, *63*, 194–201. [CrossRef]
109. Rodrigues, M.A.; Ribeiro, J.d.S.; Costa, E.d.S.; de Miranda, J.L.; Ferraz, H.C. Nanostructured Membranes Containing UiO-66 (Zr) and MIL-101 (Cr) for O_2/N_2 and CO_2/N_2 Separation. *Sep. Purif. Technol.* **2018**, *192*, 491–500. [CrossRef]
110. Ying, Y.; Cheng, Y.; Peh, S.B.; Liu, G.; Shah, B.B.; Zhai, L.; Zhao, D. Plasticization Resistance-Enhanced CO_2 Separation at Elevated Pressures by Mixed Matrix Membranes Containing Flexible Metal-Organic Framework Fillers. *J. Memb. Sci.* **2019**, *582*, 103–110. [CrossRef]
111. Chen, T.; Yu, B.; Zhao, Y.; Li, Y.; Lin, Y.S. Carbon Dioxide Permeation through Ceramic-Carbonate Dual-Phase Membrane-Effects of Sulfur Dioxide. *J. Memb. Sci.* **2017**, *540*, 477–484. [CrossRef]
112. Eiberger, J.; Wilkner, K.; Reetz, C.; Sebold, D.; Jordan, N.; De Graaff, M.; Meulenberg, W.A.; Stöver, D.; Bram, M. Influence of Coal Power Plant Exhaust Gas on the Structure and Performance of Ceramic Nanostructured Gas Separation Membranes. *Int. J. Greenh. Gas Control* **2015**, *43*, 46–56. [CrossRef]
113. Kim, K.; Hong, S.; Kim, J.; Lee, H. Preparation and Performance Evaluation of Composite Hollow Fiber Membrane for SO_2 Separation. *AIChE J.* **2014**, *60*, 2298–2306. [CrossRef]
114. Zhao, L.; Chen, Y.; Wang, B.; Sun, C.; Chakraborty, S.; Ramasubramanian, K.; Dutta, P.K.; Ho, W.S.W. Multilayer Polymer/Zeolite Y Composite Membrane Structure for CO_2 Capture from Flue Gas. *J. Memb. Sci.* **2016**, *498*, 1–13. [CrossRef]
115. Choi, S.H.; Kim, J.H.; Lee, Y. Pilot-Scale Multistage Membrane Process for the Separation of CO_2 from LNG-Fired Flue Gas. *Sep. Purif. Technol.* **2013**, *110*, 170–180. [CrossRef]

116. Krull, F.F.; Fritzmann, C.; Melin, T. Liquid Membranes for Gas/Vapor Separations. *J. Memb. Sci.* **2008**, *325*, 509–519. [CrossRef]
117. Chen, G.; Wang, T.; Zhang, G.; Liu, G.; Jin, W. Membrane Materials Targeting Carbon Capture and Utilization. *Adv. Membr.* **2022**, *2*, 100025. [CrossRef]
118. Hou, R.; Fong, C.; Freeman, B.D.; Hill, M.R.; Xie, Z. Current Status and Advances in Membrane Technology for Carbon Capture. *Sep. Purif. Technol.* **2022**, *300*, 121863. [CrossRef]
119. Sreedhar, I.; Vaidhiswaran, R.; Kamani, B.M.; Venugopal, A. Process and Engineering Trends in Membrane Based Carbon Capture. *Renew. Sustain. Energy Rev.* **2017**, *68*, 659–684. [CrossRef]
120. Siagian, U.W.R.; Raksajati, A.; Himma, N.F.; Khoiruddin, K.; Wenten, I.G. Membrane-Based Carbon Capture Technologies: Membrane Gas Separation vs. Membrane Contactor. *J. Nat. Gas Sci. Eng.* **2019**, *67*, 172–195. [CrossRef]
121. Zhao, S.; Feron, P.H.M.; Deng, L.; Favre, E.; Chabanon, E.; Yan, S.; Hou, J.; Chen, V.; Qi, H. Status and Progress of Membrane Contactors in Post-Combustion Carbon Capture: A State-of-the-Art Review of New Developments. *J. Memb. Sci.* **2016**, *511*, 180–206. [CrossRef]
122. Scholes, C.A. Challenges for CO_2 Capture by Membranes. In *Advances in Carbon Capture*; Elsevier: Amsterdam, The Netherlands, 2020; pp. 357–377.
123. Adhikari, B.; Orme, C.J.; Stetson, C.; Klaehn, J.R. Techno-Economic Analysis of Carbon Dioxide Capture from Low Concentration Sources Using Membranes. *Chem. Eng. J.* **2023**, *474*, 145876. [CrossRef]
124. National Energy Technology Laboratory NETL. Carbon Capture and Storage Database. Available online: https://netl.doe.gov/carbon-management/carbon-storage/worldwide-ccs-database (accessed on 22 November 2024).
125. MTR Carbon Capture. Available online: https://mtrccs.com/news/completion-of-the-worlds-largest-membrane-based-carbon-capture-plant/ (accessed on 22 November 2024).
126. Font-Palma, C.; Cann, D.; Udemu, C. Review of Cryogenic Carbon Capture Innovations and Their Potential Applications. *J. Carbon Res.* **2021**, *7*, 58. [CrossRef]
127. McGlashan, N.R.; Marquis, A.J. Simultaneous Removal of CO_2, SO2, and NOx from Flue Gas by Liquid Phase Dehumidification at Cryogenic Temperatures and Low Pressure. *Proc. Inst. Mech. Eng. Part A J. Power Energy* **2008**, *222*, 31–45. [CrossRef]
128. Ali, A.; Maqsood, K.; Syahera, N.; Shariff, A.B.M.; Ganguly, S. Energy Minimization in Cryogenic Packed Beds during Purification of Natural Gas with High CO_2 Content. *Chem. Eng. Technol.* **2014**, *37*, 1675–1685. [CrossRef]
129. Göttlicher, G.; Pruschek, R. Comparison of CO_2 Removal Systems for Fossil-Fuelled Power Plant Processes. *Energy Convers. Manag.* **1997**, *38*, 173–178. [CrossRef]
130. Clodic, D.; Younes, M. Procede D'extraction du Dioxyde de Carbone par Anti-Sublimation en vue de son Stockage. France Patent FR2820052A1, 2 August 2002.
131. Pan, X.; Clodic, D.; Toubassy, J. CO_2 capture by antisublimation process and its technical economic analysis. *Greenh. Gases Sci. Technol.* **2013**, *3*, 8–20. [CrossRef]
132. Wilson, M.; Morris, T.; Gale, J.; Eds, K.T. Test Results of CO_2 Capture by Anti-Sublimation Capture Efficiency and Energy Consumption for Boiler Plants. *Greenh. Gas Control. Technol.* **2005**, *II*, 1775–1780.
133. Song, C.F.; Kitamura, Y.; Li, S.H.; Ogasawara, K. Design of a Cryogenic CO_2 Capture System Based on Stirling Coolers. *Int. J. Greenh. Gas Control* **2012**, *7*, 107–114. [CrossRef]
134. Song, C.; Liu, Q.; Ji, N.; Deng, S.; Zhao, J.; Kitamura, Y. Advanced Cryogenic CO_2 capture Process Based on Stirling Coolers by Heat Integration. *Appl. Therm. Eng.* **2017**, *114*, 887–895. [CrossRef]
135. Hu, J.Y.; Dai, W.; Luo, E.C.; Wang, X.T.; Huang, Y. Development of High Efficiency Stirling-Type Pulse Tube Cryocoolers. *Cryogenics (Guildf).* **2010**, *50*, 603–607. [CrossRef]
136. Fazlollahi, F.; Saeidi, S.; Safdari, M.S.; Sarkari, M.; Klemeš, J.J.; Baxter, L.L. Effect of Operating Conditions on Cryogenic Carbon Dioxide Removal. *Energy Technol.* **2017**, *5*, 1588–1598. [CrossRef]
137. Jensen, M.J.; Russell, C.S.; Bergeson, D.; Hoeger, C.D.; Frankman, D.J.; Bence, C.S.; Baxter, L.L. Prediction and Validation of External Cooling Loop Cryogenic Carbon Capture (CCC-ECL) for Full-Scale Coal-Fired Power Plant Retrofit. *Int. J. Greenh. Gas Control* **2015**, *42*, 200–212. [CrossRef]
138. Tuinier, M.J.; van Sint Annaland, M.; Kramer, G.J.; Kuipers, J.A.M. Cryogenic CO_2 Capture Using Dynamically Operated Packed Beds. *Chem. Eng. Sci.* **2010**, *65*, 114–119. [CrossRef]
139. Willson, P.; Lychnos, G.; Clements, A.; Michailos, S.; Font-Palma, C.; Diego, M.E.; Pourkashanian, M.; Howe, J. Evaluation of the Performance and Economic Viability of a Novel Low Temperature Carbon Capture Process. *Int. J. Greenh. Gas Control* **2019**, *86*, 1–9. [CrossRef]
140. Tuinier, M.J.; Van Sint Annaland, M.; Kuipers, J.A.M. A Novel Process for Cryogenic CO_2 Capture Using Dynamically Operated Packed Beds-An Experimental and Numerical Study. *Int. J. Greenh. Gas Control* **2011**, *5*, 694–701. [CrossRef]
141. National Carbon Capture Center. Available online: https://nationalcarboncapturecenter.com/2024/03/21/national-carbon-capture-center-announces-first-cryogenic-carbon-capture-testing/ (accessed on 3 November 2024).

142. Jansen, D.; Gazzani, M.; Manzolini, G.; Dijk, E.V.; Carbo, M. Pre-Combustion CO_2 Capture. *Int. J. Greenh. Gas Control* **2015**, *40*, 167–187. [CrossRef]
143. Osman, A.I.; Abu-Dahrieh, J.K.; Cherkasov, N.; Fernandez-Garcia, J.; Walker, D.; Walton, R.I.; Rooney, D.W.; Rebrov, E. A Highly Active and Synergistic Pt/Mo2C/Al2O3 Catalyst for Water-Gas Shift Reaction. *Mol. Catal.* **2018**, *455*, 38–47. [CrossRef]
144. Cao, M.; Zhao, L.; Xu, D.; Ciora, R.; Liu, P.K.T.; Manousiouthakis, V.I.; Tsotsis, T.T. A Carbon Molecular Sieve Membrane-Based Reactive Separation Process for Pre-Combustion CO_2 Capture. *J. Memb. Sci.* **2020**, *605*, 118028. [CrossRef]
145. Osman, A.I.; Deka, T.J.; Baruah, D.C.; Rooney, D.W. Critical Challenges in Biohydrogen Production Processes from the Organic Feedstocks. *Biomass Convers. Biorefinery* **2023**, *13*, 8383–8401. [CrossRef]
146. Wang, T. An Overview of IGCC Systems. In *Integrated Gasification Combined Cycle (IGCC) Technologies*; Elsevier: Amsterdam, The Netherlands, 2017; pp. 1–80.
147. Joshi, M.M.; Lee, S. Integrated Gasification Combined Cycle—A Review of IGCC Technology. *Energy Sources* **1996**, *18*, 537–568. [CrossRef]
148. Batorshin, V.A.; Suchkov, S.I.; Tugov, A.N. Integrated Gasification Combined Cycle (IGCC) Units: History, State-of-the Art, Development Prospects (Review). *Therm. Eng.* **2023**, *70*, 418–429. [CrossRef]
149. Hossein Sahraei, M.; McCalden, D.; Hughes, R.; Ricardez-Sandoval, L.A. A Survey on Current Advanced IGCC Power Plant Technologies, Sensors and Control Systems. *Fuel* **2014**, *137*, 245–259. [CrossRef]
150. IGCC Power Plant Nakoso Power Station. Available online: https://power.mhi.com/products/igcc (accessed on 28 October 2024).
151. Jazan IGCC. Available online: https://acwapower.com/en/projects/jazan-igcc/ (accessed on 17 October 2024).
152. Stanger, R.; Wall, T.; Spörl, R.; Paneru, M.; Grathwohl, S.; Weidmann, M.; Scheffknecht, G.; McDonald, D.; Myöhänen, K.; Ritvanen, J.; et al. Oxyfuel Combustion for CO_2 Capture in Power Plants. *Int. J. Greenh. Gas Control* **2015**, *40*, 55–125. [CrossRef]
153. Richter, W.F.; Chen, S.L. CO_2 Recovery via Coal Combustion in Mixtures of Oxygen and Recycled Flue Gas. *Combust. Sci. Technol.* **1989**, *67*, 1–16. [CrossRef]
154. Chen, L.; Yong, S.Z.; Ghoniem, A.F. Oxy-Fuel Combustion of Pulverized Coal: Characterization, Fundamentals, Stabilization and CFD Modeling. *Prog. Energy Combust. Sci.* **2012**, *38*, 156–214. [CrossRef]
155. Yang, C.; Kim, Y.; Bang, B.; Jeong, S.; Moon, J.; Mun, T.Y.; Jo, S.; Lee, J.; Lee, U. Oxy-CFB Combustion Technology for Use in Power-Generation Applications. *Fuel* **2020**, *267*, 117206. [CrossRef]
156. Wall, T.; Liu, Y.; Spero, C.; Elliott, L.; Khare, S.; Rathnam, R.; Zeenathal, F.; Moghtaderi, B.; Buhre, B.; Sheng, C.; et al. An Overview on Oxyfuel Coal Combustion-State of the Art Research and Technology Development. *Chem. Eng. Res. Des.* **2009**, *87*, 1003–1016. [CrossRef]
157. Toftegaard, M.B.; Brix, J.; Jensen, P.A.; Glarborg, P.; Jensen, A.D. Oxy-Fuel Combustion of Solid Fuels. *Prog. Energy Combust. Sci.* **2010**, *36*, 581–625. [CrossRef]
158. Buhre, B.J.P.; Elliott, L.K.; Sheng, C.D.; Gupta, R.P.; Wall, T.F. Oxy-Fuel Combustion Technology For Coal-Fired Power Generation. *Prog. Energy Combust. Sci.* **2005**, *31*, 283–307. [CrossRef]
159. Go, E.S.; Kim, B.-S.; Ling, J.L.J.; Oh, S.S.; Park, H.J.; Lee, S.H. In-Situ Desulfurization Using Porous Ca-Based Materials for the Oxy-CFB Process: A Computational Study. *Environ. Res.* **2023**, *225*, 115582. [CrossRef] [PubMed]
160. Duan, L.; Zhou, W.; Li, H.; Chen, X.; Zhao, C. Sulfur Fate during Bituminous Coal Combustion in an Oxy-Fired Circulating Fluidized Bed Combustor. *Korean J. Chem. Eng.* **2011**, *28*, 1952–1955. [CrossRef]
161. Ndibe, C.; Spörl, R.; Maier, J.; Scheffknecht, G. Experimental Study of NO and NO2 Formation in a PF Oxy-Fuel Firing System. *Fuel* **2013**, *107*, 749–756. [CrossRef]
162. Yuan, Y.; Wang, L.; Zhuang, Y.; Wu, Y.; Bi, X. Energy and Economic Assessment of Oxy-Fuel Combustion CO_2 Capture in Coal-Fired Power Plants. *Energies* **2024**, *17*, 4626. [CrossRef]
163. Wohlthan, M.; Thaler, B.; Helf, A.; Keller, F.; Kaub, V.; Span, R.; Gräbner, M.; Pirker, G. Oxyfuel Combustion Based Carbon Capture Onboard Ships. *Int. J. Greenh. Gas Control* **2024**, *137*, 104234. [CrossRef]
164. NET Power, La Porte Scale-up Project. Available online: https://netpower.com/first-utility-scale-project/ (accessed on 29 December 2024).
165. NET Power, La Porte Test Facility. Available online: https://netpower.com/la-porte-test-facility/ (accessed on 16 November 2024).
166. Global CCS Institute. Available online: https://www.globalccsinstitute.com/news-media/insights/china-begins-operations-at-the-worlds-largest-oxy-fuel-combustion-ccus-project-in-cement-sector/ (accessed on 22 November 2024).
167. Frauenhofer UMSICHT Institute. Available online: https://www.umsicht-suro.fraunhofer.de/en/press-and-media/press-releases/2022/oxyfuel-combustion-biogas-CO2-separation.html (accessed on 20 November 2024).
168. Lewis, W.K.; Gilliland, E.R. Production of Pure Carbon Dioxide. U.S. Patent US2665972A, 12 January 1954.
169. Nandy, A.; Loha, C.; Gu, S.; Sarkar, P.; Karmakar, M.K.; Chatterjee, P.K. Present Status and Overview of Chemical Looping Combustion Technology. *Renew. Sustain. Energy Rev.* **2016**, *59*, 597–619. [CrossRef]

170. Daneshmand-Jahromi, S.; Sedghkerdar, M.H.; Mahinpey, N. A Review of Chemical Looping Combustion Technology: Fundamentals, and Development of Natural, Industrial Waste, and Synthetic Oxygen Carriers. *Fuel* **2023**, *341*, 127626. [CrossRef]
171. Marx, F.; Dieringer, P.; Ströhle, J.; Epple, B. Design of a 1 MWth Pilot Plant for Chemical Looping Gasification of Biogenic Residues. *Energies* **2021**, *14*, 2581. [CrossRef]
172. Alalwan, H.A.; Augustine, L.J.; Hudson, B.G.; Abeysinghe, J.P.; Gillan, E.G.; Mason, S.E.; Grassian, V.H.; Cwiertny, D.M. Linking Solid-State Reduction Mechanisms to Size-Dependent Reactivity of Metal Oxide Oxygen Carriers for Chemical Looping Combustion. *ACS Appl. Energy Mater.* **2021**, *4*, 1163–1172. [CrossRef]
173. Jin, H.; Okamoto, T.; Ishida, M. Development of a Novel Chemical-Looping Combustion: Synthesis of a Looping Material with a Double Metal Oxide of CoO-NiO. *Energy Fuels* **1998**, *12*, 1272–1277. [CrossRef]
174. Feng, X.; Li, Z.; Lin, S.; Tian, S.; Li, K. Enhanced Performance of Red Mud for Chemical-Looping Combustion of Coal by the Modification of Transition Metal Oxides. *J. Energy Inst.* **2022**, *102*, 22–31. [CrossRef]
175. Ma, J.; Hu, J.; Kang, H.; Han, Z.; Guo, Q. Behavior of Selenium during Chemical-Looping Gasification of Coal Using Copper-Based Oxygen Carrier. *Atmosphere* **2022**, *13*, 547. [CrossRef]
176. Rasi, N.M.; Hyla, A.S.; Ponnurangam, S.; Mahinpey, N. Effects of Support and Oxygen Vacancies on the Energetics of NiO Reduction with H_2 for the Chemical Looping Combustion (CLC) Reaction; A DFT Study. *Phys. Chem. Chem. Phys.* **2021**, *23*, 12795–12806. [CrossRef]
177. Daneshmand-Jahromi, S.; Karami, D.; Mahinpey, N. Novel Synthesis of High-Surface-Area Alumina Using Toluene-Dimethylformamide as Synthetic Media. *J. Environ. Chem. Eng.* **2022**, *10*, 107204. [CrossRef]
178. Czakiert, T.; Krzywanski, J.; Zylka, A.; Nowak, W. Chemical Looping Combustion: A Brief Overview. *Energies* **2022**, *15*, 1563. [CrossRef]
179. Thorne, R.J.; Bouman, E.A.; Sundseth, K.; Aranda, A.; Czakiert, T.; Pacyna, J.M.; Pacyna, E.G.; Krauz, M.; Celińska, A. Environmental Impacts of a Chemical Looping Combustion Power Plant. *Int. J. Greenh. Gas Control* **2019**, *86*, 101–111. [CrossRef]
180. Haugen, N.E.L.; Li, Z.; Gouraud, V.; Bertholin, S.; Li, W.; Larring, Y.; Luo, K.; Szlęk, A.; Flach, T.A.; Langørgen, Ø.; et al. Building the World's Largest Chemical Looping Combustion (CLC) Unit. *Int. J. Greenh. Gas Control* **2023**, *129*, 103975. [CrossRef]
181. Baek, J.-I.; Choun, M.; Kim, U.; Lee, G.; Kim, D.; Ryu, H.-J. The K-CLC Project—Development of 3 MWth CLC Steam Generation System towards a Clean Power Plant. In Proceedings of the 16th International Conference on Greenhouse Gas Control Technologies, GHGT-16, Lyon, France, 23 October 2022.
182. Zhao, H.; Tian, X.; Ma, J.; Chen, X.; Su, M.; Zheng, C.; Wang, Y. Chemical Looping Combustion of Coal in China: Comprehensive Progress, Remaining Challenges, and Potential Opportunities. *Energy Fuels* **2020**, *34*, 6696–6734. [CrossRef]
183. Trends in CO2. Available online: https://gml.noaa.gov/ccgg/trends/ (accessed on 29 November 2024).
184. House, K.Z.; Baclig, A.C.; Ranjan, M.; Van Nierop, E.A.; Wilcox, J.; Herzog, H.J. Economic and Energetic Analysis of Capturing CO_2 from Ambient Air. *Proc. Natl. Acad. Sci. USA* **1998**, *108*, 20428–20433. [CrossRef]
185. Sodiq, A.; Abdullatif, Y.; Aissa, B.; Ostovar, A.; Nassar, N.; El-Naas, M.; Amhamed, A. A Review on Progress Made in Direct Air Capture of CO_2. *Environ. Technol. Innov.* **2023**, *29*, 102991. [CrossRef]
186. Sinha, A.; Realff, M.J. A Parametric Study of the Techno-Economics of Direct CO_2 Air Capture Systems Using Solid Adsorbents. *AIChE J.* **2019**, *65*, 16607. [CrossRef]
187. Zeeshan, M.; Kidder, M.K.; Pentzer, E.; Getman, R.B.; Gurkan, B. Direct Air Capture of CO_2: From Insights into the Current and Emerging Approaches to Future Opportunities. *Front. Sustain.* **2023**, *4*, 1167713. [CrossRef]
188. Beaumont, M.L.L. Making Direct Air Capture Affordable; Technology, Market and Regulatory Approaches. *Front. Clim.* **2022**, *4*, 756013. [CrossRef]
189. Sweden Government Carbon Tax Policy. Available online: https://www.government.se/government-policy/taxes-and-tariffs/swedens-carbon-tax (accessed on 31 December 2024).
190. Brown, M.A.; Li, Y.; Soni, A. Are All Jobs Created Equal? Regional Employment Impacts of a U.S. Carbon Tax. *Appl. Energy* **2020**, *262*, 114354. [CrossRef]
191. Chai, S.; Huo, W.; Li, Q.; Ji, Q.; Shi, X. Effects of Carbon Tax on Energy Transition, Emissions and Economy amid Technological Progress. *Appl. Energy* **2025**, *377*, 124578. [CrossRef]
192. McQueen, N.; Gomes, K.V.; McCormick, C.; Blumanthal, K.; Pisciotta, M.; Wilcox, J. A Review of Direct Air Capture (DAC): Scaling up Commercial Technologies and Innovating for the Future. *Prog. Energy* **2021**, *3*, 032001. [CrossRef]
193. Lively, R.P.; Realff, M.J. On Thermodynamic Separation Efficiency: Adsorption Processes. *AIChE J.* **2016**, *62*, 3699–3705. [CrossRef]
194. Wilson, S.M.W.; Tezel, F.H. Direct Dry Air Capture of CO_2 Using VTSA with Faujasite Zeolites. *Ind. Eng. Chem. Res.* **2020**, *59*, 8783–8794. [CrossRef]
195. Climeworks Orca Plant. Available online: https://climeworks.com/plant-orca (accessed on 1 November 2024).
196. Climeworks Mammoth Plant. Available online: https://climeworks.com/plant-mammoth (accessed on 1 November 2024).
197. Sanz-Pérez, E.S.; Murdock, C.R.; Didas, S.A.; Jones, C.W. Direct Capture of CO_2 from Ambient Air. *Chem. Rev.* **2016**, *116*, 11840–11876. [CrossRef] [PubMed]

198. Veltman, K.; Singh, B.; Hertwich, E.G. Human and Environmental Impact Assessment of Postcombustion CO_2 Capture Focusing on Emissions from Amine-Based Scrubbing Solvents to Air. *Environ. Sci. Technol.* **2010**, *44*, 1496–1502. [CrossRef]
199. Zeman, F.S.; Lackner, K.S. Capturing Carbon Dioxide Directly from the Atmosphere. *World Resour. Rev.* **2004**, *16*, 157–172.
200. Bachu, S. Sequestration of CO_2 in Geological Media: Criteria and Approach for Site Selection in Response to Climate Change. *Energy Convers. Manag.* **2000**, *41*, 953–970. [CrossRef]
201. Michael, K.; Golab, A.; Shulakova, V.; Ennis-King, J.; Allinson, G.; Sharma, S.; Aiken, T. Geological Storage of CO_2 in Saline Aquifers-A Review of the Experience from Existing Storage Operations. *Int. J. Greenh. Gas Control* **2010**, *4*, 659–667. [CrossRef]
202. Hamza, A.; Hussein, I.A.; Al-Marri, M.J.; Mahmoud, M.; Shawabkeh, R.; Aparicio, S. CO_2 Enhanced Gas Recovery and Sequestration in Depleted Gas Reservoirs: A Review. *J. Pet. Sci. Eng.* **2021**, *196*, 107685. [CrossRef]
203. Godec, M.L.; Kuuskraa, V.A.; Dipietro, P. Opportunities for Using Anthropogenic CO_2 for Enhanced Oil Recovery and CO_2 Storage. *Energy Fuels* **2013**, *27*, 4183–4189. [CrossRef]
204. Zhang, S.; DePaolo, D.J. Rates of CO_2 Mineralization in Geological Carbon Storage. *Acc. Chem. Res.* **2017**, *50*, 2075–2084. [CrossRef]
205. Rigby, S.P.; Alsayah, A. Storage Sites for Carbon Dioxide in the North Sea and Their Particular Characteristics. *Energies* **2023**, *17*, 211. [CrossRef]
206. Briones-Hidrovo, A.; Copa Rey, J.R.; Cláudia Dias, A.; Tarelho, L.A.C.; Beauchet, S. Assessing a Bio-Energy System with Carbon Capture and Storage (BECCS) through Dynamic Life Cycle Assessment and Land-Water-Energy Nexus. *Energy Convers. Manag.* **2022**, *268*, 116014. [CrossRef]
207. Shukla, P.R.; Skea, J.; Slade, R.; van Diemen, R.; Haughey, E.; Malley, J.; Pathak, M.; Pereira, J.P. (Eds.) Technical Summary. In *Climate Change and Land: An IPCC Special Report on Climate Change, Desertification, Land Degradation, Sustainable Land Management, Food Security, and Greenhouse Gas Fluxes in Terrestrial Ecosystems*; Cambridge University Press: Cambridge, UK, 2019. [CrossRef]
208. Snæbjörnsdóttir, S.Ó.; Sigfússon, B.; Marieni, C.; Goldberg, D.; Gislason, S.R.; Oelkers, E.H. Carbon Dioxide Storage through Mineral Carbonation. *Nat. Rev. Earth Environ.* **2020**, *1*, 90–102. [CrossRef]
209. De Silva, G.P.D.; Ranjith, P.G.; Perera, M.S.A. Geochemical Aspects of CO_2 Sequestration in Deep Saline Aquifers: A Review. *Fuel* **2015**, *155*, 128–143. [CrossRef]
210. Wang, Y.; Vuik, C.; Hajibeygi, H. Analysis of Hydrodynamic Trapping Interactions during Full-Cycle Injection and Migration of CO_2 in Deep Saline Aquifers. *Adv. Water Resour.* **2022**, *159*, 104073. [CrossRef]
211. Szulczewski, M.L.; Hesse, M.A.; Juanes, R. Carbon Dioxide Dissolution in Structural and Stratigraphic Traps. *J. Fluid Mech.* **2013**, *736*, 287–315. [CrossRef]
212. Liu, L.C.; Li, Q.; Zhang, J.T.; Cao, D. Toward a Framework of Environmental Risk Management for CO_2 Geological Storage in China: Gaps and Suggestions for Future Regulations. *Mitig. Adapt. Strateg. Glob. Chang.* **2016**, *21*, 191–207. [CrossRef]
213. Carles, P.; Bachaud, P.; Lasseur, E.; Berne, P.; Bretonnier, P. Confining Properties of Carbonated Dogger Caprocks (Parisian Basin) for CO_2 Storage Purpose. *Oil Gas Sci. Technol.—Rev. l'Institut Français du Pétrole* **2010**, *65*, 461–472. [CrossRef]
214. Rathnaweera, T.D.; Ranjith, P.G.; Perera, M.S.A.; Lashin, A.; Al Arifi, N. Non-Linear Stress-Strain Behaviour of Reservoir Rock under Brine Saturation: An Experimental Study. *Meas. J. Int. Meas. Confed.* **2015**, *71*, 56–72. [CrossRef]
215. Rathnaweera, T.D.; Ranjith, P.G.; Perera, M.S.A.; Yang, S.Q. Determination of Effective Stress Parameters for Effective CO_2 Permeability in Deep Saline Aquifers: An Experimental Study. *J. Nat. Gas Sci. Eng.* **2015**, *24*, 64–79. [CrossRef]
216. Burnside, N.M.; Naylor, M. Review and Implications of Relative Permeability of CO_2/Brine Systems and Residual Trapping of CO_2. *Int. J. Greenh. Gas Control* **2014**, *23*, 1–11. [CrossRef]
217. Altman, S.J.; Aminzadeh, B.; Balhoff, M.T.; Bennett, P.C.; Bryant, S.L.; Cardenas, M.B.; Chaudhary, K.; Cygan, R.T.; Deng, W.; Dewers, T.; et al. Chemical and Hydrodynamic Mechanisms for Long-Term Geological Carbon Storage. *J. Phys. Chem. C* **2014**, *118*, 15103–15113. [CrossRef]
218. Kumar, S.; Foroozesh, J.; Edlmann, K.; Rezk, M.G.; Lim, C.Y. A Comprehensive Review of Value-Added CO_2 Sequestration in Subsurface Saline Aquifers. *J. Nat. Gas Sci. Eng.* **2020**, *81*, 103437. [CrossRef]
219. Holtz, M.H. Residual Gas Saturation to Aquifer Influx: A Calculation Method for 3-D Computer Reservoir Model Construction. In Proceedings of the SPE Gas Technology Symposium, SPE, Calgary, AB, Canada, 30 April 2002.
220. Zhang, D.; Song, J. Mechanisms for Geological Carbon Sequestration. *Procedia IUTAM* **2014**, *10*, 319–327. [CrossRef]
221. Wilkin, R.T.; Digiulio, D.C. Geochemical Impacts to Groundwater from Geologic Carbon Sequestration: Controls on PH and Inorganic Carbon Concentrations from Reaction Path and Kinetic Modeling. *Environ. Sci. Technol.* **2010**, *44*, 4821–4827. [CrossRef]
222. Bachu, S.; Gunter, W.D. Acid-Gas Injection in the Alberta Basin, Canada: A CO_2-Storage Experience. *Geol. Soc. London Spec. Publ.* **2004**, *233*, 225–234. [CrossRef]
223. Equinor | Deutschland | CO2 Management—Equinor.de. Available online: https://www.equinor.de/co2-management (accessed on 15 October 2024).
224. Carbon Capture and Sequestration Technologies @ MIT. Available online: https://sequestration.mit.edu/tools/projects/sleipner.html (accessed on 28 October 2024).

225. Wei, B.; Wang, B.; Li, X.; Aishan, M.; Ju, Y. CO_2 Storage in Depleted Oil and Gas Reservoirs: A Review. *Adv. Geo-Energy Res.* **2023**, *9*, 76–93. [CrossRef]
226. Carroll, S.A.; Iyer, J.; Walsh, S.D.C. Influence of Chemical, Mechanical, and Transport Processes on Wellbore Leakage from Geologic CO_2 Storage Reservoirs. *Acc. Chem. Res.* **2017**, *50*, 1829–1837. [CrossRef] [PubMed]
227. Bai, M.; Song, K.; Li, Y.; Sun, J.; Reinicke, K.M. Development of a Novel Method to Evaluate Well Integrity during CO_2 Underground Storage. *SPE J.* **2015**, *20*, 628–641. [CrossRef]
228. Ozotta, O.; Ostadhassan, M.; Liu, K.; Liu, B.; Kolawole, O.; Hadavimoghaddam, F. Reassessment of CO_2 Sequestration in Tight Reservoirs and Associated Formations. *J. Pet. Sci. Eng.* **2021**, *206*, 109071. [CrossRef]
229. Offshore CCS Planned by 2025 at Project Greensand. Available online: https://wintershalldea.com/en/newsroom/offshore-ccs-planned-2025-project-greensand (accessed on 22 November 2024).
230. Moomba CCS | Beach Energy. Available online: https://beachenergy.com.au/moomba-ccs/ (accessed on 22 November 2024).
231. Alvarado, V.; Manrique, E. Enhanced Oil Recovery: An Update Review. *Energies* **2010**, *3*, 1529–1575. [CrossRef]
232. Gbadamosi, A.O.; Kiwalabye, J.; Junin, R.; Augustine, A. A Review of Gas Enhanced Oil Recovery Schemes Used in the North Sea. *J. Pet. Explor. Prod. Technol.* **2018**, *8*, 1373–1387. [CrossRef]
233. Hascakir, B. Introduction to Thermal Enhanced Oil Recovery (EOR) Special Issue. *J. Pet. Sci. Eng.* **2017**, *154*, 438–441. [CrossRef]
234. Pothula, G.K.; Vij, R.K.; Bera, A. An Overview of Chemical Enhanced Oil Recovery and Its Status in India. *Pet. Sci.* **2023**, *20*, 2305–2323. [CrossRef]
235. Rellegadla, S.; Prajapat, G.; Agrawal, A. Polymers for Enhanced Oil Recovery: Fundamentals and Selection Criteria. *Appl. Microbiol. Biotechnol.* **2017**, *101*, 4387–4402. [CrossRef]
236. Negin, C.; Ali, S.; Xie, Q. Most Common Surfactants Employed in Chemical Enhanced Oil Recovery. *Petroleum* **2017**, *3*, 197–211. [CrossRef]
237. Lee, H.S.; Cho, J.H.; Lee, Y.W.; Lee, K.S. Compositional Modeling of Impure CO_2 Injection for Enhanced Oil Recovery and CO_2 Storage. *Appl. Sci.* **2021**, *11*, 7907. [CrossRef]
238. Claridge, E.L. Prediction of Recovery in Unstable Miscible Flooding. *Soc. Pet. Eng. J.* **1972**, *12*, 143–155. [CrossRef]
239. Kumar, N.; Augusto Sampaio, M.; Ojha, K.; Hoteit, H.; Mandal, A. Fundamental Aspects, Mechanisms and Emerging Possibilities of CO_2 Miscible Flooding in Enhanced Oil Recovery: A Review. *Fuel* **2022**, *330*, 125633. [CrossRef]
240. Zhang, N.; Wei, M.; Bai, B. Statistical and Analytical Review of Worldwide CO_2 Immiscible Field Applications. *Fuel* **2018**, *220*, 89–100. [CrossRef]
241. Coutinho, R. Petrobras Santos Basin-Salt Oil Field CCUS. In Proceedings of the CSLF: 2023 Technical Group Mid-Year Meeting, Warsaw, Poland, 13–14 June 2023.
242. Gunter, W.; Perkins, E.; Hutcheon, I. Aquifer Disposal of Acid Gases: Modelling of Water–Rock Reactions for Trapping of Acid Wastes. *Appl. Geochem.* **2000**, *15*, 1085–1095. [CrossRef]
243. Cross, W.; Iddings, J.P.; Pirsson, L.V.; Washington, H.S. Modifications of the" Quantitative System of Classification of Igneous Rocks. *J. Geol.* **1912**, *20*, 550–561. [CrossRef]
244. Kim, K.; Kim, D.; Na, Y.; Song, Y.; Wang, J. A Review of Carbon Mineralization Mechanism during Geological CO_2 Storage. *Heliyon* **2023**, *9*, e23135. [CrossRef] [PubMed]
245. Carbon Removal Solutions to Fight Climate Change. Available online: https://climeworks.com/blog/carbon-removal-solutions (accessed on 1 November 2024).
246. Metz, B.; Davidson, O.; De Coninck, H.C.; Loos, M.; Meyer, L. *IPCC Special Report on Carbon Dioxide Capture and Storage*; Cambridge University Press: Cambridge, UK, 2005.
247. Shi, J.Q.; Durucan, S. CO_2 Storage in Deep Unminable Coal Seams. *Oil Gas Sci. Technol.* **2005**, *60*, 547–558. [CrossRef]
248. Stanton, R.; Flores, R.; Warwick, P.D.; Gluskoter, H.; Stricker, G.D. Coal Bed Sequestration of Carbon Dioxide. In Proceedings of the 1st National Conference on Carbon Sequestration, Washington, DC, USA, 14–17 May 2001; pp. 14–17.
249. Fujioka, M.; Yamaguchi, S.; Nako, M. CO_2-ECBM Field Tests in the Ishikari Coal Basin of Japan. *Int. J. Coal Geol.* **2010**, *82*, 287–298. [CrossRef]
250. Hadi Mosleh, M.; Turner, M.; Sedighi, M.; Vardon, P.J. Carbon Dioxide Flow and Interactions in a High Rank Coal: Permeability Evolution and Reversibility of Reactive Processes. *Int. J. Greenh. Gas Control* **2018**, *70*, 57–67. [CrossRef]
251. Zhang, Y.; Lebedev, M.; Al-Yaseri, A.; Yu, H.; Xu, X.; Iglauer, S. Characterization of Nanoscale Rockmechanical Properties and Microstructures of a Chinese Sub-Bituminous Coal. *J. Nat. Gas Sci. Eng.* **2018**, *52*, 106–116. [CrossRef]
252. Li, L.; Yang, D.; Liu, W.; Zhang, X.; Zhao, L.; Zhu, X. Experimental Study on the Porosity and Permeability Change of High-Rank Coal under Cyclic Loading and Unloading. *ACS Omega* **2022**, *7*, 30197–30207. [CrossRef] [PubMed]
253. Niu, Q.; Cao, L.; Sang, S.; Zhou, X.; Wang, Z. Anisotropic Adsorption Swelling and Permeability Characteristics with Injecting CO_2 in Coal. *Energy Fuels* **2018**, *32*, 1979–1991. [CrossRef]
254. Global Status of CCS 2024 Report at a Glance. Available online: https://www.globalccsinstitute.com/resources/global-status-report (accessed on 28 November 2024).

255. Bradshaw, J.; Bachu, S.; Bonijoly, D.; Burruss, R.; Holloway, S.; Christensen, N.P.; Mathiassen, O.M. CO_2 Storage Capacity Estimation: Issues and Development of Standards. *Int. J. Greenh. Gas Control* **2007**, *1*, 62–68. [CrossRef]
256. Bachu, S.; Bonijoly, D.; Bradshaw, J.; Burruss, R.; Holloway, S.; Christensen, N.P.; Mathiassen, O.M. CO_2 Storage Capacity Estimation: Methodology and Gaps. *Int. J. Greenh. Gas Control* **2007**, *1*, 430–443. [CrossRef]
257. IEA, Putting CO2 to Use Creating Value from Emissions. Available online: https://www.iea.org/reports/putting-co2-to-use (accessed on 20 November 2024).
258. Dziejarski, B.; Krzyżyńska, R.; Andersson, K. Current Status of Carbon Capture, Utilization, and Storage Technologies in the Global Economy: A Survey of Technical Assessment. *Fuel* **2023**, *342*, 127776. [CrossRef]
259. Theofanidis, S.A.; Antzaras, A.N.; Lemonidou, A.A. CO_2 as a Building Block: From Capture to Utilization. *Curr. Opin. Chem. Eng.* **2023**, *39*, 100902. [CrossRef]
260. Chauvy, R.; De Weireld, G. CO_2 Utilization Technologies in Europe: A Short Review. *Energy Technol.* **2020**, *8*, 627. [CrossRef]
261. Rumayor, M.; Dominguez-Ramos, A.; Irabien, A. Environmental and Economic Assessment of the Formic Acid Electrochemical Manufacture Using Carbon Dioxide: Influence of the Electrode Lifetime. *Sustain. Prod. Consum.* **2019**, *18*, 72–82. [CrossRef]
262. Meunier, N.; Chauvy, R.; Mouhoubi, S.; Thomas, D.; De Weireld, G. Alternative Production of Methanol from Industrial CO_2. *Renew. Energy* **2020**, *146*, 1192–1203. [CrossRef]
263. Bellotti, D.; Rivarolo, M.; Magistri, L.; Massardo, A.F. Feasibility Study of Methanol Production Plant from Hydrogen and Captured Carbon Dioxide. *J. CO2 Util.* **2017**, *21*, 132–138. [CrossRef]
264. Chauvy, R.; Dubois, L.; Lybaert, P.; Thomas, D.; De Weireld, G. Production of Synthetic Natural Gas from Industrial Carbon Dioxide. *Appl. Energy* **2020**, *260*, 114249. [CrossRef]
265. Xu, Y.; Lin, L.; Xiao, M.; Wang, S.; Smith, A.T.; Sun, L.; Meng, Y. Synthesis and Properties of CO_2-Based Plastics: Environmentally-Friendly, Energy-Saving and Biomedical Polymeric Materials. *Prog. Polym. Sci.* **2018**, *80*, 163–182. [CrossRef]
266. Inoue, S.; Koinuma, H.; Tsuruta, T. Copolymerization of Carbon Dioxide and Epoxide. *J. Polym. Sci. Part B Polym. Lett.* **1969**, *7*, 287–292. [CrossRef]
267. Artz, J.; Müller, T.E.; Thenert, K.; Kleinekorte, J.; Meys, R.; Sternberg, A.; Bardow, A.; Leitner, W. Sustainable Conversion of Carbon Dioxide: An Integrated Review of Catalysis and Life Cycle Assessment. *Chem. Rev.* **2018**, *118*, 434–504. [CrossRef] [PubMed]
268. Devkota, S.; Karmacharya, P.; Maharjan, S.; Khatiwada, D.; Uprety, B. Decarbonizing Urea: Techno-Economic and Environmental Analysis of a Model Hydroelectricity and Carbon Capture Based Green Urea Production. *Appl. Energy* **2024**, *372*, 123789. [CrossRef]
269. Fu, L.; Ren, Z.; Si, W.; Ma, Q.; Huang, W.; Liao, K.; Huang, Z.; Wang, Y.; Li, J.; Xu, P. Research Progress on CO_2 Capture and Utilization Technology. *J. CO2 Util.* **2022**, *66*, 102260. [CrossRef]
270. Wang, W.; Rao, L.; Wu, X.; Wang, Y.; Zhao, L.; Liao, X. Supercritical Carbon Dioxide Applications in Food Processing. *Food Eng. Rev.* **2021**, *13*, 570–591. [CrossRef]
271. Market Insights—Bottled Water Market. Available online: https://www.marketresearchstore.com/market-insights/bottled-water-market-z39681 (accessed on 2 January 2025).
272. Grand View Research—Sparkling Water Market. Available online: https://www.grandviewresearch.com/industry-analysis/sparkling-water-market (accessed on 2 January 2025).
273. Recode Project. Available online: https://recodeh2020.eu (accessed on 11 November 2024).
274. Rumayor, M.; Dominguez-Ramos, A.; Irabien, A. Formic Acid Manufacture: Carbon Dioxide Utilization Alternatives. *Appl. Sci.* **2018**, *8*, 914. [CrossRef]
275. Mishra, A.; Kim, D.; Altahtamouni, T.; Kasak, P.; Popelka, A.; Park, H.; Han, D.S. A Comparative Study on Carbon Neutral Hydrogen Carrier Production: Formic Acid from CO_2 vs. Ammonia. *J. CO2 Util.* **2024**, *82*, 102756. [CrossRef]
276. Pérez-Fortes, M.; Schöneberger, J.C.; Boulamanti, A.; Harrison, G.; Tzimas, E. Formic Acid Synthesis Using CO_2 as Raw Material: Techno-Economic and Environmental Evaluation and Market Potential. *Int. J. Hydrogen Energy* **2016**, *41*, 16444–16462. [CrossRef]
277. Jogunola, O.; Salmi, T.; Waärnå, J.; Mikkola, J.P.; Tirronen, E. Kinetics of Methyl Formate Hydrolysis in the Absence and Presence of a Complexing Agent. *Ind. Eng. Chem. Res.* **2011**, *50*, 267–276. [CrossRef]
278. Onishi, N.; Himeda, Y. Toward Methanol Production by CO_2 Hydrogenation beyond Formic Acid Formation. *Acc. Chem. Res.* **2024**, *57*, 2816–2825. [CrossRef] [PubMed]
279. Gorbunov, D.N.; Nenasheva, M.V.; Terenina, M.V.; Kardasheva, Y.S.; Kardashev, S.V.; Naranov, E.R.; Bugaev, A.L.; Soldatov, A.V.; Maximov, A.L.; Karakhanov, E.A. Transformations of Carbon Dioxide under Homogeneous Catalysis Conditions (A Review). *Pet. Chem.* **2022**, *62*, 1–39. [CrossRef]
280. Inoue, Y.; Izumida, H.; Sasaki, Y.; Hashimoto, H. Catalytic Fixation of Carbon Dioxide to Formic Acid by Transition-Metal Complexes Under Mild Conditions. *Chem. Lett.* **1976**, *5*, 863–864. [CrossRef]
281. Tanaka, R.; Yamashita, M.; Nozaki, K. Catalytic Hydrogenation of Carbon Dioxide Using Ir(III)-Pincer Complexes. *J. Am. Chem. Soc.* **2009**, *131*, 14168–14169. [CrossRef] [PubMed]

282. COufourier, S.; Gaignard Gaillard, Q.; Lohier, J.F.; Poater, A.; Gaillard, S.; Renaud, J.L. Hydrogenation of CO_2, Hydrogenocarbonate, and Carbonate to Formate in Water Using Phosphine Free Bifunctional Iron Complexes. *ACS Catal.* **2020**, *10*, 2108–2116. [CrossRef]
283. Jeletic, M.S.; Mock, M.T.; Appel, A.M.; Linehan, J.C. A Cobalt-Based Catalyst for the Hydrogenation of CO_2 under Ambient Conditions. *J. Am. Chem. Soc.* **2013**, *135*, 11533–11536. [CrossRef] [PubMed]
284. Ogo, S.; Kabe, R.; Hayashi, H.; Harada, R.; Fukuzumi, S. Mechanistic Investigation of CO_2 Hydrogenation by Ru(Ii) and Ir(Iii) Aqua Complexes under Acidic Conditions: Two Catalytic Systems Differing in the Nature of the Rate Determining Step. *Dalt. Trans.* **2006**, *39*, 4657–4663. [CrossRef]
285. Hayashi, H.; Ogo, S.; Fukuzumi, S. Aqueous Hydrogenation of Carbon Dioxide Catalysed by Water-Soluble Ruthenium Aqua Complexes under Acidic Conditions. *Chem. Commun.* **2004**, *23*, 2714–2715. [CrossRef] [PubMed]
286. Alfian, M.; Purwanto, W.W. Multi-Objective Optimization of Green Urea Production. *Energy Sci. Eng.* **2019**, *7*, 292–304. [CrossRef]
287. Driver, J.G.; Owen, R.E.; Makanyire, T.; Lake, J.A.; McGregor, J.; Styring, P. Blue Urea: Fertilizer with Reduced Environmental Impact. *Front. Energy Res.* **2019**, *7*, 88. [CrossRef]
288. Milani, D.; Kiani, A.; Haque, N.; Giddey, S.; Feron, P. Green Pathways for Urea Synthesis: A Review from Australia's Perspective. *Sustain. Chem. Clim. Action* **2022**, *1*, 100008. [CrossRef]
289. SUN to LIQUID Project. Available online: https://www.sun-to-liquid.eu (accessed on 11 November 2024).
290. Methanol Institute. Available online: https://www.methanol.org/methanol-price-supply-demand/ (accessed on 1 January 2025).
291. Godin, J.; Liu, W.; Ren, S.; Xu, C.C. Advances in Recovery and Utilization of Carbon Dioxide: A Brief Review. *J. Environ. Chem. Eng.* **2021**, *9*, 105644. [CrossRef]
292. Khadzhiev, S.N.; Kolesnichenko, N.V.; Ezhova, N.N. Slurry Technology in Methanol Synthesis (Review). *Pet. Chem.* **2016**, *56*, 77–95. [CrossRef]
293. Bozzano, G.; Manenti, F. Efficient Methanol Synthesis: Perspectives, Technologies and Optimization Strategies. *Prog. Energy Combust. Sci.* **2016**, *56*, 71–105. [CrossRef]
294. Dalena, F.; Senatore, A.; Marino, A.; Gordano, A.; Basile, M.; Basile, A. Methanol Production and Applications: An Overview. In *Methanol*; Elsevier: Amsterdam, The Netherlands, 2018; pp. 3–28.
295. Saeidi, S.; Najari, S.; Hessel, V.; Wilson, K.; Keil, F.J.; Concepción, P.; Suib, S.L.; Rodrigues, A.E. Recent Advances in CO_2 Hydrogenation to Value-Added Products—Current Challenges and Future Directions. *Prog. Energy Combust. Sci.* **2021**, *85*, 100905. [CrossRef]
296. Ren, M.; Zhang, Y.; Wang, X.; Qiu, H. Catalytic Hydrogenation of CO_2 to Methanol: A Review. *Catalysts* **2022**, *12*, 403. [CrossRef]
297. Turner, J.; Sverdrup, G.; Mann, M.K.; Maness, P.C.; Kroposki, B.; Ghirardi, M.; Evans, R.J.; Blake, D. Renewable Hydrogen Production. *Int. J. Energy Res.* **2008**, *32*, 379–407. [CrossRef]
298. Carbon Recycling International. Available online: https://carbonrecycling.com (accessed on 11 November 2024).
299. Liu, H.; Xu, S.; Zhou, G.; Huang, G.; Huang, S.; Xiong, K. CO_2 Hydrogenation to Methane over Co/KIT-6 Catalyst: Effect of Reduction Temperature. *Chem. Eng. J.* **2018**, *351*, 65–73. [CrossRef]
300. Garcia, J.A.; Villen-Guzman, M.; Rodriguez-Maroto, J.M.; Paz-Garcia, J.M. Comparing CO_2 Storage and Utilization: Enhancing Sustainability through Renewable Energy Integration. *Sustainability* **2024**, *16*, 6639. [CrossRef]
301. Krótki, A.; Chwoła, T.; Więcław-Solny, L.; Tatarczuk, A.; Spietz, T.; Dobras, S.; Zdeb, J. Advancements in CO_2 Hydrogenation—Investigating a CNG Pilot Plant in Poland. *Fuel* **2025**, *381*, 133599. [CrossRef]
302. Alami, A.H.; Alasad, S.; Ali, M.; Alshamsi, M. Investigating Algae for CO_2 Capture and Accumulation and Simultaneous Production of Biomass for Biodiesel Production. *Sci. Total Environ.* **2021**, *759*, 143529. [CrossRef] [PubMed]
303. Moreira, D.; Pires, J.C.M. Atmospheric CO_2 Capture by Algae: Negative Carbon Dioxide Emission Path. *Bioresour. Technol.* **2016**, *215*, 371–379. [CrossRef] [PubMed]
304. Iglina, T.; Iglin, P.; Pashchenko, D. Industrial CO_2 Capture by Algae: A Review and Recent Advances. *Sustainability* **2022**, *14*, 3801. [CrossRef]
305. Doucha, J.; Straka, F.; Lívanský, K. Utilization of Flue Gas for Cultivation of Microalgae (*Chlorella* sp.) in an Outdoor Open Thin-Layer Photobioreactor. *J. Appl. Phycol.* **2005**, *17*, 403–412. [CrossRef]
306. AlgaeCytes Homepage. Available online: https://algaecytes.com (accessed on 5 January 2025).

MDPI AG
Grosspeteranlage 5
4052 Basel
Switzerland
Tel.: +41 61 683 77 34

Processes Editorial Office
E-mail: processes@mdpi.com
www.mdpi.com/journal/processes

www.ingramcontent.com/pod-product-compliance
Lightning Source LLC
Chambersburg PA
CBHW061503210926
51850CB00008B/1802
9783725878055